# Advances in Passive Cooling

BUILDINGS | ENERGY | SOLAR TECHNOLOGY

# Advances in Passive Cooling

### Edited by Mat Santamouris

Series Editor M. Santamouris

First published by Earthscan in the UK and USA in 2007

For a full list of publications please contact:

Earthscan
2 Park Square, Milton Park, Abingdon, Oxfordshire OX14 4RN
711 Third Avenue, New York, NY 10017

First issued in paperback 2015

*Earthscan is an imprint of the Taylor & Francis Group, an informa business*

ISBN: 978-1-138-96608-6 (pbk)
ISBN: 978-1-84407-263-7 (hbk)


Typeset by MapSet Ltd, Gateshead, UK
Cover design by Paul Cooper

A catalogue record for this book is available from the British Library

Library of Congress Cataloging-in-Publication Data

Advances in passive cooling / edited by Mat Santamouris
    p. cm.
  ISBN-13: 978-1-84407-263-7 (hardback)
  ISBN-10: 1-84407-263-0 (hardback)
  1. Solar air conditioning–Passive systems. I. Santamouris, M. (Matheos), 1956–
TH7687.9.A38 2007
  697.9'3–dc22

                                                        2007004085

# Contents

# List of Figures, Tables and Boxes

## FIGURES

# TABLES

# BOXES

# List of Contributors

**Hashem Akbari** is a group leader, staff scientist and principal investigator in the Environmental Energy Technologies Division at Lawrence Berkeley National Laboratory. He obtained his PhD in engineering from the University of California, Berkeley. He is the leader of the Heat Island Group and has been instrumental in the organization and initiation of the group at LBNL, funded by the US Department of Energy, Environmental Protection Agency and several other sources. He has led LBNL's efforts to investigate the energy conservation potential and environment impacts of increased tree planting and modifications of surface albedo. His research has identified the attributes of these energy efficient strategies to mitigate the heat island effect. In addition, he has performed research in energy use and conservation in buildings; advanced energy technologies; utility energy forecasting; advanced utility-customer communication, computation and control systems; energy-efficient environment; air pollution control; and environmental simulation and modelling. Dr Akbari is the author of more than 200 articles and is co-author of four books. Finally, he is an active member of ASME, ASTN, Cool Roof Rating Council and ASHRAE, and serves on several technical committees with each society.

**Evyatar Erell**, PhD, was educated as an architect and town planner, and is now an associate professor at the Desert Architecture and Urban Planning Unit of the Jacob Blaustein Institutes for Desert Research at Ben-Gurion University of the Negev, Israel. He was first employed at Ben-Gurion University in 1986, as an architect in the design of innovative energy-saving buildings. He has since participated in a series of studies on glazing systems and on passive cooling techniques for buildings. Professor Erell's research has also included studies of the urban microclimate, such as an investigation of the effect of buildings on the deposition of dust in a desert city, the development of a computer tool for modelling air temperature in urban street canyons, and a field study of the intra-urban temperature differences in a mid-latitude city. He is co-author of several books, including one on *Roof Cooling Techniques*, and has published numerous scientific papers in journals and conferences. He is a member of several expert committees at the Israel Institute of Standards, and has contributed to drafting national standards for thermal insulation and energy certification of buildings.

**Dr Bengt Hellström** is a researcher at the Division of Energy and Building Design, Lund University, Sweden. He has been engaged in the field of energy building simulation with modelling and development of building energy simulation programs for several years.

**Sebastian Herkel** is a researcher at the Fraunhofer Institute for Solar Energy Systems, Freiburg. Currently he is head of the solar building group. He has worked in the field of energy-efficient buildings, indoor climate and building simulation for 15 years.

**Maria Kolokotroni** is a Reader at the School of Engineering and Design, Brunel University, UK. She studied at the National Technical University of Athens and University College London. She carried out post-doctoral studies in the field of ventilation and low-energy cooling at the University of Westminster, followed by five years with the Indoor Environment Group at the Building Research Establishment, UK. She has taken part in a number of European and international projects on energy and ventilation. She has published extensively in scientific journals and presented results to international conferences related to energy use in buildings, indoor environmental quality, climatic design and ventilation.

**Tilmann E. Kuhn** is a researcher at the Fraunhofer Institute for Solar Energy Systems, Freiburg. Currently he is head of the group for solar facades and head of the laboratory for thermal and optical testing. He is a member of the European standardization body CEN in the group CEN/TC33/WG3/TG5, which is dealing with thermal and visual comfort.

**Professor Fergus Nicol** is best known for his work in the science of human thermal comfort, principally in 'adaptive' approaches to thermal comfort. He has run a number of major projects funded by UK and European funding agencies including the EU project Smart Controls and Thermal Comfort (SCATS). He works at Oxford Brookes and London Metropolitan Universities. At both, Fergus was responsible for developing multi-disciplinary Masters courses in energy-efficient and sustainable buildings. Fergus was recently awarded professorships by both universities and is deputy director of the Low Energy Architecture Research Unit (LEARN). He is the author of numerous journal articles and other publications. Fergus is the current convenor of the Network for Comfort and Energy Use in Buildings and organized the conference of the same name that took place in April 2006. He is involved in a number of ongoing projects in thermal comfort and human behaviour in buildings, funded by the Engineering and Physical Sciences Research Council (EPSRC) and the European Union. He has also been closely involved in developing CIBSE advice on comfort and in developing European standards for thermal comfort.

**Dr Peter Nitz** has been a researcher at the Fraunhofer Institute for Solar Energy Systems, Freiburg, since 1994. After working on switchable glazing for several years, his current focus is optical design, simulation and characterization of diffracting and refracting surface structures for daylighting applications, solar control and solar concentration.

**Dr Jens Pfafferott** is an associate researcher at the Fraunhofer Institute for Solar Energy Systems in Freiburg, Germany. His main research direction is the integration of renewable energy and passive cooling concepts in new and retrofitted non-residential buildings. He is a lecturer at the University of Applied Science, Biberach, and the University of Koblenz-Landau and member of the German Guideline Committee VDI 6018 'Thermal Comfort'.

**Susan Roaf** was born in Malaysia, and educated in Australia and England, she has degrees from the University of Manchester, the Architectural Association and Oxford Brookes University. She completed her Doctorate on the Windcatchers of Yazd, and spent ten years in the Middle East, studying ancient technologies, living with nomads and on archaeological excavations. Her first book was on the *Ice-houses of Britain*, followed by books on *Energy Efficient Buildings, Ecohouse Design, Benchmarks for Sustainable Buildings* and *Adapting Buildings and Cities for Climate Change*. She is currently a low carbon design consultant with the UK Carbon Trust and engaged in research and teaching on a wide range of subjects related to renewable energy and passive buildings and communities. She is also a visiting professor at Arizona State University and the Open University, a practising architect and an Oxford city councillor.

**Dr Burkhard Sanner** has specialized in the field of geothermal energy since 1985. He is President of the European Geothermal Energy Council, a member of the Board of Directors of the International Geothermal Association (IGA) and is Vice Chairman of the European Branch Forum of IGA.

**Mat Santamouris** is an associate professor of energy physics at the University of Athens. He is the editor in chief of the *Journal of Advances in Building Energy Research*, associate editor of the *Solar Energy Journal* and member of the editorial boards of the *International Journal of Solar Energy, Journal of Energy and Buildings, Journal of Sustainable Energy*, and of the *Journal of Ventilation*. He is editor of the series of book on Buildings, Energy and Solar Technologies (BEST) published by Earthscan. He has published nine international books on topics related to solar energy and energy conservation in buildings. Professor Santamouris has coordinated many international research programmes and is author of almost 135 scientific papers published in international scientific journals. He is visiting professor at the Metropolitan University of London.

**Professor Karsten Voss** studied mechanical engineering at the Technical University Karlsruhe, Germany. He achieved his PhD at the faculty of architecture of the Technical University of Lausanne, Switzerland, in the field of advanced low-energy housing in connection with advanced energy supply technologies such as solar and hydrogen systems. After 3 years of employment at an energy consulting company he was with the Fraunhofer Institute for Solar Energy Systems, Freiburg, for 12 years until being appointed in 2003 as professor for building physics and technical building service of the University of Wuppertal, School of Architecture. He was national expert in working groups of the International Energy Agency (IEA) and is scientific coordinator of large German programmes on solar optimized buildings.

**Dr Simone Walker-Hertkorn** is head of Systherma GmbH, Starzach-Felldorf, Germany. She is a consulting engineer for and a planner of geothermal energy systems. She is a member of the steering committee of the German Association for Heat Pumps and is a member of the German standardization body DIN in the group 'Bore-hole Heat Exchangers'.

**Dr Maria Wall** is head of the Division of Energy and Building Design, Lund University, Sweden. She has been an architect and researcher in energy-efficient buildings for more than 20 years.

# Preface: Why Passive Cooling?

## *Mat Santamouris*

The use of air conditioning in the buildings sector is increasing rapidly. Almost 46 per cent of houses in the Organisation for Economic Co-operation and Development (OECD) countries have air conditioning, and this level has been rising by 7 per cent each year. In addition, energy consumption for residential cooling in the OECD countries grew by close to 13 per cent between 1990 and 2000, and accounted for 6.4 per cent of the total electricity requirements of these countries in 2000 (IEA, 2003). In Japan, as mentioned by Waide (2006), the use of air conditioning in the service sector is estimated to be close to 100 per cent, compared with 63 per cent in the US and 27 per cent in Europe.

Intensive use of air conditioning is the result of many processes, in particular:

- adoption of a universal style of buildings that does not consider climatic issues and results in increasing energy demands during the summer period;
- increase of ambient temperature, particularly in the urban environment, owing to the heat island phenomenon, which exacerbates cooling demand in buildings;
- changes in comfort culture, consumer behaviour and expectations;
- improvement of living standards and increased affluence of consumers; and
- increase in buildings' internal loads.

## UNIVERSAL BUILDING STYLES

Adopting a universal style of buildings may have a very important impact on the cooling demand of buildings. Modern architectural styles that are poorly adapted to local climatic conditions do not allow efficient solar and thermal control, while the potential for efficient natural ventilation is seriously limited.

It is characteristic that universal-style glazed office buildings in Greece have a cooling consumption close to 200 kilowatt hours per square metre per

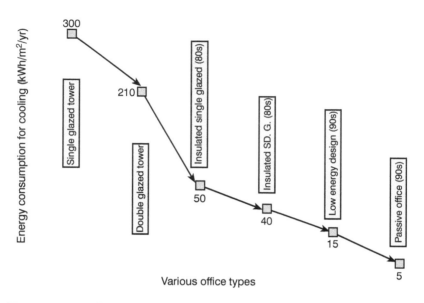

**Figure 0.1** *Cooling energy requirements of various office types in Greece*

year (kWh/m$^2$/yr), while conventional well-insulated buildings have a consumption close to 40kWh/m$^2$/yr and optimized bioclimatic offices have a consumption of about 5kWh/m$^2$/yr (see Figure 0.1).

It has to be pointed out that comfort surveys carried out in all types of offices have shown that, despite high energy consumption, comfort conditions in glazed buildings are much worse than in conventional and bioclimatic buildings. In addition, the concentration of indoor air pollutants has been found to be much higher, primarily because of inappropriate ventilation rates.

## CLIMATIC CHANGE AND THE HEAT ISLAND EFFECT

The heat balance in densely built urban environments is positive; thus, the air temperature of cities is higher than in the surrounding countryside. This phenomenon is known as the 'heat island' effect and has an adverse impact on the energy consumption of buildings for cooling. The heat island effect is the most documented phenomenon of climate change.

In Europe, important research has been carried out to document the amplitude, as well as the impact, of the heat island effect (Santamouris, 2007a). Studies have been performed almost everywhere in Europe and it has been documented that the intensity of a heat island may exceed 10° Celsius (C). Figure 0.2 shows the measured intensity of heat islands in selected cities all around Europe (Santamouris, 2007a).

Similar trends have been recorded in many cities in the rest of the world. Murakami (2006) has compiled data on the variation of the mean annual

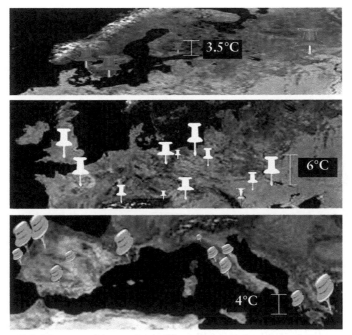

**Figure 0.2** *Measured intensity of heat islands in selected European cities*

temperature for Tokyo, New York and Paris, and found a very important increasing trend (see Figure 0.3). In parallel, he reported a very high increase in the number of days in which the daily maximum temperature exceeds 30° C (tropical day), and the number of days in which the daily minimum temperature at night exceeds 25°C (tropical night) (see Figure 0.4).

The heat island has a tremendous impact on air-conditioning demand, as well as on peak power supply. As reported by Murakami (2006), an increase by 1°C of the ambient temperature in Tokyo increases peak electricity demand by 1.8 gigawatts (GW), which is equivalent to two medium-sized nuclear plants, presenting a cost close to US$2.5 billion. Other studies on the Tokyo area reported by Ojima (1990/1991) conclude that during the period of 1965 to 2000, the cooling load of existing buildings increased by up to 50 per cent, on average, because of the heat island phenomenon.

Hassid et al (2000) and Santamouris et al (2001), using measured data on the heat island in Athens, Greece, calculated the spatial distribution of the cooling load of buildings in the city. As reported, the cooling load, as well as the peak electricity demand at the centre of Athens, is about double the figure for the surrounding region. It has been found that the difference between cooling-energy consumption in the larger municipality of western Athens (Aigaleo) and cooling-energy consumption for the reference station is equal to 180 gigawatt hours (GWh per year).

Using measured data from almost 80 surface stations on the heat island in London, Watkins et al (2002) found that the energy demand for cooling always

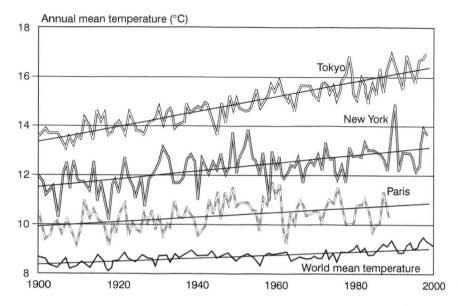

*Source:* Murakami (2006)

**Figure 0.3** *Variation in mean annual temperature for Tokyo,
New York and Paris*

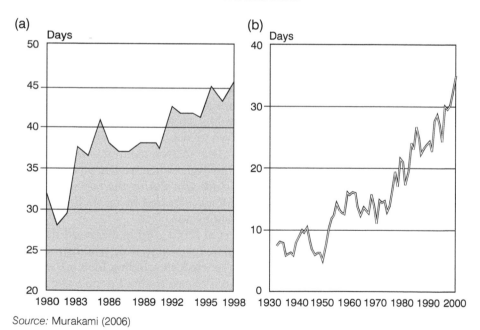

*Source:* Murakami (2006)

**Figure 0.4** *(a) Number of days in which daily maximum temperature
exceeds 30°C in central Tokyo; (b) number of days in which
daily minimum temperature at night exceeds 25°C in central Tokyo*

exceeds the need for heat, wherever the building is located. Watkins et al reported that because the urban cooling load is up to 25 per cent higher over the year, the annual heating load has been reduced by 22 per cent.

In the US, particularly in Los Angeles, it has been found that the electricity demand increases by almost 540 megawatts (MW) per 1°C increase in ambient temperature (Akbari et al, 1992). A peak temperature increase of 2.8°C since 1940 is reported for Los Angeles, resulting in an additional 1.5GW of electricity demand due to the heat island effect. In addition, it has been calculated that summer electricity costs for the US due to the heat island alone could be as much as US$1 million per hour, or over US$1 billion per year. It is estimated that 3 to 8 per cent of the current urban electricity demand is used to compensate for the heat island effect alone.

## THE IMPACT OF HEAT WAVES

Recent heat waves accelerated the demand for cooling equipment in buildings. Sales of room air conditioners have typically increased by up to 50 per cent following the European summer heat waves of 2003 (Waide, 2006). In parallel, heat waves cause important energy impacts. As reported by Meier (2006), during the heat wave of 2003 in France, there was reduced electricity supply, overheating of power plant control rooms and overheating in nuclear plants, while thermal pollution limits were exceeded in rivers. This resulted in an increase of electricity demand by 10 per cent compared to 2002; the spot market price also increased to 1000 Euros per megawatt hour (MWh) on 11 August 2003, while all interruptible contracts were terminated.

In parallel, as reported by Meier (2006), the 2003 heat wave in Italy resulted in decreased energy imports, less hydropower because of the drought, and increased line losses. High demand from air-conditioning and refrigeration equipment was registered, and the electricity shortfall exceeded 5GW. This has resulted in widespread blackouts around the country.

It is important to underline that high ambient temperatures and heat waves cause dramatic problems, especially for vulnerable people living in overheated households. DESMIE (2005) has estimated that 22,080 excess deaths occurred in England and Wales, France, Italy and Portugal during and immediately after the summer heat waves of 2003. In addition, about 7500 excess deaths were registered in Spain, and 1400 to 2200 in The Netherlands. In tandem, approximately 1250 heat-related deaths occurred in Belgium during the summer of 2003, almost 975 excess deaths in Switzerland and 1410 in Baden-Württemberg, Germany. Studies in Europe (Michelozzi et al, 1999, 2005) show that the greatest excess in mortality was registered in those populations with low socio-economic status, living in buildings with improper heat protection and ventilation.

In the US, Klinenberg (2002) reported that during the July 1995 heat wave in Chicago, almost 485 to 740 Chicagoans were killed. Most of the victims lived in single room occupancy (SRO) hotels, while only about half of the

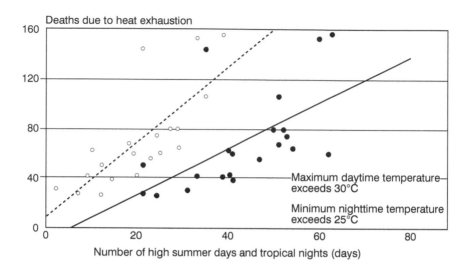

*Source:* Murakami (2006)

**Figure 0.5** *Relationship between deaths due to heat exhaustion and daytime/night-time temperatures in Tokyo*

residents had fans, and many lived in rooms with sealed windows that they could not open. As reported, some of the elderly victims had air conditioners in their homes; but many did not use them because of their more pressing fear that they would not be able to pay their utility bills, which would result in complete loss of their power. Even in Toronto, Canada, during 2005, at least six residents living in poor rooming houses died during a heat wave (Crowe, 2005).

Murakami (2006) has reported a very clear correlation between high day- and night-time ambient temperatures and mortality (see Figure 0.5). It is obvious that increased ambient temperatures have a serious human penalty.

Passive cooling techniques may help to efficiently amortize the need for air conditioning during heat waves. The city of Philadelphia in the US runs a Cool Homes Program for elderly low-income residents. The programme provides passive cooling measures such as a whole house fan, interior air sealing and an elastomeric roof coating to decrease roof temperature. It is reported that the employed measures decrease the solar gains by 80 per cent and reduce the indoor temperatures by 2.5°C. The estimated energy offer was equivalent to the energy delivered by a conventional air conditioner of 8000 British thermal units per hour (btu/h), running for four hours per day (Santamouris, 2007b).

## CHANGE OF COMFORT CULTURE, CONSUMER BEHAVIOUR AND EXPECTATIONS

Consumer behaviour has a very important impact on the use of air conditioning. Comfort is not only a function of indoor temperature, but also of clothing preferences. Parker (2004) and Waide (2006) (see Figure 0.6) provide measured data of the energy consumption of a building in Florida as a function of the ambient temperature. As shown, there is almost no dead band between the deactivation of the heating system and the activation of the air-conditioning system, which means that consumers, regardless of climatic conditions, are wearing the same clothes. This shows a very clear change of consumers' behaviour since thermal comfort could be achieved by wearing appropriate clothing, rather than by using mechanical cooling systems.

Such a development has to be seriously considered by the scientific energy efficiency community. It should be underlined that during the summer of 2005, Japanese Prime Minister Koizumi proposed a very intensive campaign to encourage Japanese office workers to take off their ties and raise set-point temperatures in order to save air-conditioning loads.

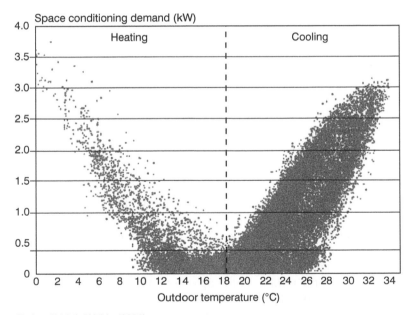

*Source:* Parker (2004); Waide (2006)

**Figure 0.6** *Electrical energy for heating or cooling as a function of external ambient temperature for a sample of Floridian households*

**Table 0.1** *Average number of people per household in selected European countries (1980 to 2003)*

| | 1980 | 1985 | 1990 | 1995 | 2000 | 2003 | Change 1980–2000 |
|---|---|---|---|---|---|---|---|
| Austria | 2.8 | 2.7 | 2.6 | 2.5 | 2.4 | 2.4 | –0.4 |
| Belgium | 2.7 | na | na | na | 2.4 | 2.4 | –0.3 |
| Cyprus | 3.5 | na | 3.2 | 3.2 | 3.1 | 3.0 | –0.4 |
| Czech Republic | 2.7 | na | 2.6 | na | 2.4 | na | –0.3 |
| Denmark | 2.5 | 2.4 | 2.3 | 2.2 | 2.2 | 2.2 | –0.3 |
| Estonia | na | na | na | 2.4 | 2.4 | 2.4 | na |
| Finland | 2.6 | 2.6 | 2.4 | 2.3 | 2.2 | 2.2 | –0.4 |
| France | 2.7 | 2.7 | 2.6 | 2.5 | 2.4 | na | –0.3 |
| Germany | 2.5 | 2.3 | 2.3 | 2.2 | 2.2 | 2.1 | –0.3 |
| Greece | 3.1 | na | 3.0 | na | 2.8 | na | –0.3 |
| Hungary | 2.8 | na | 2.6 | na | 2.6 | na | –0.2 |
| Ireland | 3.7 | 2.5 | na | 3.3 | 3.0 | 2.9 | –0.7 |
| Italy | 3.0 | 3.0 | 2.8 | 2.6 | 2.6 | 2.6 | –0.4 |
| Latvia | na | na | na | na | 2.5 | 2.5 | na |
| Lithuania | na | na | na | 2.8 | 2.6 | 2.6 | na |
| Luxembourg | 2.8 | na | 2.6 | na | 2.5 | na | –0.3 |
| Malta | na | 3.2 | na | 3.1 | 3.0 | 3.0 | na |
| Netherlands | 2.8 | 2.5 | 2.4 | 2.4 | 2.3 | 2.3 | –0.5 |
| Poland | 3.1 | 3.3 | 3.1 | 3.1 | na | 2.8 | na |
| Portugal | 3.3 | 3.3 | 3.1 | na | 2.8 | na | –0.5 |
| Slovak Republic | 3.0 | na | 2.9 | na | 2.6 | na | –0.4 |
| Slovenia | 3.2 | na | 3.0 | na | 2.8 | na | –0.4 |
| Spain | 3.5 | 3.5 | 3.4 | 3.2 | 3.1 | 2.9 | –0.4 |
| Sweden | 2.3 | 2.2 | 2.1 | 2.0 | 2.0 | 1.9 | –0.3 |
| United Kingdom | 2.7 | 2.6 | 2.5 | 2.4 | 2.4 | na | –0.3 |

*Note:* na = not available.
*Source:* National Board of Housing (2004)

## INCREASE IN LIVING STANDARDS

Europe's substantial economic growth, particularly in Southern Europe, has resulted in a considerable increase in the size of dwellings per family. In parallel, the occupied space per person has increased as well. Table 0.1 provides figures on the numbers of people per household in selected European countries for the period of 1980 to 2003.

An increase in the occupied space per family has an important impact on the energy consumption of buildings since more space per person has to be cooled. In addition, the mean size of dwellings varies considerably between various countries as a function of national economic standards (see Figure 0.7) (United Nations Economic Commission for Europe, 2005). Given that the economic distance between Southern and Northern European countries is

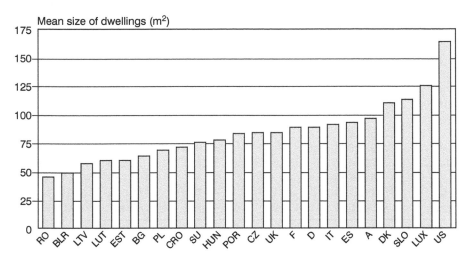

*Source:* United Nations Economic Commission for Europe (2005)

**Figure 0.7** *Mean size of dwellings per country*

considerable, there is a high potential for further social improvements. Thus, it is expected that in most of the Southern European countries, cooling consumption in the building sector will continue to increase, at least for social and economic reasons.

## EXPANSION OF AIR CONDITIONING

The air-conditioning market is expanding continuously. Based on the data given by JARN and JRAIA (2002), the total annual number of air-conditioning sales in Japan in 1998 comprised close to 35,188,000 units, while in 2000 they increased to 41,874,000 units, in 2002 to 44,614,000 units, and the predicted sales for 2006 were 52,287,000 units. In Europe, an increase close to 22 per cent on air-conditioning sales was observed between 2002 and 2006, while the corresponding increase of sales was 39.2 per cent for Asia, excluding Japan; 23.2 per cent for Oceania; 13.6 per cent for Africa; 13.3 per cent for South America; and 10.5 per cent for the Middle East. As expected, most of the cooling systems in Europe are installed in southern countries. This represents almost 74 per cent of the total number of systems (see Table 0.2).

Because of the rapid expansion of air conditioning in Europe, corresponding energy consumption has increased substantially. During 1990, the energy consumption for cooling in European Union (EU) countries was close to 1900GWh, and it is expected to exceed 44,430GWh in 2020 (Adnot, 1999). Corresponding carbon dioxide emissions have also increased from 516 kilotonnes (kt) in 1990 to 18.1 million metric tonnes (Mt) in 2020 (see Table 0.3).

**Table 0.2** *Different room air-conditioning systems in Europe (1996)*

|          | Split     | Multi-split | Windows   | Single duct | Total     |
|----------|-----------|-------------|-----------|-------------|-----------|
| Austria  | 33,400    | 21,300      | 16,600    | 7700        | 79,000    |
| France   | 752,000   | 183,850     | 106,500   | 216,750     | 1,259,100 |
| Germany  | 198,600   | 59,600      | 74,500    | 193,400     | 526,100   |
| Greece   | 138,000   | 51,830      | 555,000   |             | 744,830   |
| Italy    | 1,504,697 | 90,177      | 134,860   | 382,006     | 2,111,740 |
| Spain    | 972,000   |             | 245,000   | 152,000     | 1,369,000 |
| Portugal | 267,157   | 30,143      | 17,720    | 7800        | 322,820   |
| UK       | 516,690   |             | 54,867    | 107,755     | 674,412   |
| Others   | 119,160   | 31,100      | 44,700    | 116,040     | 315,660   |
| Total EU | 4,501,534 | 468,000     | 1,249,747 | 1,183,451   | 7,402,662 |

*Source:* Adnot (1999)

# PROBLEMS WITH AIR CONDITIONING

There are considerable problems associated with the use of air conditioning. Apart from the serious increase in the absolute energy consumption of buildings, other drawbacks include:

- an increase in the peak electricity load;
- environmental problems associated with ozone depletion and global warming; and
- indoor air-quality problems.

The use of air conditioners has a serious impact on electricity demand. Air

**Table 0.3** *Carbon dioxide ($CO_2$) emissions in selected European Union countries due to air conditioning*

| Unit: tonnes $CO_2$ | 1990                     | 1996      | 2010       | 2020       |
|---------------------|--------------------------|-----------|------------|------------|
| Austria             | 157                      | 1603      | 15,748     | 31,467     |
| France              | 26,860                   | 87,377    | 285,231    | 468,957    |
| Germany             | 7845                     | 25,615    | 139,241    | 265,983    |
| Greece              | 99,235                   | 959,939   | 2,387,187  | 3,737,087  |
| Italy               | 182,591                  | 2,247,038 | 2,923,568  | 3,623,486  |
| Portugal            | 147,358                  | 358,099   | 1,038,841  | 1,519,546  |
| Spain               | na<br>(around 90,000)    | 1,124,255 | 4,381,826  | 7,130,489  |
| UK                  | 47,710                   | 219,640   | 704,204    | 1,165,583  |
| Other EU            | 4694                     | 15,369    | 83,545     | 159,590    |
| Total EU            | 516,451<br>(606,451)     | 5,038,935 | 11,959,391 | 18,102,187 |

*Note:* na = not available.
*Source:* Adnot (1999)

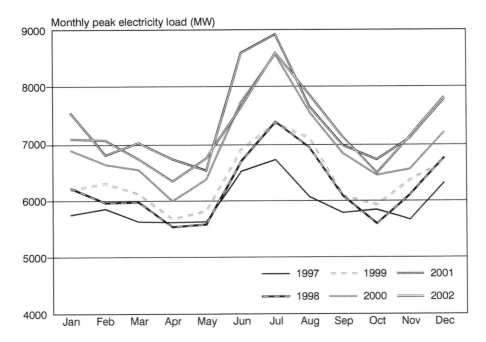

*Source:* DESMIE (2004)

**Figure 0.8** *Monthly peak electricity load in Greece for the period 1997–2002*

conditioning increases peak electricity demand and obliges utilities to build additional power plants to satisfy increasing needs. As reported by Waide (2006): 'AC [air conditioning] is already responsible for over half of peak power demand in many regions of Japan, the United States and Australia, while its contribution to peak is growing everywhere.' Southern European countries face a serious increase in their peak electricity load. Figure 0.8 depicts the recent increase in peak electricity demand in Greece, while Figure 0.9 shows the variation in peak electricity demand in Tokyo.

In addition, because of the limited use of the new plants, the cost of peak electricity increases considerably. According to OFFER and the National Audit Office, the mean European cost of 1kWh off peak is close to 3.9 Euro cents, while the mean cost during peak is 10.2 Euro cents. In parallel, the average cost of a saved kilowatt hour is 2.6 Euro cents.

## THE SOLUTION: PASSIVE COOLING

Energy conservation and passive cooling are the most efficient and cheap alternatives to conventional energy sources. As shown by Holt (2005) (see Figure 0.10), energy efficiency is the lowest-cost energy measure.

Passive cooling systems and techniques have received considerable attention recently, while scientific and technological progress has made

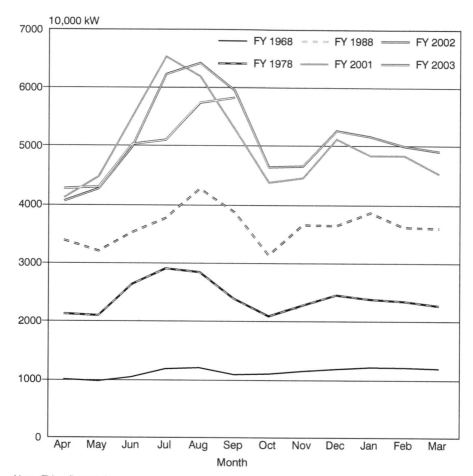

*Note:* FY = financial year
*Source:* Waide (2006)

**Figure 0.9** *Evolution of peak demand for Tepco in Tokyo, Japan*

substantial bounds.

In 1996, together with my colleagues from the University of Athens, we prepared and published our book on the *Passive Cooling of Buildings* (Santamouris and Assimakopoulos, 1996). The book has been extremely well received by an international readership and can be found in most scientific and technical libraries.

Since the publication of the book, considerable progress has been reported. Important developments have been achieved on the topic of solar and heat control, microclimate, thermal comfort, heat dissipation and natural ventilation. Thus, there was a need to update the information given in this first book. Together with some of the more respected scientists and engineers in the field, we prepared this volume. It completes the information given in the original book, while including the most important recent developments.

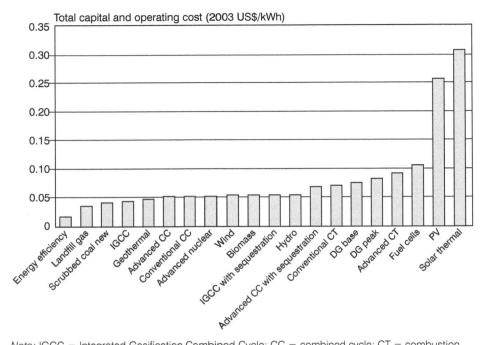

*Note:* IGCC = Integrated Gasification Combined Cycle; CC = combined cycle; CT = combustion turbine; DG = distributed generation; PV = photovoltaics.
*Source:* Holt (2005)

**Figure 0.10** *Projected capital and operating costs of energy service options*

I would like to thank all of the authors and reviewers of this book for the excellent work that they have done. I hope that the international readership will find the book useful.

## REFERENCES

Adnot, J. (ed) (1999) *Energy Efficiency of Room Air-Conditioners (EERAC)*, Study for the Directorate General for Energy (DG XVII) of the Commission of the European Communities, Luxembourg

Akbari, H., Davis, S., Dorsano, S., Huang, J. and Winett, S. (1992) *Cooling our Communities: A Guidebook on Tree Planting and Light Colored Surfacing*, US Environmental Protection Agency, Office of Policy Analysis, Climate Change Division, Washington, DC

Crowe, R. N. (2005) *Anatomy of a Heat Wave: People Die of Exposure to Heat, as Well as to Cold*, http://tdrc.net/CathyCrowe.htm

DESMIE (2005) 'Electricity load in Athens, Greece, 2004', *Eurosurveillance*, vol 10, pp7–9

Hassid S., Santamouris, M., Papanikolaou, N., Linardi, A. and Klitsikas, N. (2000) 'The effect of the heat island on air conditioning load', *Journal of Energy and Buildings*, vol 32, no 2, pp131–141

Holt, S. (2005) 'Australia: A case study', Australian Greenhouse Gas Office, Presented at the UNDP Side Event on Energy Efficiency Standards and Labelling, held during COP-11, UNFCCC, Montreal, Canada, 1 December, www.unfccc.org

IEA (International Energy Agency) (2003) *Cool Appliances*, International Energy Agency, Paris

JARN (Japan Air Conditioning and Refrigeration News) and JRAIA (Japan Refrigeration and Air Conditioning Industry Association) (2002) *Air Conditioning Market*, JARN and JRAIA

Klinenberg, E. (2002) *Heat Wave: A Social Autopsy of Disaster in Chicago*, University of Chicago Press, Chicago

Meier, A. (2006) *Europe's 2003 Heat Wave and Energy Supplies*, International Workshop on Countermeasures to Urban Heat-Islands, Tokyo, 3 August

Michelozzi, P., de'Donato, F., Bisanti, L., Russo, A., Cadum, E., DeMaria, M., D'Ovidio, M., Costa, G. and Perucci, C. A. (2005) 'The impact of the summer 2003 heat waves on mortality in four Italian cities', *European Surveillance*, vol 10, no 7, pp161–165

Michelozzi P., Perucci, C. A., Forastiere, F., Fusco, D., Ancona, C. and Dell'Orco, V. (1999) 'Inequality in health: Socio-economic differentials in mortality in Rome, 1990–1995', *Journal of Epidemiology and Community Health*, November, vol 53, no 11:6, pp87–93

Murakami, S (2006) *Technology and Policy Instruments for Mitigating the Heat-island Effect*, International Workshop on Countermeasures to Urban Heat-Islands, Tokyo, 3 August

National Board of Housing, Building and Planning of Sweden (2004) *Housing Statistics in the European Union*, National Board of Housing, Karlskroner

Ojima, T. (1990/1991) 'Changing Tokyo metropolitan area and its heat island model', *Energy and Buildings*, vol 15–16, pp191–203

Parker, D. (2004) 'How much cooling can be avoided by envelope measures?', Presented at the Conference on Cooling Buildings in a Warming Climate, International Energy Agency, Sophia Antipolis, France, 21–22 June, www.iea.org/Textbase/work/workshopdetail.asp?WS_ID=176

Santamouris, M. (2007a) 'Heat island research in Europe: The state of the art', *Journal of Advances in Building Energy Research (ABER)*, vol 1, pp123–150

Santamouris, M. (2007b) *Energy and Environmental Quality in Low Income Households*, AIVC Technical Note, in press

Santamouris M. and Assimakopoulos, D. N. (eds) (1996) *Passive Cooling of Buildings*, James and James, London

Santamouris, M., Papanikolaou, N., Livada, I., Koronakis, I., Georgakis, C. and Assimakopoulos, D. N. (2001) 'On the impact of urban climate on the energy consumption of buildings', *Solar Energy*, vol 70, no 3, pp201–216

United Nations Economic Commission for Europe (2005) *Housing Statistics*, UNECE, Geneva

Waide, P. (2006) 'Climate and comfort: International energy efficiency policy and air conditioning', in *Proceedings of the Conference on Innovative Equipment and Systems for Comfort and Food Preservation*, Auckland, 16–18 February

Watkins, J., Palmer, J., Kolokotroni, M. and Littlefair, P. (2002) 'The balance of the annual heating and cooling demand within the London urban heat island', *Building Service Engineering Research and Technology*, vol 23, no 4, pp207–213

# List of Acronyms and Abbreviations

| | |
|---|---|
| ACH | air changes per hour |
| AHU | air handling unit |
| ASHRAE | American Society of Heating, Refrigerating and Air-Conditioning Engineers |
| AST | adiabatic saturation temperature |
| ASTM | American Society for Testing of Materials |
| BBCC | building bioclimatic chart |
| BHE | borehole heat exchanger |
| BMS | building management system |
| btu | British thermal units per hour |
| °C | degrees Celsius |
| CaO | lime |
| CaCO$_3$ | calcium carbonate |
| cd | candela |
| CFD | computational fluid dynamic(s) |
| CIBSE | Chartered Institute of Building Services Engineering |
| CIT | California Institute of Technology |
| *clo* | average clothing |
| CLTC | climatic limit of thermal comfort |
| cm | centimetre |
| CO | carbon monoxide |
| COP | coefficient of performance |
| CRRC | Cool Roof Rating Council |
| CSUMM | Colorado State University Mesoscale Model |
| 3D | three dimensional |
| dbh | diameter at breast height |
| DECT | downdraught evaporative cool tower |
| dm$^3$/s | cubic decimetres per second |
| DOE | UK Department of the Environment |
| DTM | dynamic thermal model |
| EAHX | earth-to-air heat exchanger |
| ECSBC | Energy Conservation in Buildings and Community Systems |
| EED | Earth Energy Designer |
| EnREI | Energy Related Environmental Issues in Buildings programme |
| EPA | US Environmental Protection Agency |
| EPSRC | Engineering and Physical Sciences Research Council |
| EU | European Union |

| | |
|---|---|
| °F | degrees Fahrenheit |
| g | gram |
| Gbtu | giga ($10^9$) British thermal units |
| g/m²/yr | grams per square metre per year |
| GtC | metric gigatonnes of carbon |
| GW | gigawatt |
| GWh | gigawatt hours |
| HIR | heat island reduction |
| hPa | hectopascal |
| HVAC | heating, ventilating and air conditioning |
| *H/W* | height/width |
| HYSWIM | hydrogen switchable mirror |
| IAQ | indoor air quality |
| IEA | International Energy Agency |
| J | joule |
| K | Kelvin |
| kgC | kilograms carbon |
| kt | kilo tonne |
| kg | kilogram |
| kJ | kilojoule |
| km | kilometre |
| km² | square kilometres |
| Km²/W | Kelvin square metres per watt |
| kPa | kilopascal |
| kW | kilowatt |
| kWh | kilowatt hours |
| kWh/yr | kilowatt hours per annum |
| kWh/day | kilowatt hours per day |
| kWh/m²/yr | kilowatt hours per square metre per year |
| LA | Los Angeles |
| LBNL | Lawrence Berkeley National Laboratory |
| LDPE | low-density polyethylene |
| LiF | lithium fluoride |
| LLTD | lower limit of thermal discomfort |
| lm/W | lumen per watt |
| LSC | Liants Synthetiques Clairs |
| LULC | land use/land cover |
| μm | micrometre (micron) |
| m | metre |
| m² | square metres |
| m³ | cubic metres |
| m³/h | cubic metres per hour |
| MgO | magnesium oxide |
| MJ | megajoule |
| MJ/kg | megajoules per kilogram |
| mm | millimetre |

| | |
|---|---|
| m/s | metres per second |
| MSA | metropolitan statistical area |
| Mt | million metric tonnes |
| MtC | million metric tonnes of carbon |
| MW | megawatt |
| MWh | megawatt hours |
| NCAR | National Center for Atmospheric Research |
| NIR | near-infrared |
| nm | nanometre ($10^{-9}$ metres) |
| $N_2O$ | nitrous oxide |
| $NO_x$ | nitrogen oxide |
| *NTU* | number of transfer unit(s) |
| $O_3$ | ozone |
| OECD | Organisation for Economic Co-operation and Development |
| Pa | pascal |
| PbS | lead sulphide |
| PbSe | lead selenide |
| PC | polycarbonate |
| PDEC | passive downdraught evaporative cooling |
| PDLC | polymer-dispersed liquid crystal |
| PE | polyethylene |
| PG | performance grade |
| PIR | passive infrared |
| PM10 | particulate matter with a diameter less than or equal to 10 microns |
| PMV | predicted mean vote |
| PNLC | polymer network cholesteric liquid crystal |
| ppb | parts per billion |
| PPO | polyphenylenoxid |
| PSU | Pennsylvania State University |
| PV | photovoltaics |
| *PV* | present value |
| PVC | polyvinyl chloride |
| PVDF | polyvinylidene fluoride |
| Re | Reynolds number |
| ROOFSOL | Roof Solutions for Passive Cooling |
| $SF_6$ | sulphur-hexafluoride |
| SHRP | Strategic Highway Research Program |
| SI | International System of Units |
| $SO_2$ | sulphur dioxide |
| SMUD | Sacramento Municipal Utility District |
| SnO | tin oxide |
| SPD | suspended particle device |
| SRO | single room occupancy |
| TABS | thermally activated building system(s) |
| tC | metric tonnes of carbon |

| | |
|---|---|
| TCF | terrain correction factor |
| $TiO_2$ | titanium dioxide |
| TMY | Typical Meteorological Year |
| TNLC | twisted nematic liquid crystal |
| TRT | thermal response test |
| TWh | terawatt hours ($10^{12}$ watt hours) |
| UAM | Urban Air Shed Model |
| USGS | United States Geological Survey |
| UV | ultraviolet |
| VAV | variable air volume flow rate |
| VOC | volatile organic compound |
| Wh | watt hours |
| $Wh/m^2$ | watt hours per square metre |
| $Wh/m^2/d$ | watt hours per square metre per day |
| WIS | Window Information System |
| ZnO | zinc oxide |
| ZnS | zinc sulphide |
| ZnSe | zinc selenide |

# 1

# Progress on Passive Cooling: Adaptive Thermal Comfort and Passive Architecture

*Fergus Nicol and Susan Roaf*

## INTRODUCTION

This chapter seeks to show how the decision to air condition a building does not only have a fundamental influence on the comfort of the building's occupants, but also profoundly influences its sustainability and even that of the other buildings within the community, city and culture of which it is a part. The chapter also shows that the philosophy of comfort embodied in international standards promotes mechanically cooled environments, as do the assumptions that the standards incorporate and the solutions they impose on building designers. Existing international standards are derived from laboratory experiments and tend towards static solutions to preferred indoor conditions. This approach is contrasted below with the essentially dynamic 'adaptive' approach to designing for comfort, based on observations of human behaviour in real buildings, in a wide range of climates, cultures and economic environments. This approach is shown to be more compatible with reducing energy use in buildings and, in particular, with *passive* approaches to the design of buildings. The chapter ends by suggesting that there is now an environmental imperative to develop a generation of 'modern vernacular' buildings that combine the lessons of traditional buildings with the benefits of appropriate modern technology to provide truly sustainable buildings for the 21st century.

## Our thermal experience

In our daily lives most people will have a pattern of 'normal' thermal experience and behaviour that reflects their own personal circumstances and the culture and climate in which they live. Examples are given by Roaf (1988) and by Nicol (1974).

In her thesis on the wind catchers of Yazd in Iran, Roaf (1988) describes the daily thermal routine of the local adapted population occupying traditional mud brick houses, with their wind catchers set in the one- to three-storied buildings over deep cellars:

> *In contrast to the Western approach to comfort and design in which the individual chooses the climate for a room, the Yazdi living in a traditional house selects a room for its climate. Such choice and movement around a house during a day constitutes a behavioural adjustment that has been an essential adaptive strategy evolved by the people of such hot desert regions, enabling them to inhabit a seemingly hostile environment with some degree of comfort. In the heat of the Yazdi summer, starting out from sleeping on the roof, they will migrate to the courtyard, which provides shade and relative cool in the morning and, thence, to the cellar to rest through the hottest hours of the day. Towards evening they will come out into the relative heat of the courtyard, which may initially be cooled a little by water thrown on to the hot surfaces, and will then grow cooler as night draws near. In late autumn, a different migration occurs, horizontally from the shaded north-facing summer wing, to the south-facing winter rooms of the courtyard, deliberately warmed by the sun. The consequence of this daily movement is that by recording climate in one or two spaces, one does not cover the diurnal range in climate experienced by the occupants of the houses. In Yazd, it has been necessary to follow the occupants around the house, climatically, in order to record and, in turn, understand, the nature of the 'occupied' summer climate in the houses of Yazd.*

In a different context Nicol (1974) quotes a description by M. R. Sharma of the daily routine in laboratories and offices in the Central Building Research Institute in Roorkee, India:

> *The room is full of warm air in the mornings. The windows are opened and the fans run at full speed to churn cool air into the room. Within half or three quarters of an hour the air is cool enough for work to begin. Conditions remain comfortable with fans running throughout the forenoon.*

These two thermal 'diaries' share one characteristic: in each case the building occupants are using their building to make themselves comfortable – in the case of Yazd, by internal vertical and horizontal migration using the different temperatures in different parts of the building; in Roorkee, by using the ceiling fans to offset the increasing temperature as the day wears on. This *interaction* between the buildings and their occupants is crucial to the approach to thermal comfort that is presented in this chapter. It arises from observations of people in their normal environment and separates what might be called the 'traditional' approach to thermal comfort (concerned mainly with physics and physiology) and the 'adaptive' approach that is being increasingly used to inform building design.

What we try to show in this chapter is that the 'traditional' approach to thermal comfort arises from the needs of the heating and cooling industry predicated on their need to provide a predefined set of indoor conditions that are calculated to be optimal, with a minimum of active participation by occupants. Most passive buildings, on the other hand, require their occupants to take an active role in controlling the indoor environment. This makes the adaptive approach, developed from observations of user behaviour, better suited to the needs of designers who use passive cooling and heating in their buildings. The chapter introduces the thinking behind the adaptive model of comfort and shows how the choice of comfort approach has deep and far-reaching implications for the design of buildings and cities in the coming decades in which all buildings will increasingly need to rely on passive methods for their thermal performance.

## THE HEAT BALANCE APPROACH TO DEFINING COMFORT IN BUILDINGS

### Underlying assumptions and methodology

Thermal comfort is famously defined as 'that state of mind which expresses satisfaction with the thermal environment' (ASHRAE Standard 55, 2004). Despite the fact that this is a psychological definition (a state of *mind*), it has usually been modelled in terms of physiology and physics. The aim of those investigating comfort has been to define the thermal environment that a heating, ventilating and air-conditioning (HVAC) system has to provide in order to 'ensure' a comfortable environment. Much of the research on which such standards are based has been done in climate chambers – thermally controlled laboratories. Subjects' reactions are typically monitored over a period of three hours in any one set of thermal conditions. The final response of subjects is measured using the American Society of Heating, Refrigerating and Air-Conditioning Engineers (ASHRAE) scale shown in Table 1.1. In order to generalize the results, the responses have been related to a heat balance model – the assumption being that a pre-condition of thermal *neutrality* (a 0 response on the ASHRAE scale) will be a (steady state) balance between metabolic heat production and overall heat losses to the environment.

**Table 1.1** *American Society of Heating, Refrigerating and Air-Conditioning Engineers (ASHRAE) and Bedford scales of user response*

| ASHRAE descriptor | Numerical equivalent | Bedford descriptor |
|---|---|---|
| Hot | 3 | Much too hot |
| Warm | 2 | Too hot |
| Slightly warm | 1 | Comfortably warm |
| Neutral | 0 | Comfortable |
| Slightly cool | −1 | Comfortably cool |
| Cool | −2 | Too cool |
| Cold | −3 | Much too cool |

## Fanger's predicted mean vote (PMV)

The best known of such heat balance models is the predicted mean vote (PMV) of Fanger (1970), based on experiments in US and Danish universities during the 1960s. Fanger proposed values of skin temperature and sweat secretion for thermal comfort. He obtained data from climate chamber experiments, in which sweat rate and skin temperature were measured for people who were comfortable at various metabolic rates. Optimal sets of environmental conditions for thermal comfort were deduced from the metabolic rate and the clothing insulation.

Fanger extended his work by proposing a method by which the mean thermal sensation of a group of people (on the ASHRAE scale in Table 1.1) could be predicted. His assumption for this was that the sensation experienced was a function of the physiological strain imposed by the environment. This he defined as 'the difference between the internal heat production and the heat loss to the actual environment for a man kept at the comfort values for skin temperature and sweat production at the actual activity level' (Fanger, 1970). He calculated this thermal load for people involved in climate chamber experiments and used it to predict their comfort vote.

The final equations for optimal thermal comfort and for PMV are presented in textbooks (e.g. McIntyre, 1980; Parsons, 2002). Fanger presented the results in the form of diagrams and tables from which optimal comfort conditions can be read, and CEN-ISO 7730 (ISO, 1994) and ASHRAE 55 (2004) include a computer program that can be used to calculate PMV for a group of people in a particular environment given a knowledge of their mean metabolic rate and clothing insulation.

## Strengths of the heat balance model in defining standards for highly serviced buildings

*Creating thermal comfort for man is a primary purpose of the heating and air-conditioning industry, and this has had a radical influence ... on the whole building industry ... thermal comfort is the 'product' which is produced and sold to the customer.* (Fanger, 1970, pp14, 15)

Fanger's approach is justified by his aim to define the 'product' (comfort) that the HVAC industry is 'selling' to the customer. The requirement is to define conditions for comfort in a building serviced by heating or air conditioning. From this flows the assumption that a stable and closely controlled indoor environment is required, as it is assumed that any deviation from optimal conditions will increase the risk that the subjects will become uncomfortable. This assumption is embodied in international and European standards, such as ISO 7730 (ISO, 1994) and CEN 15251 (CEN, 2007). International standards are now dividing buildings into A (best), B or C categories solely on how closely their indoor environment is controlled ($\pm$0.2PMV, 0.5PMV or 0.7PMV, respectively).

## Methodological and philosophical weakness of the heat balance approach in a real situation

One obvious strength of this approach that helps to explain why PMV is so widely used is that it provides a 'number' that engineers, facilities managers or building occupants can set the thermostat to, only changing it (if at all) at the transition between winter and summer. The difference between indoor and outdoor temperatures will consequently be wide when the weather is particularly cold or hot, causing more energy to be used to run the systems than would be the case if the indoor temperature was to a greater extent linked to the outdoor temperature. When the method was being developed it may have been necessary to manually set the thermostat of the system. Today the technology exists to enable the indoor temperatures to track outdoor weather conditions in such a way as to maintain indoor comfort while reducing energy use (Nicol and McCartney, 2001).

For environmental designers, the calculation of PMV poses other problems (Nicol et al, 1995):

- First, we must assume that conditions in the building approach those of the steady state in the climate chamber.
- We must also know the mean clothing insulation of the building occupants and their mean metabolic rate (and there is an additional problem for buildings where a number of activities are taking place in the same space). Both of these variables are difficult to measure and, in the absence of accurate knowledge, the tendency has been to make an assumption (e.g. a clothing insulation of 0.5 in summer and 1.0 *clo* in winter) (CIBSE, 2006).

A further fundamental question has also been raised by Humphreys and Nicol (1996) about the assumption behind the PMV equation itself. The PMV equation is based on the assumption that the thermal sensation of a person (away from neutral) is a function of the thermal load on the body, which is then expressed as a deviation from a state of thermal neutrality. This is different from the comfort criteria in the equation for comfort expressed in terms of heat balance, mean skin temperature and sweat secretion (Fanger, 1970). This leads to an internal contradiction when applied to people not in thermal neutrality who are wearing different levels of clothing so that for a single load, different levels of thermal sensation and physiological response (skin temperature and sweat rate) would occur. In addition, if there is a net thermal load, then theoretically the body will either warm up or cool down and this is incompatible with a model that assumes thermal balance. Experimental investigations by Parsons et al (1997) tended to confirm these doubts about PMV.

## The heat balance method applied to passively cooled buildings

Conditions in passively cooled buildings often cannot be controlled to the same extent as in buildings with mechanical air conditioning. Using natural phenomena such as the wind, the sun and the outdoor temperature, such buildings cannot be closely regulated to a single temperature in the same way as those with fully mechanical systems. If the 'highest' level of indoor environment is defined in terms of the *constancy* of the indoor environment, then it is difficult for passive buildings to achieve this. Does this mean that the passive cooling systems of buildings are second-rate technologies? Many well-known buildings use these technologies and do so very successfully.

# THERMAL COMFORT SURVEYS AND THE ADAPTIVE APPROACH

## Basic field survey methodologies and outputs

The adaptive model of thermal comfort is based on the field survey, an alternative approach to understanding and deriving the conditions that people find comfortable. In a field survey, participants are asked to take part in the survey to assess their thermal sensation on a subjective scale (see Table 1.1). The environmental variables are measured at the same time as the subjective reactions are recorded. Because the aim is to obtain a reaction to typical conditions, there is no attempt to interfere with the environmental conditions, the activity or the modes of dress, and thus the full complexity of the context is included in the responses of the participants. Although the clothing insulation and metabolic rate have often been included in the measurements, they are not necessary to the derivation of comfort conditions and the purpose of recording them has usually been to test the results against theoretical models. The outdoor climate, the availability and use of environmental controls, the building type and function, and the social milieu can all be recorded as potential influences on the participants' response.

The underlying assumption of the field survey is that people are able to act as 'meters' of their environment. In effect, the participant is used as a 'comfort meter', not of temperature alone, but of all the environmental and social variables simultaneously. One aim can be to discover what combination of environmental variables best describes people's subjective responses; another is to predict which environment people in the survey will find most comfortable. To do this, the researcher performs a statistical analysis of comfort vote to show how it depends upon different aspects of the physical environment. A number of such 'comfort indices' have been put forward over the years. Some examples are Bedford's (1936) equivalent temperature, which was developed from a survey of British factory workers, Webb's (1959) Equatorial Comfort (or Singapore) Index from office workers in Singapore and, more recently, Sharma and Ali's (1986) Tropical Summer Index from a survey of Indian office workers. The changing effect of time has generally been ignored in the analysis, although some surveys (e.g. Humphreys, 1979; McCartney and Nicol, 2002; Morgan et al, 2002) have tried to estimate the effect of recent thermal history on the response. Current conditions for comfort may relate to the conditions at some previous time, or to some more complex time series of the people's experience.

One output from these statistical analyses is the so-called 'comfort temperature'. This is a theoretical temperature at which it is calculated a particular individual or group will be most comfortable (or most likely will be comfortable). In a continually changing environment, it is subject to gradual change. It also takes no direct account of differences between individual people. It is, however, a convenient concept when talking of how people respond to buildings, climate and culture and can have a reasonably definite value at any particular time.

The individual response to temperature is influenced by the person's thermal history, and the same is true of groups of people. Hence, it is important to understand not only the short-term thermal history of the person or group, but also their seasonal experience in terms of the drift in their comfort conditions over the year. The upshot is that the temperature which feels comfortable changes with time and can be expressed in terms of an amalgamation over time, such as the running mean of the temperature they experienced (McCartney and Nicol, 2002) (see below).

Individual differences occur between people and have been investigated for relationship to gender, age, race, nationality and so on. While there is some evidence for a difference between men and women in the temperature that they prefer, the difference between individuals within any sub-group is generally far greater than the difference between groups. Group differences are related more to climate and culture than to physiological variation.

## Interpretation of field study output: The adaptive approach

Adaptive thermal comfort is based on the theory of Nicol and Humphreys (1973) that thermal sensation is part of the feedback system by which the human body is kept in thermal equilibrium, and deep-body temperature is controlled within narrow limits. Discomfort (or the perceived danger of discomfort) is the trigger for behavioural responses to the thermal environment. This insight has since been expressed in the form of an adaptive principle: *if a change occurs such as to produce discomfort, people react in ways that tend to restore their comfort* (Humphreys and Nicol, 1998).

People react in one of two principal ways:

1   They change 'themselves' to avoid discomfort in the prevailing conditions: in many cases, this is through clothing adjustments, but also by other means – for instance, changes in posture (Raja and Nicol, 1997) or activity.
2   They change the environment to suit their needs: adjustments such as opening windows (to change temperature and air movement), drawing blinds (to reduce incoming radiation) or changing their location to a more comfortable spot in the room. The use of mechanical systems such as heating, cooling or fans can also be seen as examples of adaptive behaviour (Nicol and Humphreys, 2004).

Changes in air movement are often achieved through the use of controls (but also by moving to a new place), although the physical effect is to change the comfort temperature.

These responses are far from random. People evolve, perhaps over millennia, a locally appropriate type of building, clothing, diet and way of life that enable them to be generally comfortable within their normal environment. These learned patterns of behaviour are an integral part of the local culture and ensure that individuals do not risk losing control over their internal thermal equilibrium despite the fact that at some points in the day or night the thermal environment they experience may be hostile. In such behavioural cycles they can experience conditions that outside the dynamic thermal cycle of their day would be considered uncomfortable, but in the context of their lifestyles are not experienced as such.

Assuming that these responses are, on the whole, successful, the outcome of adaptive behaviour is that subjects report themselves to be comfortable at a temperature which is typical of what they would expect to experience in their normal environment – their 'customary' temperature, within a known behavioural lifestyle. However, such a customary temperature is not static, but can change with, for instance, season and the weather outside, or the location of a person's work.

'Adaptive comfort' has been characterized as applying only to the case of naturally ventilated buildings by ASHRAE (2004) and many others, but Humphreys and Nicol (2002) have suggested that adaptation is a characteristic of human behaviour in all environments. It is patently 'unnatural' for human beings to pass their days in environments with only a single temperature, as there are few if any places where such conditions exist naturally. Hence, it could be argued that the only reason air-conditioned offices are considered tolerable in hot climates (Busch, 1992) is because the building occupants have adapted to these unnatural conditions.

The significant distinction between buildings in this context is the extent to which the internal climate reflects that outdoors. If people are taking action to be comfortable at normally occurring conditions, then in a building whose indoor environment is permanently decoupled from outdoor conditions (e.g. because it is air conditioned), the indoor comfort temperature will also be decoupled. In a building whose indoor conditions are related to those outdoors (e.g. a building that is free running – neither heated nor cooled), the customary temperature (and, consequently, the comfort temperature) will track outdoor conditions. A naturally ventilated building that is heated is decoupled in winter and effectively free running in summer. Comfort temperatures in such buildings will follow different 'modes' at different times.

Humphreys (1981) and de Dear and Brager (2002) performed meta-analyses of numerous field surveys. In both cases, the dependence of indoor comfort temperature upon mean outdoor temperature was a feature of buildings where indoor and outdoor temperatures were linked (free-running buildings in the case of Humphreys and naturally ventilated buildings in de Dear and Brager). The dependence was much less marked (though still there) in buildings where the indoor climate was controlled.

## Using the heat balance approach to predict the results of comfort surveys

Conditions in real buildings are continually changing. Even in centrally air-conditioned buildings, the continual change in heat load from changing population, outdoor air temperatures and the ingress of solar radiation will be reflected in temperature variations indoors unless the mechanical system is exceptionally responsive and controlled. There will also inevitably be variations in the thermal conditions in different parts of a room related to height, room configuration and proximity to openings and external walls. People will also change their clothing as seasons change (Nicol and McCartney, 2001; Morgan et al, 2002) and their metabolic rate as they go about their everyday tasks. On top of this there will be small changes in posture and numerous other minor adjustments. A comfort temperature based on steady-state heat transfer and the controlled conditions of a climate chamber may be able to account for changes in the mean value of the thermal sensation over long periods, but cannot adequately account for the multitude of adaptive adjustments. If, for instance, some of the adjustments are aimed at reducing discomfort, then the steady-state model may predict more discomfort than actually occurs.

In the case of naturally ventilated buildings where occupants are continually interacting with the building (Nicol and Humphreys, 2004), the mismatch between prediction and actual comfort can become important, as has been shown by de Dear and Brager (2002) in Figure 1.1. This figure shows that the difference between the predicted comfort temperature and the actual comfort temperature is negative in cold conditions (outdoor temperature of less than 20°C) and positive in hot conditions. PMV tends to prefer a more

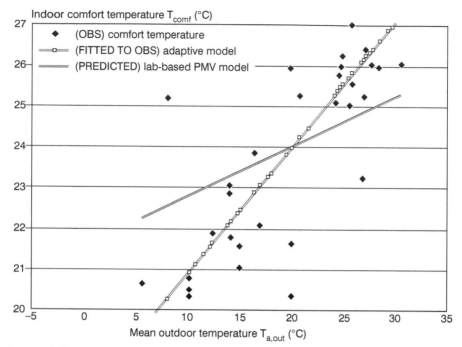

*Source:* de Dear and Brager (2002)

**Figure 1.1** *Mismatch between actual comfort temperature and that predicted using the predicted mean vote (PMV) index among naturally ventilated buildings in the ASHRAE database of comfort surveys*

controlled environment, favouring an indoor temperature that may be needlessly warm in cold conditions and cooler than necessary in hot conditions.

## LESSONS OF THE ADAPTIVE APPROACH

### Comfort is a goal to achieve, not a product to define

Thermal adaptation is essentially dynamic. Comfort is not a 'product' that is provided for building occupants, it is a goal which they achieve provided they are able to exert the necessary control over their environment. The control that they can exert over the environment will partly be decided by the building they occupy and its services. The environmental conditions needed to achieve this comfort goal also change with time and there are undoubtedly limits to the range of indoor climates that any group of people can adapt to, related as much to their thermal experience and their climatic, social, economic and cultural position as to their physiology (Nicol and Roaf, 1996).

This dynamic model for comfort requires a different approach to providing comfort than one that assumes a single temperature is acceptable. Change and movement, typically within the context of well-understood patterns of behaviour, is the essence of the adaptive approach: stasis, the existence of a static relationship between occupant and environment, is only achieved in very specific circumstances (Nicol and Roaf, 2005).

## Buildings need to allow people opportunities for adjustment, in an understandable way, in a variable and varying context

### Occupant control

Adaptation is assisted by providing thermal control, and convenient and effective means of control should be provided so that the occupants can adjust the thermal environment to their own requirements. This 'adaptive opportunity' (Baker and Standeven, 1996) may be provided – for instance, by ceiling fans and openable windows in summertime, or by local temperature controls in winter. A control band of ±2K (Nicol and Humphreys, 2006) (or an equivalent band of air speed) should be sufficient to accommodate the great majority of people. Individual control is more effective than group control.

### Customary thermal environments and comfort

People adapt more readily to thermal environments with which they are familiar. The building should therefore be designed to provide a thermal environment that is within the customary range for the particular type of accommodation, according to climate, season and cultural context. Such 'customary' temperatures (usually expressed in terms of operative temperature, which combines the effects of radiant and air temperature; CIBSE, 2006) can often be found in guides and textbooks, established from the experience of building services professionals, but may need to be modified in the light of climate change.

### Drift of comfort conditions

These customary temperatures are not fixed, but are subject to gradual drift in response to changes in conditions, both outdoors and indoors, and are modified by climate and social custom. A sudden departure from the current customary temperature is likely to provoke discomfort and complaint, while a similar change, occurring gradually over several days, may not be so likely to do so.

### Dress codes

The extent of seasonal variation in indoor temperature that is consistent with comfort depends upon the occupants' ability to wear cool clothing in warm conditions and warm clothing when it is cool. Strict dress codes can therefore affect thermal comfort in offices and any code should incorporate adequate seasonal flexibility and personal choice.

## Temperature drift during the day

Field studies have found that in offices and schools, people adjust their clothing relatively seldom during the working day (Humphreys, 1979), so the temperature during occupied hours in any day should not vary much from the comfort temperature. Temperature drifts within ±1K of the customary temperature would attract little notice. ±2K would be likely to attract attention and could result in mild discomfort among a small proportion of the occupants. In this respect, changes that arise from the building or its services are different from adaptive changes made by the occupants to increase their comfort.

## Temperature drift over several days

Clothing and other adjustments in response to day-to-day changes in weather and season occur quite gradually (Humphreys, 1979; Nicol and Raja, 1996; Morgan et al, 2002), and take a week or so to complete. Therefore, it is desirable that the day-to-day change in mean indoor operative temperature during occupied hours should not normally exceed about 1K, nor should the cumulative change over a week exceed about 3K. These figures apply to sedentary or lightly active people (see also the section on 'Exponentially weighted running mean outdoor temperatures'). If these simple suggestions are followed, people can be comfortable in naturally ventilated buildings in many climates during the whole (or part of the) year, thus reducing the need for the use of fossil fuel-driven air conditioning. These simple rules about acceptable temperature change can be formalized into a mathematical relationship, as is explained in the section on 'Time in the relationship of comfort temperatures to climate'.

# Predicting the most likely customary temperature from the outdoor temperature

In his survey of data from all over the world, Humphreys (1981) found that the relationship between indoor and outdoor temperatures in free-running buildings was strong and linear. Because the subjects were, over time, adapted to their mean indoor temperature, their comfort temperature also followed a linear relationship with outdoor temperature.

The mean indoor and comfort temperature followed the relationships:

$$T_{in} = 0.55T_{out} + 14.1 \qquad \text{standard error} = 1.2K \qquad R = 0.96 \qquad [1]$$
$$T_{comf} = 0.534T_{out} + 11.9 \quad \text{standard error} = 1.0K \qquad R = 0.97 \qquad [2]$$

where $T_{in}$ is the mean indoor temperature, $T_{comf}$ is the comfort temperature and $T_{out}$ is the monthly mean outdoor temperature calculated from the monthly mean maximum $(T_{max})$ and the monthly mean minimum $(T_{min})$ temperatures for the site and the months of the survey using the relationship:

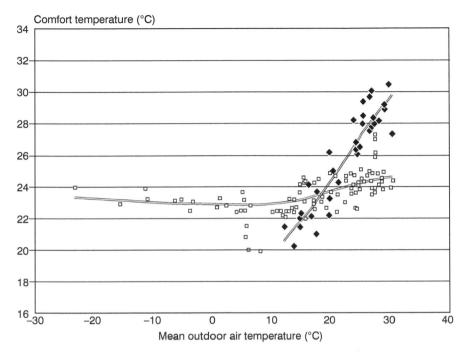

*Source:* data from the ASHRAE database from Humphreys and Nicol (2000)

**Figure 1.2** *Relationship between comfort temperature and mean outdoor temperature for buildings that are heated or cooled (curve, open symbols) or free running (straight line, filled symbols) at the time of the survey*

$$T_{out} = (T_{max} + T_{min})/2 \qquad\qquad [3]$$

A subsequent analysis of the data from the ASHRAE database (de Dear, 1998) has shown that this relationship for free-running buildings has changed little in 25 years (Humphreys and Nicol, 2000).

These are results from the 'average' free-running building. The fact that all buildings or occupants are not identical is reflected in the range (standard deviation of approximately 1K) of comfort temperatures at any one value of outdoor temperature. Some of the buildings measured may have incorporated some passive cooling techniques; but, on the whole, these were typical buildings for their location.

If we assume that during hot spells a well-designed passively cooled building provides indoor conditions which, while cooler than a building without passive cooling, are nonetheless not as cool as would be achieved by a mechanical system, we might expect that in Figure 1.2 the comfort temperature would lie below that of a free-running building, but above that of a mechanically cooled building. We can assume that our passively cooled building will provide acceptable indoor conditions if indoor temperatures lie between the two at any given outdoor temperature.

## Time in the relationship of comfort temperatures to climate

The relationship between indoor comfort and outdoor temperature has usually been expressed in terms of the monthly mean of the outdoor temperature (Humphreys, 1981; ASHRAE, 2004). This is because the monthly mean outdoor temperature is generally available from meteorological records. Important variations of outdoor temperature do, however, occur at much shorter intervals. Adaptive theory suggests that people respond on the basis of their thermal experience, with more recent experience being more important. A running mean of outdoor temperatures, weighted according to their distance in the past, is therefore more appropriate than a monthly mean.

### Exponentially weighted running mean outdoor temperatures

An exponentially weighted running mean of the daily mean outdoor air temperature $t_{rm}$ is an appropriate expression of the outdoor temperature, and is calculated from the series:

$$t_{rm} = (1 - \alpha) \cdot \{t_{od-1} + \alpha \cdot t_{od-2} + d^2 t_{od-3} \cdots\} \qquad [4]$$

where $t_{od-1}$ is the daily mean outdoor temperature for the previous day, $t_{od-2}$ is the daily mean outdoor temperature for the day before, and so on. $\alpha$ is a constant between 0 and 1 that defines the speed at which the running mean responds to the outdoor temperature. The use of an infinite series would be impracticable were not Equation 4 reducible to the form:

$$_n t_{rm} = (1 - \alpha) \cdot t_{od-1} + \alpha \cdot _{n-1} t_{rm} \qquad [5]$$

where $_n t_{rm}$ is the running mean temperature for day $n$ and $_{n-1} t_{rm}$ for the previous day.

So if the running mean has been calculated (or assumed) for one day, then it can be readily calculated for the next day, and so on.

## Defining an adaptive comfort standard for European offices, incorporating time and context

Data applicable to Europe are available from extensive surveys of office workers (Nicol and McCartney 2001, McCartney and Nicol, 2002). A value in the region of 0.8 was found to be suitable for $\alpha$ in the running mean temperature, a value previously found suitable for data from the UK (Nicol and Raja, 1996). This value suggests that the characteristic time that subjects take to adjust fully to a change in the outdoor temperature is about a week.

Bands within which comfortable conditions have been found to lie are shown in relation to the running mean outdoor temperature in Figure 1.3, both for the free-running mode of operation and for the heated or cooled mode. Comfortable conditions for mixed-mode operation lie within and between these bands. The bands indicate the indoor temperatures within which people readily adapt in relation to the outdoor temperature. A thermally successful building is one whose indoor temperatures change only gradually in response to changes in the outdoor temperature (see above) and rarely stray beyond these bands. The limits of the comfort bands derived from new European data (Nicol and McCartney, 2001) are given by the equations.

For free-running operation:

upper margin:   $t_{comf} = 0.33t_{rm} + 20.8$ [6]

lower margin:   $t_{comf} = 0.33t_{rm} + 16.8$ [7]

For heated or cooled operation:

upper margin:   $t_{comf} = 0.09t_{rm} + 24.6$ [8]

lower margin:   $t_{comf} = 0.09t_{rm} + 20.6$ [9]

*Example: naturally ventilated office in summer.* For the assessment of the adequacy of the building in summer, the upper margin of the free-running zone is examined. This line may be used to indicate the probable upper limit of the comfort temperature (Nicol and Humphreys, 2006). In the UK, the running mean outdoor temperature rarely exceeds 20°C. At this temperature the upper limit of the band is 27.4°C. So the temperature during occupied hours should

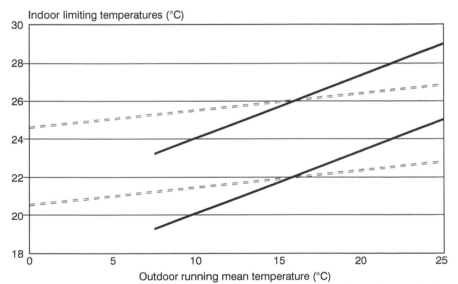

*Note:* Separate bands are shown for buildings in the free-running and the heated and cooled modes from field surveys in Europe.
*Source:* McCartney and Nicol (2002); CIBSE (2006)

**Figure 1.3** *Bands of comfort temperatures in offices related to the running mean of the outdoor temperature*

preferably not exceed this value. Operative temperatures drifting a little above this value might attract little notice; but temperatures 2K or more above it would be likely to attract increasing complaint. On a more normal summer day, the running mean outdoor temperature would be about 15°C. This would give a value of 25.8°C, and the indoor temperature should preferably not be higher than this. Again, temperatures a little above this value would attract little notice, while temperatures more than about 2K above the line would be likely to attract increasing complaint.

Expected percentages of occupants experiencing discomfort have sometimes been estimated (ASHRAE, 2004); but the percentage varies from building to building, depending upon where its comfort temperature lies within the band and upon the adaptive opportunity it affords (Baker and Standeven, 1996). Temperatures below these values would be found satisfactory provided the advice on within-day and day-on-day temperature changes (see above) is observed.

## PROBLEMS OF THE ADAPTIVE APPROACH

### Adaptive comfort and the challenge of climate change

The changing climate provides designers with the challenge of keeping temperatures within safe ranges. We need to be able to respond adequately to climate change. An important factor in achieving the goal of comfort is that local populations understand the thermal conditions that are a risk to their health or survival. An important question is whether a population can adapt to conditions beyond their normal experience and to what extent existing buildings will be adequate when conditions change. The need for adequate occupant control is particularly crucial in a changing climate; yet, most air-conditioning systems are currently not designed to deal with change. Given the opportunity, many people are capable of adapting to changes in their own local environment. Problems of comfort may become problems of health if changes become more extreme, and people are without the means or the understanding needed to combat these adverse effects. This was tragically demonstrated in 2003 when 35,000 deaths occurred in Europe due to unexpected hot weather.

At the same time, lack of adaptability of centrally controlled systems can lead to complete building and system failure in extreme temperatures. At the recently refurbished Her Majesty's Treasury building in London, the entire staff were sent home at lunchtime on 8 August 2003 during the heat wave because the building was simply too hot to occupy. It is unclear if this was the fault of the design of the refurbishment, during which an inner court was completely covered over, or whether the air conditioning was not properly commissioned. The centrally controlled system provided no opportunity for local system overrides or even to simply open a window, and this was in a building that had previously been naturally ventilated successfully for nearly a century (Roaf et al, 2005, p90).

## Comfort standards for passively cooled or heated buildings

If a passively cooled building were to deliver very steady indoor conditions, the occupants' response would probably be little different from their response in a mechanically cooled building, providing that the constant temperature that it supplied was within the local cultural norms for indoor temperature. The mean indoor temperature would be decided not by the building engineer, but by the physics of the building form, the amount and location of its mass, the size of its windows, the extent of their shading in summer, the insulation provided by its walls, floor and roof, and, of course, the effects of any passive technology.

In reality, passively cooled buildings are less completely controlled. The mean temperature in the building will change from season to season, and unless it is a zero-energy building, it will often be heated for at least part of the year. In a well-designed passively cooled (or heated) building, the building itself, as a result of its form and fabric, will moderate both the diurnal swings and seasonal changes in indoor temperature. A passive building may not behave like a 'normal' free-running building, but neither will it produce the controlled environment typical of full air conditioning. As suggested above, the resulting 'comfort zone' for such buildings will lie between that of heated or cooled buildings and that of free-running buildings (see Figure 1.3), and such a comfort zone is incorporated in Dutch guidelines (van der Linden et al, 2006).

## Limits of simulation in a shifting environment: Predicting the future performance of buildings

There are now a number of dynamic physical/physiological simulation models of the human body – for example, those developed by Fiala (Fiala et al, 1999). These are beginning to address the problem of providing a psychophysical simulation of thermal comfort responses. There are also increasingly sophisticated dynamic thermal simulation packages for buildings. A weakness of both types of simulation, from the perspective of adaptive comfort, is that they simplify or eliminate the parts of the model dealing with the building/occupant interaction: physiological models oversimplify the building, while the building simulations, at best, use steady-state models for human response. Behaviour is also treated in a very simplistic way. In order to realistically include the actual dynamic relationship between people and buildings, consideration must be given to the integration of these two very different types of dynamic model in order to build a single dynamic model of the complete occupant–building system. This integration would need to be informed by appropriate experimental data.

The ways in which people use building controls is stochastic rather than precise. Thus, there is not a precise temperature at which people switch on a fan or open a window; but there is an increasing likelihood that they will do so as the temperature rises. Using these relationships to inform simulations assumes that the aim of the simulation is to predict a distribution of indoor temperatures, rather than a precise value. More discussion of the approach and problems associated with it are presented in Nicol (2003).

## BUILDING DESIGN, AIR CONDITIONING AND THE ADAPTIVE APPROACH

### The fundamental problems of the mechanical approach

Air-conditioned buildings use cooled and, possibly, dehumidified air, circulated around buildings in systems of ducts, using mainly electrical power. This means that temperature, air movement and relative humidity are the key variables used to define the comfort 'standards' because they are relatively easily controlled by the air-conditioning system (Shove, 2003). Furthermore, because of the way in which comfort standards have been formulated, it is assumed that an accurately controlled environment is needed. Many things follow from this assumption, and the fuller implications are discussed below.

Air conditioning was originally used in commercial buildings, where it was considered to increase satisfaction and productivity. Mechanical systems of environmental control in buildings became pervasive in the US as the technologies spread from workplace to the home. As Cooper (1998) pointed out:

> *The plug-in (air-conditioning) appliance privileged the consumer; but its complete divorce from the building compromised its performance. The standardized installation of the tract development [in the US] provided affordability and performance; but a building that was dependant upon its mechanical services and alarmingly inefficient in energy consumption.*

Marsha Ackerman (2002) builds on this, writing:

> *The counterpart of technologically enabled control is dependency, and the history of air conditioning provides it in full measure. Air conditioning has made it possible to erect structures that must be evacuated when the power fails, to make buildings in which people get sick. It gulps electricity; roars, wheezes and whines; makes urban heat islands even hotter with the exhaust of a million air-conditioned cars and thousands of sealed buildings.*

Yet, as Ackerman (2002) wrote: 'For better and for worse, our world tomorrow will be air conditioned', not least because the 'standards' demand it even in climates where the need for air conditioning is minimal (Haves et al, 1998). The trend in the use of air conditioning to cool buildings is undermining the use of traditional building forms, materials and elements of regionally appropriate passive buildings. It has 'liberated' designers to create buildings that have increasingly become more 'fashionably' disconnected from the climate and environment in which they are found.

In *Vers une Architecture*, Le Corbusier, one of the greatest proponents of the Modern Movement made it clear why 'the engineer's aesthetic' is far more successful than architecture: engineering takes advantage of the latest and most innovative building types, technologies and construction systems, based on mass production, standardization and industrialization. Architecture, if it wanted to embrace the 'new spirit', needed to embrace these modern methods of technology and progress away from the static 'safe' traditional architecture (Le Corbusier, 1928). Le Corbusier championed the decoupling of the indoor from the outdoor climate at a time long before the problem of the finite nature of fossil fuel resources became manifest, and when air conditioning seemed to have profound advantages, particularly in regions with more extreme climates. In 1937, Markham (1944) identified the power of air conditioning to enable people, adapted to the colder climates of Europe and the US, to colonize the world.

Unseen, and often unacknowledged, thermal comfort theory, through its philosophy of the need for control and its consequent standards, has played a major role in shaping the buildings, cities and society of the 20th century. Elizabeth Shove (2003) traces the way in which air conditioning has moved from being a luxury for the few to being a necessity for the many, especially in the US. This trend is not restricted to the reliance on air conditioning and Shove cites rising demands for comfort, cleanliness and convenience as key motivators.

The problem that Shove is particularly concerned with is the irreversibility of such changes. Comfort that relies on a high level of services and control also changes the way in which buildings are designed and constructed. Buildings designed for air-conditioning systems are different from those without. The danger is not just that air-conditioned buildings use more energy than necessary, but increasingly that the energy supply will become prohibitively expensive, or will become less secure in its supply as grids are overloaded, particularly in extreme weather. In the latter case, the occupant is left in a building that is extremely uncomfortable or even dangerous when the power grid does fail.

### Air-conditioned buildings use more energy and therefore cost more to run and are more vulnerable

The Building Research Establishment (1991) has estimated that a naturally ventilated office in the UK typically uses half as much energy as an air-conditioned one. Table 1.2 suggests the reason for this. The cooling and circulation of air uses almost half of the total energy cost of an air-conditioned building. The social and environmental costs of energy use are fully expounded elsewhere (e.g. Roaf et al, 2005); but the high proportion of greenhouse gas emissions resulting from highly serviced buildings is one of the primary arguments driving the move back to the use of natural ventilation and passive

**Table 1.2** *Typical proportions of $CO_2$ production
in UK air-conditioned office buildings*

| System | Proportion of $CO_2$ produced (percentage) |
|---|---|
| Fans and pumps | 30 |
| Cooling | 13 |
| Heating and hot water | 32 |
| Lights | 21 |
| Catering | 4 |

*Source:* adapted from the data of Max Fordham, personal communication

heating and cooling of buildings (Santamouris, 2005).

As fossil fuel supply drops or becomes increasingly expensive, buildings relying on grid-supplied electricity will become more vulnerable. Air conditioning requires such high levels of energy to cool and distribute air around buildings that it cannot be run from building-integrated renewable energy systems. Robust passive buildings can be cooled using much less energy and need not fail when the grid does go down.

### Buildings have become 'faster' in their response times and therefore threaten energy supply

In air-conditioned buildings, the indoor climate has apparently been disconnected from the outdoor climate; but the often poor thermal performance of their envelopes, because of their high percentage of glazing and their low thermal capacity, means that they intrinsically respond sharply to changing weather. This is masked by using the speedy reaction of the mechanical systems to maintain a stable temperature within the building. The rapid thermal response time of air-conditioned buildings is a key concern of the growing 'slow buildings' movement. The fast-response building can result in a surge in demand for energy, whether gas or electricity, to cope with the swing in cooling or heating loads, and causes load spikes that can pull down the entire grid. This was the case in July and August 2003 when the heat wave caused grids to go down in many regions in Europe and North America (Roaf et al, 2005).

### Cost drivers and the problems of central control

The increasing reliance on machines to moderate indoor climates has led to the increased first-cost expenditure on equipment in buildings, often paid for by the reduction of spend on the fabric of the building. The heating, ventilating and air-conditioning (HVAC) engineer is typically paid according to the cost of the cooling and heating systems, not the cost of the building itself. Substantial walls and well-designed openable windows, essential to passive design, give way to thin tight envelopes. A conflict exists between the needs of the developer to justify a cheaper first-cost design and those of the eventual occupier who may have to pay the building running costs and deal with uncomfortable staff.

In order to minimize the floor area to plot ratio, deep-plan buildings are obviously an advantage for developers. It has been argued that deep-plan buildings can be efficient in reducing heating and cooling loads due to their volume-to-surface ratios. By air conditioning a building, they could be made as deep plan as possible. However, by eliminating the light wells and courts at the hearts of shallow plan buildings on deep sites, the running costs of the resultant buildings soared as natural ventilation and daylight were replaced by mechanical cooling and artificial lighting.

**Predicting building performance**

Heating and ventilating engineers are not trained to take advantage of the role that mass plays in the indoor climate of a building. Many calculations ignore the ability of mass to store heat beyond 24 hours because they are based on an assumption of a diurnal temperature swing. In order to be able to calculate how a new building will perform, building modellers have made simplified assumptions. The difficulty of predicting occupant behaviour means that it has been seen as a problem, rather than as a possible part of the solution. As a result, options for action by ordinary building occupants to ameliorate their local climate have become increasingly limited and are eventually replaced by the building management systems and facilities manager.

Calculations find it hard to take into account the variable contribution of wind speed and direction, solar energy and outdoor air temperature, so there is a tendency for natural ventilation, let alone the behaviour of occupants, to get written out. The upshot is the assumption that if mechanical systems are to work predictably in the building, the windows must not be openable.

**Architects are deskilled and buildings are disconnected from the seasons**

As a consequence of the move to air-conditioned buildings, which entails the indoor climate of a building becoming more disconnected from the outdoor climate, decreasing importance is given to regional and cultural variations in preferred temperatures. Pre-eminence is given to standards that – because they were formulated in climate chamber experiments which exclude cultural differences – conclude that people around the world have similar comfort needs, an assumption which has been proved wrong in any number of field studies of thermal comfort (Humphreys, 1981; Brager and de Dear, 1998).

A major consequence of the development of the 'international' building style has been that architects have lost much of their skill in optimizing energy use, life-cycle cost and comfort benefits in the building:

- *using thermal storage in mass to optimize the value of 'free energy'* – for instance, using coolth from the night sky or low night-time air temperatures to cool the building the following day; using the heat from the sun, people or machines; or shifting energy use away from times of high load demand;

- *using seasonal climate patterns, of winds and solar access*, to minimize the need for fossil fuel heating and cooling in buildings through climate-related design of the building form and its elements and materials.

This is despite excellent research published over the last half century on the problems that arise from reducing the mass of a building and increasing its glazed area. In particular, several studies from the 1960s by the UK Building Research Establishment clearly saw that the trend towards over-glazed, light-weight buildings was causing significant problems with overheating, discomfort and high energy running costs (Loudon and Keighley, 1964; Loudon and Danter, 1965; Black and Milroy, 1966; Loudon, 1968).

### Cultural and health impacts

Traditional industries related to buildings are often badly affected by the movement towards 'international'-style buildings. The design of such buildings is best done by large firms, so traditional patterns of usage by local culturally adapted populations are replaced by the culturally amorphous universal 'air-conditioned' lifestyle in which the time of day or year play little part. In Spain in December 2005, the afternoon siesta, a wonderful cultural adaptation to the hot Mediterranean afternoons, was officially cancelled for office workers (Drenzer, 2005).

Buildings are becoming increasingly 'sick'. Indoor air quality can be worse in an air-conditioned building than in a comparable naturally ventilated one. Worryingly, researchers are finding that the filters, ducts and plant of air-conditioning systems are often filthy, introducing air that is dirtier than if one simply opened the window, even in cities (Chen and Vine, 1998; Clausen et al, 2002). Ducts can harbour potential killer bacteria, moulds and particulates that are released from the filter. This can be exacerbated by changes in the weather – for instance, on the arrival of warmer and wetter air (Mauderly, 2002). Other hazards such as legionella are also associated with inadequate commissioning and maintenance of plant.

## Buildings and adaptive comfort

The time has come to re-evaluate the 20th-century approach to comfort standards, to identify their weaknesses and build on their strengths in order to enable passive buildings to emerge. It is passive buildings that will enable us to survive the climate change and fossil fuel exigencies of the 21st century. This chapter is an attempt to further the discussion upon which those new standards may be built.

Applying the insights of the adaptive approach to the thermal design of buildings will assist the evolution of a truly 21st-century building paradigm, enabling designers to create buildings that remain comfortable for their occupants in the increasingly extreme weather events ahead and with the decreasing availability of affordable oil and gas to help us solve our comfort problems. Only in passive, adaptive and slow buildings can we reduce fossil fuel use to the extent that a large part of the required energy to run them is supplied by free, renewable on-site energy. So what will such buildings be like? They will probably have:

- *shallow plans* for better daylight and natural ventilation;
- *opening windows* for natural ventilation;
- *adaptive skins* where elements such as shades, awnings, blinds and shutters are designed to maximize the potential to protect the building from wind and sun;
- *high levels of thermal mass* to stabilize internal temperatures in heat waves or cold snaps and to store renewable energy; the inherent response time of a building to external temperature fluctuations determines the ability of that building to ride a heat wave with acceptable indoor temperatures (Meir and Roaf, 2005);
- *more control given to occupants* in order to maximize their 'adaptive opportunities' for moderating their immediate environment;
- *sustainable building designers* – a new breed of professional who understand the performance implications of the building form and fabric, and the environmental impacts of the design decisions they make (e.g. how passive heating and cooling systems may be used in the building, how to embed renewable energy systems within the buildings, how occupants will interact with the building); building physics must grow as a discipline and incorporate HVAC engineering;
- *connected indoor and outdoor climates* where the seasons are reflected in the indoor temperatures of the building;
- *local building industries and craft inputs* as the trend towards bioregional sourcing of materials and skilled workers picks up speed in response to rapidly rising materials processing and transport costs;
- *local lifestyles* as the need to save energy kicks in and the energy advantage of reintroducing traditional patterns of building use by local culturally adapted populations become clear;
- *building contracts* that ensure that building engineers are paid according to how low the energy bills and carbon dioxide emissions are, rather than how much air-conditioning equipment goes into the building;
- *street life improved* – with the re-establishment of the connection between indoor and outdoor climates, life can spill back out of the buildings for much of the year in many parts of the world, in this way reclaiming the streets;
- *more robust buildings* that are less vulnerable to catastrophic thermal failure ceasing to be totally dependent upon electrical energy to remain habitable;
- *lower energy use* resulting in the building itself playing a greater role in indoor climate control;
- *fewer greenhouse gas emissions* as energy use is reduced and more of the energy is generated by renewable energy systems; and
- *healthier indoor environments* as fresher, cleaner air is reintroduced through open windows.

# Conclusion: Towards a 'modern vernacular' for buildings and cities

## Vernacular architecture as a paradigm for passive design

Vernacular is a term that is applied to local buildings that have evolved over time in one location to suit the local climate, culture and economy. People live traditional lifestyles in vernacular buildings in virtually every climate in the world, from the Arctic circle to the tropics, in temperatures from below zero to over 40°C, and historically without the benefit of gas or electrically driven mechanized heating and cooling systems (Meir and Roaf, 2003a).

If it is assumed, as in international standards such as CEN/ISO 7730 or ASHRAE 55, that people suffer less discomfort in very closely controlled conditions, then such vernacular buildings, along with modern passive buildings, cannot provide their occupants with 'comfortable' indoor climates. If, however, as the adaptive comfort approach suggests, the closely controlled environment is just one of many possible high-quality indoor environments, then an unvarying temperature resulting in high energy use is not necessary for comfort.

At the same time, the vernacular tradition has never been driven solely by comfort considerations, nor is it necessarily compatible with present-day lifestyles, and may require, for instance, differently sized and proportioned houses, different materials, and different dress codes, daily routines and levels of activity. Vernacular design strategies and technologies cannot be simply reused to provide cooling in modern buildings without taking care to understand the impacts on design of such local customs. At the same time, there is much to learn from the vernacular when designing the passive low-energy types of buildings upon which the 21st-century design paradigms will increasingly be styled (Meir and Roaf, 2005).

## Developing a 'modern vernacular'

Using the insights of the adaptive approach to thermal comfort in and around buildings, a 'new vernacular' can be developed harnessing the types of low-tech solutions that are familiar to most of us from the vernacular, together with modern passive and active renewable energy technologies and strategies to reflect the new cultural, climatic and economic realities of the 21st century (Roaf et al, 2003). What may prove the strongest influence on the emergence of the new vernacular is the rapidly rising price of oil and gas, associated with the 'peak oil' issue (Roaf et al, 2005).

We must sooner or later accept that buildings that rely on centralized control and conventional oil- and gas-powered systems will no longer be able to deliver adequate comfort at an affordable price for the majority of the world's population. The development of a new robust and adaptable paradigm of truly sustainable buildings is an essential precursor in the process of future-proofing our buildings. At the heart of this new paradigm lie the insights afforded by an understanding of the adaptive approach to comfort in buildings.

This adaptive approach suggests that:

- Current international standards advising on comfort in buildings are formulated to suit the needs of closely controlled air-conditioned spaces, precise simulation technologies and the professions who specialize in them.
- Provision of closely controlled static indoor conditions is one way of providing comfort for building occupants; but it is not the only way to do so.
- Comfort is a goal that building occupants seek, and the function of a good passive low-energy building is to provide the opportunity for the occupants to create their own comfort in a partnership with the building.
- The temperature that building occupants will find comfortable is generally close to the temperature which they customarily experience in buildings within the local culture.
- The changes in temperature in buildings within any day, or from day to day, should be kept within the ranges shown in the earlier section on 'Lessons of the adaptive approach'.
- If the customary temperature for a building is not known, a first approximation of it can be made from Equation 2 for free-running buildings, where the monthly mean temperature is known. In Europe, a better approximation can be made by the appropriate use of Equations 6 to 9 if the running mean temperature can be calculated (Equation 5).

We can design passive buildings for most climates – after all, we have been doing it for millennia already. New drivers in 21st-century buildings all suggest that we must move away from designs which rely solely on machines to maintain indoor climates that are disconnected from those outdoors. Perhaps the Bauhaus is on the verge of giving way to the Ecohouse. Certainly, if we are to survive the coming difficult decades with our societies intact, then it is not only the technology that must change, but our mindsets, as we pioneer the built environments of the 21st century. This chapter provides an introduction to the comfort standards that can be used to inform this new age of architecture.

## ACKNOWLEDGEMENTS

The authors are grateful for all the interactions we have had down the years with our fellow researchers, in particular Reverend Professor Michael Humphreys. The section on 'Lessons of the adaptive approach' is based on the 'adaptive' section 1.6 of CIBSE (2006).

# REFERENCES

Ackerman, M. (2002) *Cool Comfort: America's Romance with Air-conditioning*, Smithsonian Institution Press, Washington, DC

ASHRAE Standard 55 (2004) *Thermal Environment Conditions for Human Occupancy*, American Society of Heating Refrigeration and Air-Conditioning Engineers, Atlanta, GA

Baker, N. V. and Standeven, M. A. (1996) 'Thermal comfort in free-running buildings', *Energy and Buildings*, vol 23, no 3, pp175–182

Bedford, T. (1936) *The Warmth Factor in Comfort at Work*, MRC Industrial Health Board Report No 76, HMSO, London

Black, F. and Milroy, E. (1966) 'Experience of air-conditioning in offices', *Journal of the Institute of Heating and Ventilating Engineers*, September, pp188–196

Brager, G. S. and de Dear, R. J. (1998) 'Thermal adaptation in the built environment: A literature review', *Energy and Buildings*, vol 27, no 1, pp83–96

Building Research Establishment (1991) *Energy Consumption Guide 19: Energy Use in Offices*, BRECSU, Garston

Busch, J. (1992) 'A tale of two populations: Thermal comfort in air-conditioned and naturally ventilated offices in Thailand', *Energy and Buildings*, vol 18, pp235–249

CEN (2007) 'EN15251: Indoor environmental input parameters for design and assessment of energy performance of buildings addressing indoor air quality, thermal environment, lighting and acoustics', European Committee for Standardization, Brussels

CEN ISO 7730 (1994) *Moderate Thermal Environments – Determination of the PMV and PPD Indices and Specification of the Conditions for Thermal Comfort*, International Organization for Standardization, Geneva

Chen, A, and Vine, E, {1998} *A Scoping Study on the Costs of Indoor Air Quality Illnesses: An Insurance Loss Reduction Perspective*, US Department of Commerce, Washington, DC

CIBSE (Chartered Institution of Building Services Engineers) (2006) 'Environmental criteria for design', in *CIBSE Guide: An Environmental Design*, CIBSE, London, Chapter 1

Clausen, G., Olm, O. and Fanger, P. O. (2002) 'The impact of air pollution from used ventilation filters on human comfort and health', in Levin, H. (ed) *Proceedings of the 9th International Conference on Indoor Air Quality and Climate*, vol 1, Indoor Air 2002, Santa Cruz, US, pp338–343

Cooper, G. (1998) *Air-Conditioning America: Engineers and the Controlled Environment, 1900–1960*, Johns Hopkins Studies in the History of Technology, John Hopkins University, Baltimore, MD, p190

de Dear, R. J. (1998) 'A global database of thermal comfort field experiments', *ASHRAE Transactions*, vol 104, no 1b, pp1141–1152

de Dear, R. J. and Brager, G. S. (2002) 'Thermal comfort in naturally ventilated buildings: Revisions to ASHRAE Standard 55', *Energy and Buildings*, vol 34, no 6, pp549–561

Drenzer, D. (2005) 'Adios siesta', www.danieldrezner.com/archives/002491.html

Fanger, P. O. (1970) *Thermal Comfort*, Danish Technical Press, Copenhagen

Fiala D., Lomas, K. J. and Stohrer, M. A. (1999) 'Computer model of human thermoregulation for a wide range of environmental conditions: The passive system', *Journal of Applied Physiology (American Physiology Society)*, vol 87, no 5, pp1957–1972

Haves, P., Roaf, S. and Orr, J. (1998) 'Climate change and passive cooling in Europe', in *Proceedings of PLEA Conference,* Lisbon, pp463–466

Humphreys, M. A. (1979) 'The influence of season and ambient temperature on human clothing behaviour', in Fanger, P. O. and Valbjorn, O. (eds) *Indoor Climate,* Danish Building Research, Copenhagen

Humphreys, M. A. (1981) 'The dependence of comfortable temperature upon indoor and outdoor climate', in K. Cena and J. A. Clark (eds) *Bioengineering, Thermal Physiology and Comfort,* Elsevier, Oxford

Humphreys, M. A. and Nicol, J. F. (1996) 'Conflicting criteria for thermal sensation within the Fanger Predicted Mean Vote equation', in *Proceedings of the CIBSE/ASHRAE Joint National Conference 1996, Harrogate,* part 1, Chartered Institution of Building Services Engineers, London, pp153–158

Humphreys, M. A. and Nicol, J. F. (1998) 'Understanding the adaptive approach to thermal comfort', *ASHRAE Transactions,* vol 104, no 1, pp991–1004

Humphreys, M. A. and Nicol, J. F. (2000) 'Outdoor temperature and indoor thermal comfort: Raising the precision of the relationship for the 1998 ASHRAE database of field studies', *ASHRAE Transactions,* vol 106, no 2, pp485–492

Humphreys, M. A. and Nicol, J. F. (2002) 'The validity of ISO-PMV for predicting comfort votes in every-day thermal environments', *Energy and Buildings,* vol 34, no 6, pp667–684

ISO (International Organization for Standardization) (1994) 'International Standard 7730', in *Moderate Thermal Environments: Determination of PMV and PPD Indices and Specification of the Conditions for Thermal Comfort,* ISO, Geneva

Le Corbusier ([1928] 1986) *Towards a New Architecture,* Dover Publications, London, and Mineola, New York (originally published in 1928, Paris, as *Vers une Architecture*)

Loudon, A. G. (1968) *Window Design Criteria to Avoid Overheating by Excessive Solar Gains,* BRS Current Paper 4/68, Building Research Station, Garston

Loudon, F. J. and Danter, E. (1965) 'Investigations of summer overheating', *Building Science,* vol 1, pp89–94

Loudon, F. J. and Keighley, E. C. (1964) 'User research in office design', *Architect's Journal,* vol 139, no 6, pp333–339

Markham, S. F. (1944) *Climate and the Energy of Nations,* Oxford University Press, Oxford

Mauderly, J. (2002) 'Linkages between outdoor and indoor air quality issues: Pollutants and research problems crossing the threshold', in Levin, H. (ed) *Proceedings of the 9th International Conference on Indoor Air Quality and Climate,* vol 1, Indoor Air 2002, Monterey, Santa Cruz, US, pp12–13

McCartney, K. J. and Nicol, J. F. (2002) 'Developing an adaptive control algorithm for Europe: Results of the SCATs project', *Energy and Buildings,* vol 34, no 6, pp623–635

McIntyre, D. A. (1980) *Indoor Climate,* Applied Science Publishers, London

Meir, I and Roaf, S. (2003a) 'Between Scylla and Charibdis: In search of the sustainable design paradigm between vernacular and hi-tech', in *Proceedings of PLEA Conference,* Santiago, Chile

Meir, I. and Roaf, S. (2003b) 'Thermal comfort: Thermal mass housing in hot dry climates', in Levin, H. in *Proceedings of the 9th International Conference on Indoor Air Quality and Climate,* vol 1, Indoor Air 2002, Monterey, Santa Cruz, US, pp12–13

Meir, I. and Roaf, S. (2005) 'The future of the vernacular: Towards new methodologies for the understanding and optimization of the performance of vernacular buildings', in Asquith, L. and Vellinga, M. (eds) *Vernacular Architecture in the Twenty-First Century: Theory, Education and Practice,* Spon, London

Morgan, C. A., de Dear, R. and Brager, G. (2002) 'Climate clothing and adaptation in the built environment', in Levin, H. (ed) *Proceedings of the 9th International Conference on Indoor Air Quality and Climate,* vol 5, Indoor Air 2002, Santa Cruz, US, pp98–103

Nicol, J. F. (1974) 'An analysis of some observations of thermal comfort in Roorkee, India and Baghdad, Iraq', *Annals of Human Biology,* vol 1, pp411–426

Nicol, J. F (2003) 'Thermal comfort', in Santamouris, M. (ed) *Solar Thermal Technologies for Buildings,* James and James, London

Nicol, J. F. and Humphreys, M. A. (1973) 'Thermal comfort as part of a self-regulating system', *Building Research and Practice (Journal of CIB),* vol 6, no 3, pp191–197

Nicol, J. F. and Humphreys, M. A. (2004) 'A stochastic approach to thermal comfort, occupant behaviour and energy use in buildings', *ASHRAE Transactions,* vol 110, no 2, pp554–568

Nicol, J. F. and Humphreys, M. A. (2006) 'Maximum temperatures in European office buildings to avoid heat discomfort', *Solar Energy,* vol 81, no 3, pp295–304

Nicol, F., Humphreys, M., Sykes, O. and Roaf, S. (eds) (1995) *Standards for Thermal Comfort: Indoor Air Temperature Standards for the 21st Century,* E. & F. N. Spon, London

Nicol, F. and McCartney, K. (2001) *Final Report (Public) Smart Controls and Thermal Comfort (SCATs): Report to the European Commission of the Smart Controls and Thermal Comfort Project (Contract JOE3-CT97-0066),* Oxford Brookes University, Oxford

Nicol, F. and Raja, I. (1996) *Thermal Comfort, Time and Posture: Exploratory Studies in the Nature of Adaptive Thermal Comfort,* School of Architecture, Oxford Brookes University, Oxford

Nicol, F. and Roaf, S. (1996) 'Pioneering new indoor temperature standards: The Pakistan Project', *Energy and Buildings,* vol 23, pp169–174

Nicol, J. F. and Roaf, S. (2005) 'Post occupancy evaluation and thermal comfort surveys', *Building Research and Information,* vol 33, no 4, pp338–346

Parsons, K. C. (2002) *Human Thermal Environments,* second edition, Blackwell Scientific, Oxford

Parsons, K. C., Webb, L. H., McCartney, K. J., Humphreys, M. A. and Nicol, J. F. (1997) 'A climatic chamber study into the validity of Fanger's PMV/ PPD thermal comfort index for subjects wearing different levels of clothing insulation', in *Proceedings of the CIBSE National Conference 1997, London,* part 1, Chartered Institution of Building Services Engineers, London, pp193–205

Raja, I. A. and Nicol, J. F. (1997) 'A technique for postural recording and analysis for thermal comfort research', *Applied Ergonomics,* vol 27, no 3, pp221–225

Roaf, S. (1988) *The Windcatchers of Yazd,* PhD thesis, Oxford Polytechnic, Oxford, p204

Roaf, S., Crichton, D. and Nicol, F. (2005) *Adapting Buildings and Cities for Climate Change,* Architectural Press, London

Roaf, S., Fuentes, M. and Thomas, S. (2003) *Ecohouse 2: A Design Guide,* Architectural Press, Elsevier, Oxford

Santamouris, M. (ed) (2005) *Air Conditioning, Energy Consumption and Environmental Quality*, EOLSS Publishers, Oxford

Sharma, M. R. and Ali, S. (1986) 'Tropical Summer Index: A study of thermal comfort in Indian subjects', *Building and Environment*, vol 21, no 1, pp11–24

Shove, E. (2003) *Comfort Cleanliness and Convenience*, Berg Publishers, Oxford

van der Linden, A. C., Boerstra, A. C., Raue, A. K., Kurvers, S. R. and de Dear, R. J. (2006) 'Adaptive temperature limits: A new guideline in The Netherlands: A new approach for the assessment of building performance with respect to thermal indoor climate', *Energy and Buildings*, vol 38, no 1, pp8–17

Webb, C. G. (1959) 'An analysis of some observations of thermal comfort in an equatorial climate', *British Journal of Industrial Medicine*, vol 16, no 3, pp297–310

# 2

# Opportunities for Saving Energy and Improving Air Quality in Urban Heat Islands

*Hashem Akbari*

## Introduction and Background on the Urban Heat Island

World energy use is the main contributor to atmospheric carbon dioxide ($CO_2$). In 2002, about 7 metric gigatonnes of carbon (GtC) were emitted internationally by combustion of gas, liquid and solid fuels (CDIAC, 2006), two to five times the amount contributed by deforestation (Brown et al, 1988). The share of atmospheric carbon emissions for the US from fossil fuel combustion was 1.6GtC. Increasing use of fossil fuel and deforestation together has raised atmospheric $CO_2$ concentration some 25 per cent over the last 150 years. According to global climate models and preliminary measurements, these changes in the composition of the atmosphere have already begun raising the Earth's average temperature. If current energy trends continue, these changes could drastically alter the Earth's temperature, with unknown but potentially catastrophic physical and political consequences. During the last three decades, increased energy awareness has led to conservation efforts and levelling of energy consumption in the industrialized countries. An important by-product of this reduced energy use is the lowering of $CO_2$ emissions.

Of all electricity generated in the US, about one sixth is used to air condition buildings. The air-conditioning use is about 400 terawatt hours (TWh), equivalent to about 80 million metric tonnes of carbon (MtC) emissions, and translating to about US$40 billion per year. Of this US$40 billion per year, about half is used in cities that have pronounced 'heat islands'.

The contribution of the urban heat island to the air-conditioning demand has increased over the last 40 years, and it is currently at about 10 per cent. Metropolitan areas in the US (e.g. Los Angeles, Phoenix, Houston, Atlanta and New York City) have typically pronounced heat islands that warrant special attention by anyone concerned with broad-scale energy efficiency (HIG, 2006).

The ambient air is primarily heated through three processes: direct absorption of solar radiation, convection of heat from hot surfaces, and man-made heat (exhaust from cars, buildings, etc.). Air is fairly transparent to light – the direct absorption of solar radiation in atmospheric air only raises the air temperature by a small amount. Typically, about 90 per cent of solar radiation reaches the Earth's surface and then is either absorbed or reflected. The radiation absorbed on the surface increases the surface temperature. And, in turn, the hot surfaces heat the air. This convective heating is responsible for the majority of diurnal temperature ranges. The contribution of man-made heat (e.g. air conditioning and cars) is very small compared to the heating of air by hot surfaces, except for the downtown high-rise areas.

Modern urban areas have darker surfaces or lower 'effective' albedo[1] and relatively less vegetation than their more natural surroundings, which affects urban climate, energy use and thermal environmental conditions. Dark roofs, for example, heat up more than their more reflective counterparts and thus raise the summertime cooling demands of buildings. Collectively, on a neighbourhood scale, dark surfaces and reduced vegetation warm the air over urban areas, contributing to urban heat islands. Figure 2.1 shows a sketch of a typical summer afternoon urban heat island. On a clear summer afternoon, the air temperature in a typical city can be as much as 2.5 Kelvin (K) higher

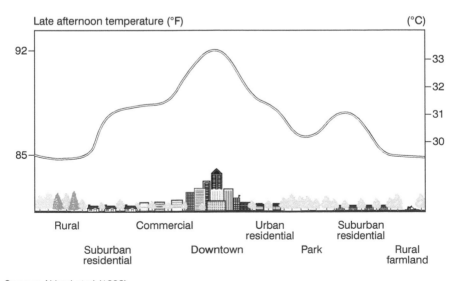

*Source:* Akbari et al (1992)

**Figure 2.1** *Sketch of a hypothetical urban heat island profile*

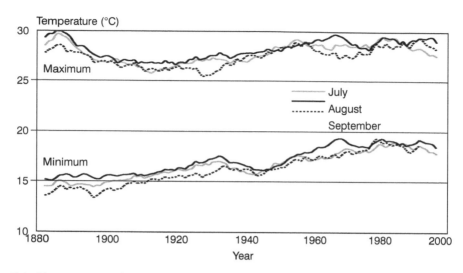

*Note:* The ten-year running average is calculated as the average temperature of the previous four years, the current year and the next five years. Note that the maximum temperatures have increased about 2.5K since 1920. During the same period, the minimum temperature also increased by about 3K.
*Source:* Akbari et al (2001); updated data to include 1998–2004

**Figure 2.2** *Ten-year running average summertime monthly maximum and minimum temperatures in Los Angeles, California (1877–2004)*

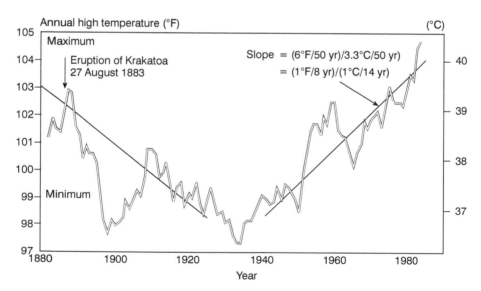

*Note:* The ten-year running average is calculated as the average temperature of the previous four years, the current year and the next five years.
*Source:* Akbari et al (1992)

**Figure 2.3** *Ten-year running average maximum annual temperatures in Los Angeles, California (1877–1997)*

than surrounding rural areas.[2] In hot cities, peak urban electric demand in the US rises by 2 to 4 per cent for each 1K rise in daily maximum temperature above ambient air temperatures of 15°C to 20°C.

Temperatures in cities are generally increasing. An analysis of summertime monthly maximum and minimum temperatures between 1877 and 1997 in downtown Los Angeles clearly indicated that maximum temperatures are now about 2.5K higher than in 1920 (Akbari et al, 2001; see Figures 2.2 and 2.3). Minimum temperatures are about 4K higher than in 1880. A California study analysing the average urban–rural temperature differences for 31 urban and 31 rural stations from 1965 to 1989 showed that urban temperatures have increased by about 1K (Goodridge, 1987, 1989; see Figure 2.4). This trend in increasing temperatures in urban areas is typical of most US metropolitan areas and is observed in many other cities across the world (Akbari et al, 1992; see Figure 2.5). Santamouris (2007) has also reviewed the existing heat-island data in Europe and noted the increasing trends in summertime temperatures in many European cities. Summertime urban heat islands can exacerbate demand for cooling energy. Note that this is above and beyond what is believed to be the global warming trend. Since most people live in cities, they would experience the effects of both global warming and urban heat islands.

Increasing urban ambient temperatures result in increased system-wide electricity use. In the Los Angeles Basin, the heat island-induced increase in power consumption of 1GW to 1.5GW can cost ratepayers US$100 million

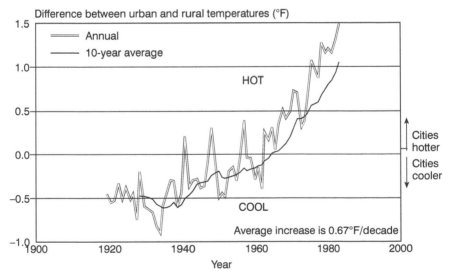

*Note:* During 1920 to 1960, cities were actually cooler than suburban areas, probably because of relatively more vegetation in urban areas.
*Source:* Akbari et al (1990), based on data from Goodridge (1989)

**Figure 2.4** *Warming trend in California urban areas; since 1940,*
*the temperature difference between urban and rural meteorological stations*
*has shown an increase of about 0.37K (0.67°F) per decade*

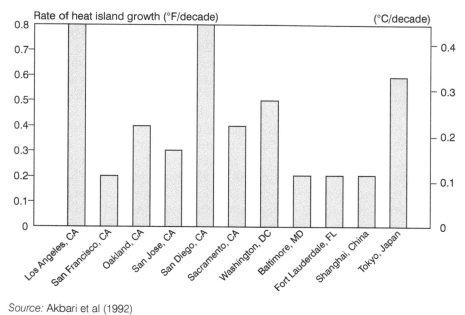

*Source:* Akbari et al (1992)

**Figure 2.5** *Trend of increasing urban temperature over the last three to eight decades in selected cities*

per year (see Figure 2.6). In the US, additional air-conditioning use from increased urban air temperature comprises 5 to 10 per cent of urban peak electric demand at a direct cost of several billion dollars per year. Since cooling demand on hot summer days is the cause of peak demand for electricity, the electric utilities have installed additional capacity to compensate for the heat island effects.

Besides increasing system-wide cooling loads, summer heat islands increase smog production. Smog production is a highly temperature-sensitive process. In the Los Angeles Basin, at daily maximum temperatures below 22°C, maximum ozone concentration is typically below the California standard (90 parts per billion, ppb); at above 32°C, practically all days are smoggy (see Figure 2.7).

The relationship between the urban heat islands and pollution has also been studied in several European cities. Sarrat et al (2006) have shown that the urban heat island has an important effect on the primary and secondary regional pollutant (nitrogen oxide, $NO_x$, and ozone) in the Paris metropolitan area. Stathopoulou et al (2007) collected air temperature and ozone concentration data from several stations in the greater Athens area and found a strong positive correlation between daytime air temperature and ozone concentration.

Summer heat islands increase citizens' discomfort and heat wave-related mortalities. According to the US Centers for Disease Control and Prevention (CDC, 2006), over the past 20 years, more Americans were killed by heat than

(a) Southern California Edison Company (SCE) 2002 system-wide load

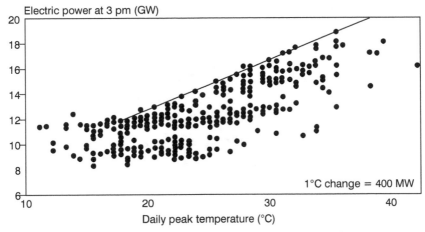

(b) Los Angeles Department of Water and Power (LADWP)
2002 system-wide load

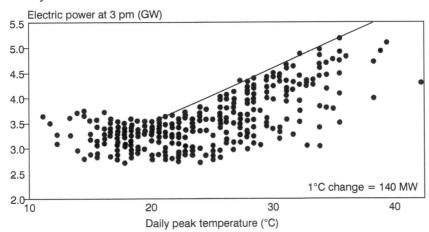

*Note:* The increased summertime temperatures cause increased cooling requirements. In the Los Angeles Basin (primarily served by Southern California Edison and Los Angeles Department of Water and Power), we estimate that about 1GW to 1.5GW of power is used to compensate the heat island effect. This increased power adds about US$100,000 per hour (US$100 million a year) during summer days to the utility customers' electricity bills.
*Source:* Southern California Edison Company (SCE) and Los Angeles Department of Water and Power (LADWP)

**Figure 2.6** *Daily peak utility electric power demand versus daily peak air temperature*

by hurricanes, lightning, tornadoes, floods and earthquakes combined. Within a five-day period, the 1995 Chicago heat wave killed between 525 and 726 people, depending upon the method used for determining which deaths were attributable to the high temperatures. In the heat wave of 1980, some 1250

(a) Ozone concentration measured at Los Angeles, North Main Street, 2002

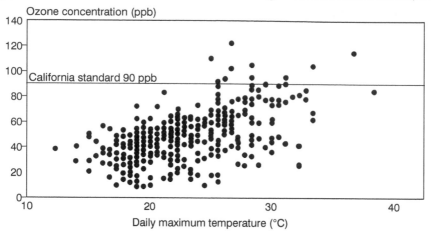

(b) Ozone concentration measured at Los Angeles, W Flint Street, 2002

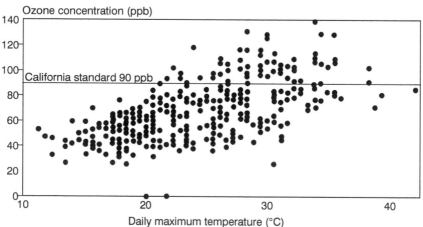

*Note:* The impact of the heat island is also seen in smog. The formation of smog is highly sensitive to temperatures; the higher the temperature, the higher the formation and, hence, the concentration of smog. In Los Angeles at temperatures below 22°C, the concentration of smog (measured as ozone) is below the California standard. At temperatures of about 32°C, practically all days are smoggy. Cooling the city by about 3°C would have a dramatic impact on smog concentration.

*Source:* South Coast Air Quality Management District, Diamond Bar, California

**Figure 2.7** *Daily maximum ozone concentration versus daily maximum temperature in two locations in Los Angeles*

Americans died. A heat wave in the summer of 2003 in India killed at least 1200 people. Most tragic is the death of between 10,000 and 15,000 people who died in France's scorching heat wave in August 2003. Many of the victims were elderly people living in poorly designed houses or apartments that were not air conditioned.

*Source:* Akbari et al (1999b)

**Figure 2.8** *Orthophoto of a typical mixed urban area in Sacramento, California*

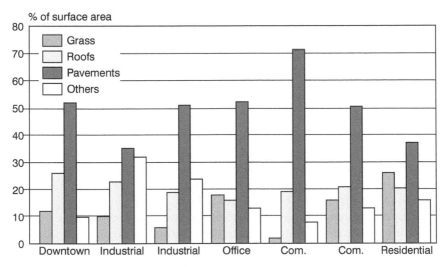

*Note:* In all areas, paved surfaces and roofs together comprise more than 55 per cent of the developed areas.
*Source:* Akbari et al (1999b)

**Figure 2.9** *Urban fabric of several residential and commercial areas in Sacramento, California*

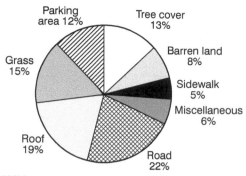

*Source:* Akbari et al (1999b)

**Figure 2.10** *The land use/land cover (LULC) percentages for
Sacramento, California*

In France, the heat wave brought temperatures of up to 40°C in the first two
weeks of August 2003 in a country where air conditioning is rare. Although
high temperatures may be the immediate cause of the higher mortality rates,
the lack of preparation to face the high temperatures is the real cause for these
'natural' disasters. In regions where higher summer temperatures are prevalent
(the Mediterranean, North Africa and the Middle East), incidents of such
disasters are far lower.

It is important to note that the heat island is a direct result of urbanization
that creates an urban fabric consisting mostly of roofs, paved surfaces (roads,
driveways and parking lots) and less vegetation (trees, lawns, bushes and
shrubs). Understanding and quantifying the fabric of a city is an important
first step in analysing and designing implementation programmes to mitigate
urban heat islands. Of particular importance is the fraction of each surface
type within an area. An accurate characterization of the urban surfaces will
also allow a better estimate of the potential for increasing surface albedo (roofs
and pavements) and urban vegetation. This would, in turn, provide more
accurate modelling of the impact of heat-island reduction measures on ambient
cooling and urban ozone air quality.

In four studies, Akbari et al (1999b), Akbari and Rose (2001a, 2001b) and
Rose et al (2003) characterized the fabric of Sacramento, California; Salt Lake
City, Utah; Chicago, Illinois; and Houston, Texas, using high-resolution aerial
digital orthophotos covering selected areas in each city (see Figure 2.8 for an
example of high-resolution orthophotos). Four major land-use types were
examined: commercial, industrial, transportation and residential. These
orthophotos were analysed to estimate the fraction of each major land-use
type (defined as urban fabric) and to estimate the land use/land cover (LULC)
in each city (see Figures 2.9 and Figure 2.10). Although there were differences
among the fabrics of these four metropolitan areas, some significant
similarities were found.

Table 2.1 shows the LULC for the four metropolitan areas based on United
States Geological Survey (USGS) data. Of approximately 800 square

**Table 2.1** *USGS land use/land cover (LULC) percentages for four cities: Sacramento, California; Salt Lake City, Utah; Chicago, Illinois; and Houston, Texas*

|  | Sacramento | Salt Lake City | Chicago | Houston |
|---|---|---|---|---|
| Total metropolitan area (km²) | 809 | 624 | 2521 | 3433 |
| **LULC (percentage)** | | | | |
| Residential | 49.3 | 59.1 | 53.5 | 56.1 |
| Commercial/service | 17.1 | 15.0 | 19.2 | 5.1 |
| Industrial | 7.2 | 4.9 | 11.5 | 9.3 |
| Transportation/communication | 11.4 | 9.8 | 7.7 | 2.9 |
| Industrial and commercial | 0.3 | 0.0 | 0.1 | 4.8 |
| Mixed urban or built-up land | 5.2 | 1.9 | 0.4 | 3.5 |
| Other | 9.5 | 9.4 | 7.6 | 18.3 |

*Note:* USGS = United States Geological Survey
*Source:* Akbari et al (1999b), Akbari and Rose (2001a, 2001b) and Rose et al (2003)

kilometres of urban area in Sacramento, about 49 per cent were residential; in Salt Lake City, about 59 per cent of the 620 square kilometres; in Chicago, about 53 per cent of the 2520 square kilometres; and in Houston, about 56 per cent of the 3430 square kilometres. The fraction of industrial, transportation and mixed urban land uses in these four cities varied only by a few per cent.

For the entire metropolitan area, the percentage of the total roof areas, as seen from above the canopy, was about 19 per cent in the Sacramento and Salt Lake City metropolitan areas; 25 per cent in the Chicago metropolitan area; and 21 per cent in the Greater Houston (see Table 2.2). The percentage of paved areas ranged from 29 to 39 per cent; vegetated areas, 29 to 41 per cent; and other areas, 10 to 40 per cent. Under the canopy, the roof area ranged from 20 to 25 per cent; paved surfaces, 29 to 45 per cent; vegetated areas, 20 to 37 per cent; and other areas, 9 to 15 per cent.

In residential areas, the percentage of the total roof areas, as seen from above the canopy, ranged from 19 to 26 per cent; paved surfaces, 25 to 26 per cent; vegetated areas, 39 to 49 per cent; and others, 4 to 16 per cent. Under the canopy, roof area ranged from 20 to 27 per cent; paved surfaces, 24 to 32 per cent; vegetated areas, 33 to 47 per cent; and other areas, 6 to 17 per cent.

Other researchers involved in the analysis of urban climate have tried to quantitatively characterize the surface-type composition of various urban areas. Myrup and Morgan (1972) conducted such work with the analysis of the urban fabric in Sacramento. They applied the strategy of examining the land-use data in progressively smaller segments of macro scale (representative areas of Sacramento), meso scale (individual communities), micro scale (land-use ordinance zones) and basic scale (city blocks). The data they used included USGS photos, parks and recreation plans, city engineering road maps, and detailed aerial photos. Their analysis covered 195 square kilometres of urban areas. The percentages of the land-use areas were calculated as follows:

**Table 2.2** *The land use/land cover (LULC) percentages for four cities:*
*Sacramento, California; Salt Lake City, Utah; Chicago, Illinois;*
*and Houston, Texas*

| City | Vegetation | Roofs | Pavements | Other |
|------|-----------|-------|-----------|-------|
| **Above the canopy** | | | | |
| Metropolitan Salt Lake City | 40.9 | 19.0 | 30.3 | 9.7 |
| Metropolitan Sacramento | 28.6 | 18.7 | 38.5 | 14.3 |
| Metropolitan Chicago | 30.5 | 24.8 | 33.7 | 11.0 |
| Greater Houston | 38.6 | 21.4 | 29.0 | 10.9 |
| Residential Salt Lake City | 46.6 | 19.7 | 25.3 | 8.5 |
| Residential Sacramento | 39.2 | 19.4 | 25.6 | 15.8 |
| Residential Chicago | 44.3 | 25.9 | 25.7 | 4.1 |
| Residential Houston | 48.9 | 20.5 | 24.7 | 6.0 |
| | | | | |
| **Under the canopy** | | | | |
| Metropolitan Salt Lake City | 33.3 | 21.9 | 36.4 | 8.5 |
| Metropolitan Sacramento | 20.3 | 19.7 | 44.5 | 15.4 |
| Metropolitan Chicago | 26.7 | 24.8 | 37.1 | 11.4 |
| Greater Houston | 37.1 | 21.3 | 29.2 | 12.4 |
| Residential Salt Lake City | 38.6 | 23.9 | 31.6 | 6.0 |
| Residential Sacramento | 32.8 | 19.8 | 30.6 | 16.8 |
| Residential Chicago | 35.8 | 26.9 | 29.2 | 8.1 |
| Residential Houston | 47.4 | 21.1 | 23.9 | 7.6 |

*Source:* Akbari et al (1999b), Akbari and Rose (2001a, 2001b) and Rose et al (2003)

residential, 35.5 per cent; commercial, 7.2 per cent; industrial, 13.5 per cent; streets and freeways, 17 per cent; institutional, 3.2 per cent; and open space and recreational, 23.6 per cent. They found the average residential area to be composed of about 22 per cent streets, 23 per cent roofs, 22 per cent other impervious surfaces and 33 per cent green areas. Overall, they found a composition of 14 per cent streets, 22 per cent roofs, 22 per cent other impervious surfaces, 36 per cent green areas and 3 per cent water surfaces. They defined 'other impervious surfaces' to include highway shoulder strips, airport runways and parking lots. Streets included curbs and sidewalks.

## HEAT ISLAND MITIGATION TECHNOLOGIES

Possible technologies used in lowering the summertime ambient temperatures and increasing comfort include use of light-coloured materials on roofs and walls; trees and vegetation to shade buildings, walkways and streets; and using light-coloured paving materials for streets, parking lots, driveways and sidewalks. Santamouris (2001) provides a thorough description of building and pavement construction materials that have been historically used as a countermeasure for urban heat islands. In addition, Doulos et al (2004) have measured and compared the thermal performance of 93 construction materials commonly used in Greece.

Use of high-albedo urban surfaces and planting of urban trees are inexpensive measures that can reduce summertime temperatures. The effects of modifying the urban environment by planting trees and increasing albedo are best quantified in terms of 'direct' and 'indirect' effects. The direct effect of planting trees around a building or using reflective materials on roofs or walls is to alter the energy balance and cooling requirements of that particular building. However, when trees are planted and the albedo of roofs and pavements is increased throughout an entire community, the energy balance of the whole community is modified, producing community-wide changes in climate. Phenomena associated with community-wide changes in climate are referred to as indirect effects because they indirectly affect the energy use in an individual building. Direct effects give immediate benefits to the building that applies them. Indirect effects achieve benefits only with widespread deployment.

When dark roofs are heated by the sun, they directly raise the demand for cooling for the buildings beneath those roofs. For highly absorptive (low albedo) roofs, the surface/ambient air temperatures difference may reach 50K, while for less absorptive (high albedo) surfaces with similar insulative properties (e.g. white-coated roofs), the difference can be only about 10K. Clearly, a cool roof reduces the cooling-energy requirements of its own building.

Hot roofs also heat the outside ambient air, thus indirectly increasing the cooling demand of neighbouring buildings. We have simulated the effect of urban-wide application of reflective roofs on cooling-energy use and smog in many metropolitan areas (Taha et al, 1995, 2000, 2001). We estimate roof albedos can realistically be raised by 0.30 on average, resulting in a 1K to 2.5K cooling at 3 pm (on a sunny August day). This temperature reduction reduces building cooling-energy use even further. Other benefits of light-coloured roofs include a potential increase in the roof's useful life.

The beneficial effects of trees are both direct in the shading of buildings and indirect in cooling the ambient air (urban forest). Trees can intercept sunlight before it warms buildings and can cool the air by evapotranspiration. In winter, trees can shield buildings from cold winds. Urban shade trees offer significant benefits by reducing building air conditioning and lowering air temperature, thus improving urban air quality (reducing smog). Savings associated with these benefits vary by climate and region and, over a tree's life, can reach up to US$200 per tree. The cost of planting and maintaining trees can vary from US$10 to $500 per tree. Tree planting programmes can be low cost, offering savings to tree-planting communities. The choice of tree species is also important. Drought-resistant trees that are low emitters of volatile organic compounds are typically recommended.

The issue of direct and indirect effects also enters into our discussion of atmospheric pollutants. Planting trees has the direct effect of reducing atmospheric $CO_2$ because each individual tree directly sequesters carbon from the atmosphere through photosynthesis. However, planting trees in cities also has an indirect effect on $CO_2$. By reducing the demand for cooling energy, urban trees indirectly reduce emission of $CO_2$ from power plants. Akbari et al (1990) showed that the amount of $CO_2$ avoided via the indirect effect is considerably greater than the amount sequestered directly. Similarly, trees directly trap ozone precursors (by dry deposition, a process in which ozone is directly absorbed by tree leaves) and indirectly reduce the emission of these precursors from power plants by reducing the combustion of fossil fuels and, hence, by reducing $NO_x$ emissions from power plants (Taha, 1996).

There are other important benefits associated with urban trees. These include improvement in environmental quality, increased property values and decreased runoff, which leads to flood protection.

Urban pavements are made predominantly of asphalt concrete. The advantages of this smooth and all-weather surface type for vehicles are obvious; but some associated problems are perhaps not so well appreciated. Sunlight on dark asphalt surfaces produces increased heating. An air temperature increase, in turn, increases cooling-energy use in buildings and can accelerate smog formation. The albedo of fresh asphalt concrete pavement is about 0.05: the relatively small amount of black asphalt coats the lighter-coloured aggregate. As an asphalt concrete pavement is worn down and the aggregate is revealed, albedo increases to about 0.10 to 0.15 (the value of ordinary aggregates). If a reflective aggregate is used, the long-term albedo can be higher.

Unlike cool roofs and urban trees, cool pavements provide only indirect effects in terms of urban cooling energy use – that is, through lowered ambient temperatures. Lower temperatures have two effects:

1   reduced demand for electricity for air conditioning; and
2   decreased rate of smog production (ozone).

Savings from reduced electricity demand and from the externalities of lower ozone concentrations can be significant. Furthermore, the temperature of a pavement affects its structural performance; cooler pavements last longer in hot climates. The reflectivity of pavements can improve visibility at night and can reduce electric street-lighting demand. Street lighting is more effective if pavements are more reflective, increasing safety as a result. Despite concerns that, in time, dirt will darken light-coloured pavements, experience with cement concrete roads suggests that the light colour of the pavement can actually persist after long usage.

We estimate that by full implementation of the above mitigation measures (cool roofs, shade trees and cool pavements), the cooling demand in the US can be decreased by 20 per cent. This equals about 40TWh per year in savings, worth over US\$4 billion per year by 2015 in cooling-electricity savings alone. If smog reduction benefits are included, savings could total over US\$10 billion per year. Achieving these potential savings is conditional on receiving the

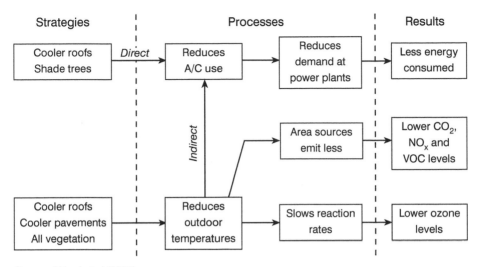

*Source:* Akbari et al (2001)

**Figure 2.11** *Methodology for energy and air quality analysis*

necessary federal, state and local community support. Scattered programmes for planting trees and increasing surface albedo already exist; but the initiation of an effective and comprehensive campaign would require an aggressive agenda.

Over the past two decades, scientists at the Lawrence Berkeley National Laboratory (LBNL) have been studying the energy savings and air quality benefits of heat-island mitigation measures. The approaches used for analysis included direct measurements of the energy savings for cool roofs and shade trees, simulations of direct and indirect energy savings of the mitigation measures (cool roofs, cool pavements and vegetation), and meteorological and air quality simulations of the mitigation measures. Figure 2.11 depicts the overall methodology used in analysing the impact of heat-island mitigation measures on energy use and urban air pollution.

## Cool roofs

At the building scale, a dark roof is heated by the sun and thus directly raises the summertime cooling demand of the building beneath it. For highly absorptive (low albedo) roofs, the difference between the surface and ambient air temperatures may be as high as 50K (Berdahl and Bretz, 1997; see Figure 2.12). For this reason, 'cool' surfaces (which absorb little solar radiation) can be effective in reducing cooling-energy use. Highly absorptive surfaces contribute to the heating of the air, and thus indirectly increase the cooling demand of, in principle, all buildings. In most applications, cool roofs incur no additional cost if colour changes are incorporated within routine re-roofing and resurfacing schedules (Bretz et al, 1997; Rosenfeld et al, 1992).

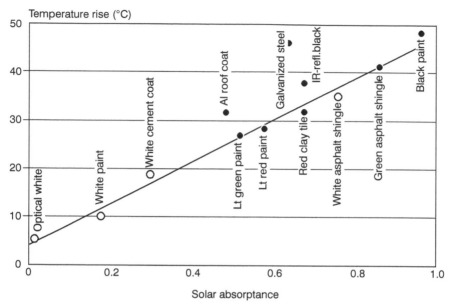

Note: All samples were insulated on the back and the measurements were made at low wind speed.
Source: Berdahl and Bretz (1997)

**Figure 2.12** *Temperature rise (surface temperature minus air temperature) of various roofing materials measured at peak solar conditions*

Most high-albedo roofing materials are light coloured, although selective surfaces that reflect a large portion of the infrared solar radiation, but absorb some visible light, can be dark coloured and yet have relatively high albedos (Berdahl and Bretz, 1997; Levinson et al, 2005a, 2005b).

### Energy, smog and other benefits of cool roofs
*Direct energy savings* Field studies in California and Florida have demonstrated cooling-energy savings in excess of 20 per cent upon raising the solar reflectance of a roof to 0.6 from a prior value of 0.1 to 0.2 (Konopacki et al, 1998; Konopacki and Akbari, 2001; Parker et al, 2002; see Table 2.3). Energy savings are particularly pronounced in older houses that have little or no attic insulation, especially if the attic contains the air distribution ducts. Akbari et al (1997a) observed cooling-energy savings of 46 per cent and peak power savings of 20 per cent achieved by increasing the roof reflectance of two identical portable classrooms in Sacramento, California. Konopacki et al (1998) documented measured energy savings of 12 to 18 per cent in two commercial buildings in California. In a large retail store in Austin, Texas, Konopacki and Akbari (2001) documented measured energy savings of 12 per cent. Akbari (2003) documented energy savings of 31 to 39Wh per square metre per day in two small commercial buildings with very high internal loads by coating roofs with a white elastomer with a reflectivity of 0.70. Parker et al

**Table 2.3** *Comparison of measured summertime air-conditioning daily energy savings from application of reflective roofs*

| Location | Building type | Roof area (m²) | R-value | Roof system duct | Δρ | Savings (Wh/m²/day) |
|---|---|---|---|---|---|---|
| **California** | | | | | | |
| Davis | Medical office | 2945 | 1.4 | Interior | 0.36 | 68 |
| Gilroy | Medical office | 2211 | 3.3 | Plenum | 0.35 | 39 |
| San José | Retail store | 3056 | RB | Plenum | 0.44 | 4.3 |
| Sacramento | School bungalow | 89 | 3.3 | Ceiling | 0.60 | 47 |
| Sacramento | Office | 2285 | 3.3 | Plenum | 0.40 | 14 |
| Sacramento | Museum | 455 | 0 | Interior | 0.40 | 20 |
| Sacramento | Hospice | 557 | 1.9 | Attic | 0.40 | 11 |
| Sacramento | Retail store | 1600 | RB | None | 0.61 | 72 |
| San Marcus | Elementary school | 570 | 5.3 | None | 0.54 | 45 |
| Reedley | Cold storage facility | | | | | |
| | Cold storage | 4900 | 5.1 | None | 0.61 | 69 |
| | Fruit conditioning | 1300 | 4.4 | None | 0.33 | |
| | Packing area | 3400 | 1.7 | None | 0.33 | Nil (open to outdoor) |
| **Florida** | | | | | | |
| Cocoa Beach | Strip mall | 1161 | 1.9 | Plenum | 0.46 | 7.5 |
| Cocoa Beach | School | 929 | 3.3 | Plenum | 0.46 | 43 |
| **Georgia** | | | | | | |
| Atlanta | Education | 1115 | 1.9 | Plenum | n/a | 75 |
| **Nevada** | | | | | | |
| Battle Mountain | Regeneration | 14.9 | 3.2 | None | 0.45 | 31 |
| Carlin | Regeneration | 14.9 | 3.2 | None | 0.45 | 39 |
| **Texas** | | | | | | |
| Austin | Retail store | 9300 | 2.1 | Plenum | 0.70 | 39 |

*Note:* Δρ is change in roof reflectivity; RB is radiant barrier; duct is the location of air-conditioning ducts; R-value is roof insulation in K·m²/W; n/a stands for not available.
Source: Akbari et al (2005)

(1998a) measured an average of 19 per cent energy savings in 11 Florida residences by applying reflective coatings on their roofs. Parker et al (1997, 1998b) also monitored seven retail stores in a strip mall in Florida before and after applying a high albedo coating to the roof and measured a 25 per cent drop in seasonal cooling-energy use. Hildebrandt et al (1998) observed daily energy savings of 17, 26 and 39 per cent in an office, a museum and a hospice, respectively, retrofitted with high albedo roofs in Sacramento. Akridge (1998) reported energy savings of 28 per cent for a school building in Georgia after an unpainted galvanized roof was coated with white acrylic. Boutwell and Salinas (1986) showed that an office building in southern Mississippi saved 22 per cent after the application of a high-reflectance coating. Simpson and McPherson (1997) measured energy savings in the range of 5 to 28 per cent in several quarter-scale models in Tucson, Arizona.

Cool roofs also significantly reduce buildings' peak electric demand in summer (Akbari et al, 1997a; Levinson et al, 2005c).

More recently, Akbari et al (2005) monitored the effects of cool roofs on energy use and environmental parameters in six California buildings at three different sites: a retail store in Sacramento; an elementary school in San Marcos (near San Diego); and a four-building cold storage facility in Reedley (near Fresno). The latter included a cold storage building, a conditioning and fruit-palletizing area, a conditioned packing area, and two unconditioned packing areas. Results showed that installing a cool roof reduced the daily peak roof-surface temperature of each building by 33K to 42K. In the retail store building in Sacramento, for the monitored period of 8 August to 30 September 2002, the estimated savings in average air-conditioning energy use was about 72Wh per square metre per day (52 per cent). In the school building in San Marcos, for the monitored period of 8 July to 20 August 2002, the estimated savings in average air-conditioning energy use was about 42 to 48Wh per square metre per day (17 to 18 per cent). In the cold storage facility in Reedley, for the monitored period of 11 July to 14 September 2002, and 11 July to 18 August 2003, the estimated savings in average chiller energy use was about 57 to 81Wh per square metre per day (3 to 4 per cent). Using the measured data and calibrated simulations, Akbari et al (2005) extrapolated the results and estimated savings for similar buildings installing cool roofs in retrofit applications for all California climate zones.

In addition to these building monitoring studies, computer simulations of cooling-energy savings from increased roof albedo in residential and commercial buildings have been documented by many studies, including Konopacki and Akbari (1998), Akbari et al (1998), Parker et al (1998b) and Gartland et al (1996). Konopacki et al (1997) estimated the direct energy savings potential from high albedo roofs in 11 US metropolitan areas (see Figure 2.13). The results showed that four major building types account for over 90 per cent of the annual electricity and monetary savings in the US: pre-1980 residences (55 per cent); post-1980 residences (15 per cent); and office buildings and retail stores together (25 per cent). Furthermore, these four building types account for 93 per cent of the total air-conditioned roof area. Regional savings were found to be a function of three factors: energy savings in the air-conditioned residential and commercial building stock; the percentage of buildings that were air conditioned; and the aggregate regional roof area. Metropolitan-wide annual savings from the application of cool roofs on residential and commercial buildings were as low as US$3 million in the heating-dominated climate of Philadelphia, and as much as US$37 million for Phoenix and US$35 milion in Los Angeles.

The results for the 11 metropolitan statistical areas (MSAs) were extrapolated to estimate the savings in the entire US. At 8 cents per kilowatt hour (kWh), the value of US potential nationwide net commercial and

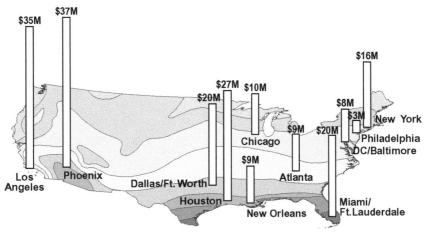

*Note:* About ten residential and commercial building prototypes in each area are simulated. Both savings in cooling and penalties in heating are considered. The estimated saving potential is about US$175 million (1997 energy prices) per year for the 11 cities. Extrapolated national energy savings are about US$0.75 billion per year.
*Source:* Konopacki et al (1997)

**Figure 2.13** *Estimated energy-saving potentials of light-coloured roofs in 11 US metropolitan areas*

residential energy savings (cooling savings minus heating penalties) exceeds US$750 million per year (Akbari et al, 1999a). The study estimates that, nationwide, light-coloured roofing could produce savings of about 10TWh per year (about 3.0 per cent of the national cooling-electricity use in residential and commercial buildings), an increase in natural gas (heating) use by 26 giga ($10^9$) British thermal units (Gbtu) per year (1.6 per cent), and a decrease in peak electrical demand of 7GW (2.5 per cent) (equivalent to 14 power plants each with a capacity of 0.5GW).

Analysis of the scale of urban energy savings potential was further refined for five cities: Baton Rouge, Los Angeles; Chicago, Illinois; Houston, Texas; Sacramento, California; and Salt Lake City, Utah; by Konopacki and Akbari (2000a, 2000b, 2002). The study included the direct and indirect effects of both cool roofs and trees. The direct saving potentials for cool roofs in these five metropolitan areas ranged from US$8 million to $38 million (see Table 2.4 and its notes for details).

*Indirect energy and smog benefits* Indirect effects require that a large fraction of the urban area be modified to produce a change in the local climate. To date, results have been attained only by computer simulations. Using the Los Angeles Basin as a case study, Taha (1996, 1997) examined the impacts of using cool surfaces (cool roofs and pavements) on urban air temperature and thus on cooling-energy use and smog. In these simulations, Taha estimates that about 50 per cent of the urbanized area in the Los Angeles Basin is covered by

Table 2.4 Metropolitan-wide estimates of annual energy savings, peak power avoided and annual carbon emissions reduction from heat-island reduction strategies for residential and commercial buildings in Baton Rouge, Chicago, Houston, Sacramento and Salt Lake City

| Metropolitan area and heat-island reduction strategy | Annual energy (US$ million) | Annual electricity (GWh) | Annual electricity (US$ million) | Annual natural gas (million therms) | Annual natural gas (US$ million) | Peak power (MW) | Annual carbon (thousand tonnes) |
|---|---|---|---|---|---|---|---|
| **Baton Rouge** | | | | | | | |
| Base Case | 114.8 | 1275 | 92.8 | 30.7 | 21.9 | 858 | 257 |
| Savings | | | | | | | |
| Direct shade trees | 5.2 | 94 | 6.9 | (2.4) | (1.7) | 62 | 12 |
| Direct high albedo | 8.0 | 120 | 8.7 | (1.0) | (0.7) | 60 | 19 |
| Indirect | 2.3 | 39 | 2.8 | (0.7) | (0.5) | 13 | 6 |
| Combined | 15.5 | 253 | 18.4 | (4.1) | (2.9) | 135 | 36 |
| **Chicago** | | | | | | | |
| Base case | 879.4 | 3505 | 293.4 | 804.3 | 586.0 | 3456 | 1749 |
| Savings | | | | | | | |
| Direct shade trees | 13.5 | 293 | 25.0 | (15.6) | (11.4) | 128 | 26 |
| Direct high albedo | 10.9 | 224 | 18.9 | (11.0) | (8.1) | 237 | 21 |
| Indirect | 5.4 | 65 | 5.6 | (0.3) | (0.2) | 33 | 10 |
| Combined | 29.8 | 582 | 49.5 | (26.9) | (19.7) | 398 | 58 |
| **Houston** | | | | | | | |
| Base case | 696.6 | 7230 | 572.0 | 169.7 | 124.7 | 5158 | 1453 |
| Savings | | | | | | | |
| Direct shade trees | 27.8 | 421 | 34.3 | (8.8) | (6.5) | 247 | 58 |
| Direct high albedo | 38.3 | 523 | 42.0 | (5.0) | (3.7) | 269 | 80 |
| Indirect | 15.6 | 236 | 19.1 | (4.7) | (3.5) | 218 | 33 |
| Combined | 81.8 | 1181 | 95.4 | (18.5) | (13.6) | 734 | 170 |

**Sacramento**

| | | | | | | | |
|---|---|---|---|---|---|---|---|
| *Base case* | *296.2* | *2238* | *185.9* | *162.2* | *110.3* | *2454* | *608* |
| Savings | | | | | | | |
| Direct shade trees | 9.8 | 247 | 20.6 | (15.8) | (10.7) | 180 | 18 |
| Direct high albedo | 14.6 | 220 | 18.3 | (5.5) | (3.8) | 163 | 29 |
| Indirect | 5.9 | 114 | 9.5 | (5.3) | (3.6) | 106 | 11 |
| Combined | 30.3 | 581 | 48.4 | (26.6) | (18.1) | 449 | 59 |

**Salt Lake City**

| | | | | | | | |
|---|---|---|---|---|---|---|---|
| *Base case* | *67.0* | *511* | *31.4* | *70.8* | *35.6* | *488* | *188* |
| Savings | | | | | | | |
| Direct shade tree | 1.1 | 52 | 3.3 | (4.2) | (2.2) | 33 | 3 |
| Direct high albedo | 1.8 | 45 | 2.8 | (2.0) | (1.0) | 32 | 5 |
| Indirect | 0.8 | 25 | 1.6 | (1.6) | (0.8) | 20 | 2 |
| Combined | 3.7 | 122 | 7.7 | (7.8) | (4.0) | 85 | 9 |

*Notes:* The methodology consisted of the following:

1 define prototypical building characteristics in detail for old and new construction;

2 simulate annual energy use and peak power demand using the DOE-2.1E model;

3 determine direct and indirect energy benefits from high albedo surfaces (roofs and pavements) and trees;

4 identify the total roof area of air-conditioned buildings in each city; and

5 calculate the metropolitan-wide impact of heat-island reduction (HIR) strategies.

Base energy expenditures and peak power demand are calculated for buildings without shade trees and with a dark roof (albedo = 0.2). Direct savings are determined for buildings with eight shade trees (retail: four) and a high-albedo roof (residential = 0.5 and commercial = 0.6), and indirect savings include the impact of reduced air temperature from urban reforestation and high albedo surfaces.

The conversion from GWh to carbon corresponds to the US mix of electricity. 1997 regional electricity and gas costs are used in the calculations (EIA, 1997). EIA (1997) shows that 3000TWh sold emitted 500 million metric tonnes of carbon (MtC); thus, 1GWh emits 167tC. The estimated carbon emission from combustion of natural gas is 1.447kgC per therm.

*Source:* Konopacki and Akbari (2000a, 2000b, 2002)

**Table 2.5** *Energy savings, ozone reduction and avoided peak power resulting from the use of cool roofs in the Los Angeles Basin*

| Benefits | Direct | Indirect | Smog | Total |
|---|---|---|---|---|
| 1   Cost savings from cool roofs | | | | |
| (US$ million per year) | 46 | 21 | 104 | 171 |
| 2   Δ Peak power (GW) | 0.4 | 0.2 | | 0.6 |
| 3   Present value per 100m² | | | | |
| of roof area (US$) | 153 | 25 | 125 | 303 |

*Source:* Rosenfeld et al (1998)

roofs and roads, the albedos of which can realistically be raised by 0.30 when they undergo normal repairs. This results in a 2K cooling at 3 pm during an August episode. This summertime temperature reduction has a significant effect on further reducing building cooling-energy use. The annual savings in Los Angeles are estimated at US$21 million (Rosenfeld et al, 1998).

Taha (1997) also simulated the impact of urban-wide cooling in Los Angeles on smog – predicting a reduction of 10 to 20 per cent in population-weighted smog (ozone). In Los Angeles, where smog is especially serious, the potential savings were valued at US$104 million per year (Rosenfeld et al, 1998; see Table 2.5). Table 2.5 also shows the present value (*PV*) of all future savings associated with installation of cool roofs. The *PV* of future savings from the installation of cool roofs is calculated using:

$$PV = a \frac{1 - (1 + d)^{-n}}{d} \qquad [1]$$

where:
- *a* = annual savings (US$);
- *d* = real discount rate (3 per cent); and
- *n* = life of the savings from cool roofs, in years.

In a more recent study, Akbari and Konopacki (2005) developed summary tables (sorted by heating and cooling degree days) to estimate the potential of heat island reduction (HIR) strategies (i.e. solar-reflective roofs, shade trees, reflective pavements and urban vegetation) to reduce cooling-energy use in buildings. Their tables provide estimates of savings for both direct effect and indirect effect (see Table 2.6 for a summary of their results). The estimated savings in Table 2.6 includes both direct and indirect effects of cool roofs, cool pavements and shade trees. About 50 per cent of the savings are the direct savings from the application of cool roofs. The estimated indirect savings from the combined effects of cool roofs, shade trees and cool pavements are about 25 per cent. The study does not address the smog benefits from HIR.

**Table 2.6** *Estimated ranges of annual base case (electricity use, gas use and peak demand) and savings from heat-island reduction measures across all climate regions*

| Prototype building | Electricity (kWh/100m²) | | Gas (therm/100m²) | | Peak power (kW/100m²) | |
|---|---|---|---|---|---|---|
| | Base use | Savings | Base use | Penalties | Base use | Savings |
| **Residential** | | | | | | |
| Pre-1980 gas heated | 1600–11,000 | 400–1200 | 0–1000 | 0–50 | 3.1–4.0 | 0.4–0.6 |
| Pre-1980 electrically heated | 8500–20,000 | 100–1200 | | | 3.1–4.0 | 0.4–0.6 |
| 1980+ gas heated | 700–7000 | 150–700 | 0–500 | 0–20 | 1.7–3.3 | 0.2–0.4 |
| 1980+ electrically heated | 5000–9000 | 50–600 | | | 1.7–3.3 | 0.2–0.4 |
| **Office** | | | | | | |
| Pre-1980 gas heated | 7000–18,700 | 1200–1400 | 0–500 | 0–20 | 6.3–8.4 | 0.5–1.0 |
| Pre-1980 electrically heated | 12,600–18,700 | 1100–1300 | | | 6.3–8.4 | 0.5–1.0 |
| 1980+ gas heated | 3500–10,800 | 500–600 | 0–300 | 0–10 | 3.5–4.6 | 0.2–0.5 |
| 1980+ electrically heated | 5700–10,800 | 300–600 | | | 3.5–4.6 | 0.2–0.5 |
| **Retail store** | | | | | | |
| Pre-1980 gas heated | 8200–15,700 | 1400–1500 | 0–200 | 0–10 | 4.5–5.7 | 0.4–0.7 |
| Pre-1980 electrically heated | 10,700–17,200 | 1300–1700 | | | 4.1–5.7 | 0.4–0.7 |
| 1980+ gas heated | 3100–8900 | 500–700 | 0–60 | 0–6 | 2.2–2.8 | 0.2–0.3 |
| 1980+ electrically heated | 4000–8900 | 300–700 | | | 2.2–2.8 | 0.2–0.3 |

*Source:* Akbari and Konopacki (2005)

*Other benefits of cool roofs* Another benefit of a light-coloured roof is a potential increase in its useful life. The diurnal temperature fluctuation and concomitant expansion and contraction of a light-coloured roof is smaller than that of a dark one. Also, the degradation of materials resulting from the absorption of ultraviolet light is a temperature-dependent process. For these reasons, cooler roofs may last longer than hot roofs of the same material.

## Potential problems with cool roofs

Several possible problems may arise from the use of reflective roofing materials (Bretz and Akbari, 1994, 1997). A drastic increase in the overall albedo of the many roofs in a city has the potential to create glare and visual discomfort. Besides being unpleasant, extreme glare could possibly increase the incidence of traffic accidents. Fortunately, the glare from roofs is not a major problem for those who are at street level.

In addition, many types of building materials, such as tar roofing, are not well adapted to painting. Although such materials could be specially designed to have a higher albedo, this would entail a greater expense than painting. Additionally, to maintain a high albedo, roofs may need to be recoated or washed on a regular basis. The cost of a regular maintenance programme could be significant.

A possible conflict of great concern is the fact that building owners and architects like to choose their preferred colour for their rooftops. This is particularly a concern for sloped roofs. The roofing industry has responded to this concern by developing and marketing cool-coloured materials for roofs (see the following section on 'Cool-coloured roofing materials').

## Cost of cool roofs

To change the albedo, the rooftops of buildings may be painted or covered with a new material. Since most roofs have regular maintenance schedules or need to be re-roofed or recoated periodically, the change in albedo should be done at these times to minimize the costs.

High albedo alternatives to conventional roofing materials are usually available, often at little or no additional cost. For example, a built-up roof typically has a coating or a protective layer of mineral granules or gravel. In such conditions, it is expected that choosing a reflective material at the time of installation should not add to the cost of the roof. Also, roofing shingles are available in a variety of colours, including white, at the same price. The incremental price premium for choosing a non-black single-ply membrane roofing material is less than 10 per cent. Cool roofing materials that require an initial investment may turn out to be more attractive in terms of life-cycle cost than conventional dark alternatives. Usually, the lower life-cycle cost results from longer roof life and/or energy savings.

## Cool-coloured roofing materials

Suitable cool white materials are available for most products, with the notable exception of asphalt shingles; cooler-coloured (non-white) materials are needed for all types of roofing, especially in the residential market. Coatings coloured with conventional pigments tend to absorb the invisible 'near-infrared' (NIR) radiation that bears more than half of the power in sunlight (see Figure 2.14). Replacing conventional pigments with 'cool' pigments that absorb less NIR radiation can yield coloured coatings that look the same to the eye, but have higher solar reflectance. These cool coatings lower roof surface temperature,

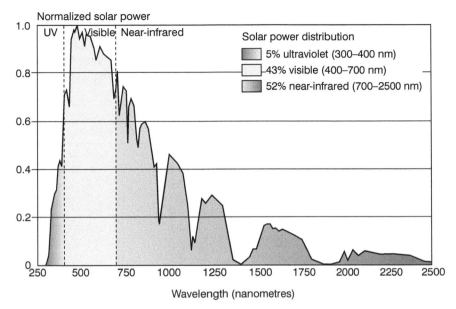

*Source:* Akbari et al (2006)

**Figure 2.14** *Peak-normalized solar spectral power; over half of all solar power arrives as invisible 'near-infrared' radiation*

reducing the need for cooling energy in conditioned buildings and making unconditioned buildings more comfortable.

According to the magazine *Western Roofing, Insulation and Siding* (2002), the total value of the 2002 projected residential roofing market in 14 western US states (Alaska, Arizona, California, Colorado, Hawaii, Idaho, Montana, Nevada, New Mexico, Oregon, Texas, Utah, Washington and Wyoming) was about US$3.6 billion. We estimate that 40 per cent (US$1.4 billion) of that amount was spent in California. The lion's share of residential roofing expenditure was for fibreglass shingle, which accounted for US$1.7 billion, or 47 per cent of sales. Concrete and clay roof tiles made up US$0.95 billion (27 per cent), while wood, metal and slate roofing collectively represented another US$0.55 billion (15 per cent). The value of all other roofing projects was about US$0.41 billion (11 per cent). We estimate that the roofing market area distribution was 54 to 58 per cent fibreglass shingle, 8 to 10 per cent concrete tile, 8 to 10 per cent clay tile, 7 per cent metal, 3 per cent wood shake, and 3 per cent slate (see Table 2.7).

Suitable cool white materials are available for most roofing products, with the notable exception (prior to March 2005)[3] of asphalt shingles. Cool non-white materials are needed for all types of roofing. Industry researchers have developed complex inorganic colour pigments that are dark in colour, but highly reflective in the near-infrared portion of the solar spectrum. The high near-infrared reflectance of coatings formulated with these and other 'cool'

**Table 2.7** *Projected residential roofing market in the US western region surveyed by* Western Roofing, Insulation and Siding Magazine *(2002); the 14 states included in the US western region are Alaska, Arizona, California, Colorado, Hawaii, Idaho, Montana, Nevada, New Mexico, Oregon, Texas, Utah, Washington and Wyoming*

| Roofing type | Market share by US$ | | Estimated market share by roofing area |
| --- | --- | --- | --- |
| | US$ billion | Percentage | Percentage |
| Fibreglass shingle | 1.70 | 47.2 | 53.6–57.5 |
| Concrete tile | 0.50 | 13.8 | 8.4–10.4 |
| Clay tile | 0.45 | 12.6 | 7.7–9.5 |
| Wood shingle/shake | 0.17 | 4.7 | 2.9–3.6 |
| Metal/architectural | 0.21 | 5.9 | 6.7–7.2 |
| Slate | 0.17 | 4.7 | 2.9–3.6 |
| Other | 0.13 | 3.6 | 4.1–4.4 |
| SBC modified | 0.08 | 2.1 | 2.4–2.6 |
| APP modified | 0.07 | 1.9 | 2.2–2.3 |
| Metal/structural | 0.07 | 1.9 | 2.2–2.3 |
| Cementitious | 0.04 | 1.1 | 1.2–1.3 |
| Organic shingles | 0.02 | 0.5 | 0.6 |
| Total | 3.60 | 100 | 100 |

*Note:* SBC = Styrene Block Copolymer;
APP = Atactic Polypropylane Polymer.
*Source: Western Roofing, Insulation and Siding* (2002)

pigments – for example, chromium oxide green, cobalt blue, phthalocyanine blue and Hansa yellow – can be exploited to manufacture roofing materials that reflect more sunlight than conventionally pigmented roofing products.

Cool-coloured roofing materials are expected to penetrate the roofing market within the next few years. Preliminary analysis suggests that they may cost up to US$1 per square metre more than conventionally coloured roofing materials. However, this would raise the total cost of a new roof (material plus labour) by only 2 to 5 per cent.

A US consortium (Cool Team) of 2 national research laboratories and 12 companies that manufacture roofing materials, including shingles, roofing granules, clay tiles, concrete tiles, tile coatings, metal panels, metal coatings and pigments, are collaborating to expedite the manufacturing of cool-coloured roofing materials (Akbari et al, 2004, 2006). The iterative development of cool-coloured materials has included selection of cool pigments, choice of base coats for the two-layer applications (discussed later in this chapter) and identification of pigments to avoid.

*Creating cool non-white coatings* In order to determine how to optimize the solar reflectance of a pigmented coating matching a particular colour, and how the performance of cool-coloured roofing products compares to those of standard materials, the Cool Team has:

- identified and characterized the optical properties of over 100 pigmented coatings;
- created a database of pigment characteristics; and
- developed a model to maximize the solar reflectance of roofing materials for a choice of visible colour.

The LBNL Cool Team measured the spectral reflectance $r$ and transmittance $t$ of a thin coating containing a single pigment or a binary mix of pigments (Levinson et al, 2005a, 2005b). These spectral, or wavelength-dependent, properties of the pigmented coating were measured at 441 evenly spaced wavelengths spanning the solar spectrum (300 to 2500 nanometres). Then, using a modified version of the Kebelka-Monk two-flux model, each sample was characterized by its computed spectral absorption coefficient, $K$, and backscattering coefficient, $S$. A cool colour is defined by a large absorption coefficient $K$ in parts of the visible spectral range in order to permit the attainment of desired colours and a small $K$ in the NIR. For cool colours, the $S$ is small (or large) in the visible spectral range for formulating dark (or light) colours and large in the NIR.

Inspection of the film's spectral absorptance (calculated as $1 - r - t$) reveals whether a pigmented coating is cool (has low NIR absorptance) or hot (has high NIR absorptance). The spectral reflectance and transmittance measurements were also used to compute spectral rates of light absorption and backscattering (reflection) per unit depth of film. The spectral reflectance of a coating coloured with a mixture of pigments can then be estimated from the spectral absorption and backscattering rates of its components. The results of these measurements and analyses are summarized in a database detailing the optical properties of the characterized pigmented coatings.

## Creating cool non-white roofing products

Roofing shingles, tiles and metal panels comprise more than 90 per cent (by roof area) of the residential roofing market in the US. The Cool Team has evaluated the best ways to increase the solar reflectance of these products and to produce cool roofing materials. As the direct result of this collaborative effort, manufactures of roofing materials have introduced cool shingles, clay tiles, concrete tiles, metal roofs and concrete tile coatings.

In addition to using NIR-reflective pigments in the manufacturing of cool roofing materials, application of novel engineering techniques can further economically enhance the solar reflectance of coloured roofing materials. Cool-coloured pigments are partly transparent to NIR light; thus, any NIR light not reflected by the cool pigment is transmitted to the layer underneath, where it can be absorbed. To increase the solar reflectance of coloured materials with cool pigments, a reflective undercoating can be used. This method is referred to as a two-layered technique.

Figure 2.15 demonstrates the application of the two-layered technique to manufacture cool-coloured materials. A thin layer of dioxazine purple (14μm to 27μm) is applied on four substrates:

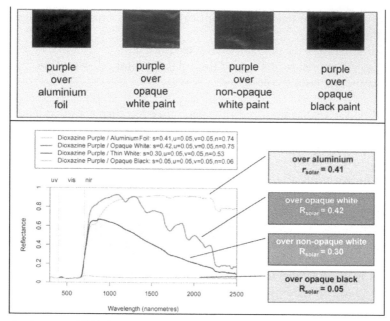

*Source:* Levinson et al (2007)

**Figure 2.15** *Application of the two-layered technique to manufacture cool-coloured materials*

1    aluminium foil (~ 25µm);
2    opaque white paint (~ 1000µm);
3    non-opaque white paint (~ 25µm); and
4    opaque black paint (~ 25µm).

It appears to the eye (and is confirmed by the visible reflectance spectrum) that the colour of the material is black even when applied to an opaque white or aluminium foil substrate. However, the solar reflectance of these samples exceed 0.4; the solar reflectance over a black substrate is only 0.05.

*Cool-coloured shingles* The solar reflectance of a new shingle, by design, is dominated by the solar reflectance of its granules, which cover over 97 per cent of its surface. Manufacturers use granules coated with titanium dioxide ($TiO_2$) rutile white to produce white (or grey) shingles. Because a thin $TiO_2$-pigmented coating is reflective but not opaque in the NIR, multiple layers are needed to obtain high solar reflectance. This technique has been used to produce 'super-white' (meaning truly white, rather than grey) granulated shingles with solar reflectances exceeding 0.5 (see Figure 2.16).

Although white roofing materials are popular in some areas (e.g. Greece; see Figure 2.17), many consumers aesthetically prefer non-white roofs. Manufacturers have also tried to produce coloured granules with high solar reflectance by using non-white pigments with high NIR reflectance. To increase

| Black Shingle | Conventional White Shingle | Advanced White Shingle |
|:---:|:---:|:---:|
|  | | |
| $r = 5\%, t = 180°F$ | $r = 29\%, t = 157°F$ | $r = 60\%, t = 128°F$ |

*Source:* Akbari's personal files

**Figure 2.16** *Development of super-white shingles*

the solar reflectance of coloured granules with cool pigments, multiple colour layers, a reflective undercoating and/or reflective aggregate should be used. Obviously, each additional coating increases the cost of production.

Several prototype cool-coloured fibreglass asphalt shingles with solar reflectances ranging from 0.28 to 0.37 were developed in 2004 and 2005. Also, in 2005, a major manufacturer of roofing shingles in California announced the availability of cool-coloured shingles in three popular colours.

*Cool-coloured tiles and tile coatings* Clay and concrete tiles are used in many areas around the world. In the US, clay and concrete tiles are especially popular in the hot climate regions. There are three ways of improving the solar

*Source:* Akbari's personal files

**Figure 2.17** *White roofs and walls are used in Santorini, Greece*

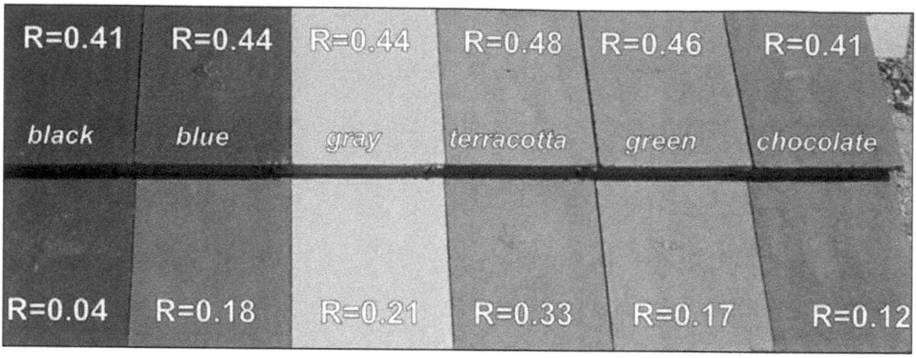

*Source:* Akbari et al (2004)

**Figure 2.18** *Palette of colour-matched cool (top row) and conventional (bottom row) roof-tile coatings developed by industrial partner American Rooftile Coatings; shown on each coated tile is its solar reflectance (R)*

reflectance of coloured tiles:

1   Use clay or concrete with low concentrations of light-absorbing impurities, such as iron oxides and elemental carbon.
2   Colour the tile with cool pigments contained in a surface coating or mixed integrally.
3   Include an NIR-reflective (e.g. white) sub-layer beneath an NIR-transmitting coloured topcoat.

Although all of these options are, in principle, easy to implement, they may require changes in the current production techniques that may add to the cost of the finished products. Colorants can be included throughout the body of the tile or used in a surface coating. Both methods need to be addressed.

The American Rooftile Coating Company has developed a palette of cool non-white coatings for concrete tiles. Each of the cool-coloured coatings shown in Figure 2.18 has a solar reflectance higher than 0.40. The solar

**Table 2.8** *Sample cool-coloured clay tiles and their solar reflectances*

| Model | Initial solar reflectance | Solar reflectance after three years |
|---|---|---|
| Weathered green blend | 0.43 | 0.49 |
| Natural red | 0.43 | 0.38 |
| Brick red | 0.42 | 0.40 |
| White buff | 0.68 | 0.56 |
| Tobacco | 0.43 | 0.41 |
| Peach buff | 0.61 | 0.48 |
| Regency blue | 0.38 | 0.34 |
| Light cactus green | 0.51 | 0.52 |

*Source:* www.MCATile.com

reflectance of each cool coating exceeds that of a colour-matched, conventionally pigmented coating by 0.15 (terracotta) to 0.37 (black). MCA-Tile manufactures clay tiles in many colours (glazed and unglazed) with solar reflectance greater than 0.4 (see Table 2.8).

Synnefa and Santamouris (2006) and Synnefa et al (2006) have also measured the solar spectral reflectance of ten prototype cool-coloured coatings, developed at the National and Kapodistrian University of Athens. These coatings are developed to be used as measures to reduce the summertime cooling-energy use in buildings and to reduce summertime urban temperatures.

*Cool-coloured metal panels* Metal roofing materials are installed on a small (but growing) fraction of US residential roofs. Historically, metal roofs have had only about 3 per cent of the residential market. However, the architectural appeal, flexibility and durability, due in part to the cool-coloured pigments, has steadily increased the sales of painted metal roofing, and as of 2002, its sales volume has increased to 8.9 per cent of the residential market, making it the fastest growing residential roofing product (F. W. Dodge, 2005). Metal roofs are available in many colours and can simulate the shape and form of many other roofing materials. Application of cool-coloured pigments in metal roofing materials may require the fewest number of changes (and, in many cases, no changes) to the existing production processes. In fact, cool pigments have been incorporated within paint systems used for metal roofing since 2002. For example, the BASF Industrial Coatings line of cool coatings for metal includes over 20 cool-coloured products. As in the cases of tile and asphalt shingle, cool pigments can be applied to metal via a single- or double-layered technique. If the metal substrate is highly reflective, a single-layered technique may suffice. The coatings for metal shingles are thin durable polymer materials. These thin layers use materials efficiently, but limit the maximum amount of pigment present. However, the metal substrate can provide some NIR reflectance if the coating is transparent in the NIR. Several manufactures have developed cool-coloured metal roof products.

Cool non-white coatings have been enthusiastically adopted by premium coil coaters and metal roofing manufacturers. Metal panels and clay tiles were the first types of roofing to be produced in cool colours. BASF Industrial Coatings (Southfield, Minnesota) has launched a line of cool-coloured siliconized-polyester coatings that is quickly replacing their conventional siliconized-polyester coatings. Steelscape Inc (Kalama, Washington) has recently introduced a cool polyvinylidene difluoride (PVDF) coating for the metal building industry. Custom-Bilt Metals (Chino, California) has switched more than 250 of its metal roofing products to cool colours. The Cool Team is currently testing a cool-coloured metal roof on a demonstration house in Sacramento.

*Durability of cool non-white coatings* The durability of cool materials has been tested in weatherometers after being exposed to 5000 hours of xenon-arc light and to about 10,000 hours of fluorescent light. Figure 2.19 compares the

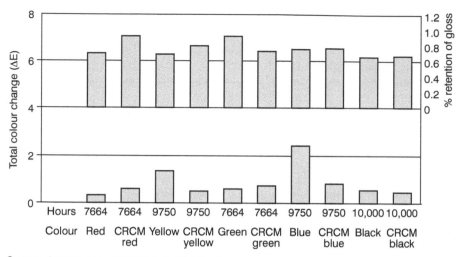

*Source:* data courtesy of BASF Industrial Coatings

**Figure 2.19** *Fade resistance and gloss retention of painted metals*

total colour change and reduction in gloss of cool-roofing coloured metals and standard coloured metals exposed to accelerated fluorescent ultraviolet (UV) light. In almost all cases, cool materials have performed better than standard materials.

## Cool pavements

The practice of widespread paving of city streets with asphalt began only during the past century. The advantages of this smooth and all-weather surface for the movement of bicycles and automobiles are obvious; but some of the associated problems are perhaps not readily noticed. One consequence of streets covered with dark asphalt surfaces is that the pavements heat the air, increasing the temperature of the city. Measured data clearly indicate that changing the pavement albedo has a significant effect on the pavement surface temperature. If urban surfaces were lighter in colour, more of the incoming light would be reflected back into space and the surfaces and the air would be cooler. This tends to reduce the need for air conditioning. Pomerantz et al (1997) present an overview of cool paving materials for urban heat-island mitigation.

Urban pavements are made predominantly of asphalt concrete that is dark in colour. The challenge is to develop cool pavements that are economical and practical. Figure 2.20 shows some measurements of the effect of albedo on pavement temperature. The data clearly indicate that significant modification of the pavement temperature can be achieved: a 10K decrease in temperature for a 0.25 increase in albedo.

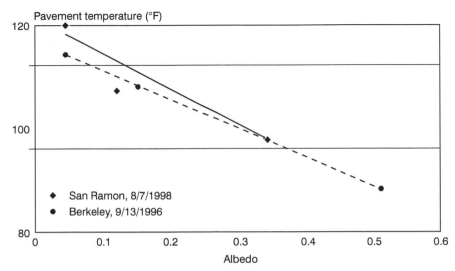

*Note:* Data in Berkeley, California, were taken at about 3 pm on new, old and light-colour coated asphalt pavements. The data from San Ramon, California, were taken at about 3 pm on four asphalt concrete and one cement concrete (albedo = 0.35) pavements.
*Source:* Pomerantz et al (1997)

**Figure 2.20** *The dependence of pavement surface temperature upon albedo*

## Energy and smog benefits of cool pavements

Cool pavements affect energy use and air quality through lowered ambient temperatures. Lower temperature has two important effects:

1   reduced demand for electricity for air conditioning; and
2   decreased production of smog (ozone).

Rosenfeld et al (1998) estimated the cost savings of reduced demand for electricity and of the externalities of lower ozone concentrations in the Los Angeles Basin. Simulations for the Los Angeles (LA) Basin indicate that a reasonable change in the albedo of the city could cause a noticeable decrease in temperature. Taha (1997) predicted a 1.5K decrease in temperature of the downtown area. The lower temperatures in the city are calculated based on the assumption that all roads and roofs are improved. From the meteorological simulations of three days in each season, the temperature changes for every day in a typical year were estimated for Burbank, typical of the hottest one third of the LA Basin. The energy consumptions of typical buildings were then simulated for the original weather and also for the modified weather. The differences are the annual energy changes due to the decrease in ambient temperature. The result is a citywide annual saving of about US$71 million, due to combined albedo and vegetation changes. The kilowatt hour savings attributable to the pavement are US$15 million per year, or US$0.012 per square metre per year. Analysis of the hourly demand indicates that cooler pavements could save an estimated 100MW of peak power in Los Angeles.

The simulations of the effects of higher albedo on smog formation indicate that an albedo change of 0.3 throughout the developed 25 per cent of the city would yield a 12 per cent decrease in the population-weighted ozone exceeding the California air-quality standard (Taha, 1997). The estimated annual cost to the residents of Los Angeles, because of air quality-related medical costs and lost work time, is about US$10 billion (Hall et al, 1992). The greater part of pollution is particulates; but the ozone contribution averages about US$3 billion per year. Assuming a proportional relationship of the cost with the amount of smog exceeding standards, the cooler-surfaced city would save 12 per cent of US$3 billion per year, or US$360 million per year. As above, we attribute about 21 per cent of the saving to pavements. Rosenfeld et al (1998) value that the benefits from smog improvement by altering the albedo of all 1250 square kilometres of pavements by 0.25 saves about US$76 million per year (about US$0.06 per square metre per year).

## Other benefits of cool pavements

It has long been known that the temperature of a pavement affects its performance (Yoder and Witzak, 1975). This has been emphasized by the new system of binder specification advocated by the Strategic Highway Research Program (SHRP). Beginning in 1987, this programme led pavement experts to carry out the task of researching and then recommending the best methods of making asphalt concrete pavements (Monismith et al, 1994). A result of this study was the issuance of specifications for the asphalt binder. The temperature range which the pavement will endure is a primary consideration (Cominsky et al, 1994). The performance grade (PG) is specified by two temperatures:

1   the average seven-day maximum temperature that the pavement will probably encounter; and
2   the minimum temperature the pavement will probably attain.

Reflectivity of pavements is also a safety factor in visibility at night and in wet weather, affecting the demand for electric street lighting. Street lighting is more effective if pavements are more reflective, which can lead to greater safety; or, alternatively, less lighting could be used to obtain the same visibility. These benefits have not yet been monetized.

## Potential problems with cool pavements

A practical drawback of high reflectivity is glare; but this does not appear to be a problem. Instead of black asphalt, with an albedo of about 0.05 to 0.12, we suggest the application of a product with an albedo of about 0.35, similar to that of cement concrete. The experiment to test whether this will be a problem has already been performed: every day millions of people drive on cement concrete roads, and we rarely hear of accidents caused by glare or of people even complaining about the glare on such roads.

There is also a concern that, after some time, light-coloured pavement will darken because of dirt. Again, experience with cement concrete roads suggests that the light colour of the pavement persists after long usage. Most drivers can see the difference in reflection between an asphalt and a cement concrete road when they drive over them, even when the roads are old.

## Cost of cool pavements

It is clear that cooler pavements will have energy, environmental and engineering benefits. The issue is, then, whether there are ways of constructing cool pavements that are feasible and economical. The economic question is whether the savings generated by a cool pavement over its lifetime are greater than its extra cost. Properly, one should distinguish between initial cost and lifetime costs (including maintenance, repair time and length of service of the road). Often the initial cost is decisive.

*Thick pavements* A typical asphalt concrete contains about 7 per cent of asphalt by weight, or about 17 per cent by volume; the remainder is rock aggregate, except for a few per cent of voids. The cost of ordinary asphalt (1998 prices) is about US$125 per tonne, and the price of aggregate is about US$20 per tonne, exclusive of transportation costs. Thus, in 1 tonne of mixed asphalt concrete the cost of materials only is about US$28 per tonne, of which about US$9 is in the binder and US$19 is in the aggregate. For a pavement about 10cm thick (4 inches), with a density of 2.1 tonnes per cubic metre, the cost of the binder alone is about US$2 per square metre and aggregate costs about US$4.2 per square metre.

Experimentally, the albedo of a fresh asphalt concrete pavement is about 0.05 (Pomerantz et al, 1997) because the relatively small amount of black asphalt coats the lighter-coloured aggregate. As an asphalt concrete pavement is worn down and the aggregate is revealed, we observed an albedo increase to about 0.10 to 0.15 for ordinary aggregate. If it were made with a reflective aggregate, we could expect the long-term albedo to approach that of the aggregate.

Using the assumptions for Los Angeles, a cooler pavement would generate a stream of savings of US$0.07 per square kilometre per year for the lifetime of the road, about 20 years. At a real interest rate of 3 per cent per year, the present value of potential savings is estimated at US$1.1 per square metre. This saving would allow for the purchase of a binder costing US$3 per square metre, instead of US$2 per square metre, or 50 per cent more expensive. Or, one could buy aggregate; instead of spending US$4.2 per square metre, one can now afford US$5.2 per square metre (a 20 per cent more expensive, whiter aggregate).

In the special case of a climate in which the pavement is subjected to wide temperature swings, additional savings accrue because higher quality binders may be avoided. Note, importantly and logically, that it is the pavement temperature and not the air temperature that is considered in specifying a binder. If an asphalt pavement may be exposed to large temperature variations over the year, the binder must be specially formulated to handle the expansion, contraction and viscosity changes between the maximum and minimum temperatures. There is a rule of thumb in the industry, 'Rule of 90', that when the difference of these temperatures is greater than 90°C, some kind of modification of the asphalt will be needed; this adds to the cost. The Rule of 90 arises because ordinary asphalt has difficulty in performing over wide temperature ranges. Additives, such as polymers, are needed to attain performance over a wide range. For example, if a binder is specified as PG 58-22, it is intended to function between 58°C and –22°C. The difference, 58 – (–22) = 80. An ordinary grade of asphalt binder will suffice; its cost is about US$125 per tonne. If, however, the pavement temperature varies between 76°C and –16°C, or PG 76-16, the difference 76 – (–16) = 92. An enhanced binder is recommended at a price of about US$165 per tonne (Bally, 1998): a 30 per cent increase in price. It may be possible to stay within the Rule of 90 and avoid the increased cost of binder if the pavement albedo is increased and the pavement does not get as hot. For a 10cm-thick new road, the cost of ordinary asphalt is US$2 per square metre, and higher grade asphalt costs US$2.6 per square metre. Instead of buying the higher grade binder, one could apply a chip seal, which costs about US$0.6 per square metre. Chip seals comprise a binder onto which aggregate is pressed. The aggregate is visible from the outset; if it is reflective, the pavement stays cooler. It might be sufficiently cool that it is unnecessary to use the higher grade binder. For example, the data show that a 0.25 increase in albedo can reduce the pavement temperature by 10K. This suggests that the maximum temperature specification for the pavements might be reduced by 10K, which means a lower grade of binder might then be acceptable. The reduced cost of the binder cancels the cost of the chip seal, and one enjoys the cooling benefit at no extra cost.

Thus, for thick pavements, the energy and smog savings may not pay directly for whiter roads. If, however, the lighter-coloured road leads to a substantially longer lifetime, the initial higher cost is offset by lifetime savings. An example of this is to be seen when a higher grade binder is replaceable by a whiter surface.

### Thin pavements

At some time in its life, a pavement needs to be maintained (i.e. resurfaced). This offers an opportunity to get cooler pavements economically. Good maintenance practice calls for resurfacing a new road within about ten years (Dunn, 1996), and the lifetime of resurfacing is only about five years. Hence, within ten years all of the asphalt concrete surfaces in a city can be made light coloured. As part of this regular maintenance, any additional cost of the whiter material will be minimized. Note also that because the lifetime of the resurfacing is only about five years, the present value of the savings is five times greater than the annual savings. Thus, for Los Angeles, the present value is about US$0.36 per square metre.

Can a pavement be resurfaced with a light colour at an added cost less than this saving? For resurfacing, there are the options of a black topping, such as a slurry seal, or a lighter-coloured surface achieved by using a chip seal. The costs of both of these are about the same: US$0.6 per square metre (Means, 2006). For a chip seal, about half the materials cost is aggregate and half is the binder. If special light-coloured aggregate is used in the chip seal, there will be an extra cost. For example, if the aggregate costs 50 per cent more, instead of US$0.3 per square metre, it will cost US$0.45 per square metre and the price of the chip seal will rise by US$0.15 per square metre. If the energy, environmental and durability benefits over the lifetime of the resurfacing exceed US$0.15 per square metre, the cooler pavement pays for itself. Again, this depends upon local circumstances: the climate and smog conditions versus the cost of light-coloured aggregate. For Los Angeles, we have estimated that energy and environmental savings alone are about US$0.36 per square metre (present value over the lifetime of five years for a resurfacing), and thus one could afford to pay twice the usual price for aggregate and still have no net increase in cost. Lifetime benefits would also accrue in addition to energy and smog benefits.

## Cool pavement materials

As stated earlier, most urban paved surfaces are either made of asphalt concrete (commonly referred to as asphalt pavements) or cement concrete (known as concrete pavements). Installing new pavements typically requires grading of the terrain and a new base course of rock. The thickness of this base and its preparation will depend upon the anticipated traffic. The topmost (wearing) course, which is relatively independent of the base, is the important part for the albedo of the pavement. A pavement is typically maintained (repaired and resurfaced several times) throughout its life. The maintenance usually involves resurfacing the topmost layer of pavement. This makes routine maintenance an ideal time for introducing light-colour surfaces to roads. The following is a brief description of various technologies used in the pavement industry.

*New pavements*
There are three main types of new pavements: asphalt concrete, cement concrete and porous paver. In general, a pavement consists of a binder (asphalt, tar or Portland cement) and aggregate (stones of various sizes down to sand). The function of the binder is to glue the aggregate together. The aggregate provides the strength, friction and resistance to wear, and the binder keeps the stones from dispersing under the forces of the traffic and weather.

*Asphalt concrete in new pavements.* Asphalt or bituminous materials are the most common binders of road surfaces (Asphalt Institute, 1989). The relative amount of asphalt and aggregate is about 1 part in 10 (typically about 7 per cent asphalt by weight, or 17 per cent by volume). This type of pavement is properly called 'asphalt concrete', suggestive of its composite nature. The fact that about 80 per cent of roads now in service are made of asphalt concrete is a result of its relatively low initial cost and ease of repair.

Asphalt is derived from petroleum. It is often the residue after lighter components, such as gasoline and kerosene, are fractioned from crude oil.

As such, it varies in composition depending upon the reservoir of origin and upon the fractionating process to which it is subjected. Compared to the Portland cement concrete, bituminous concrete is more flexible. This has the advantage that the wearing surface tends to conform to any movements of the sub-grade with less cracking; but too much softness can lead to spreading or rutting of the road. In particular, asphalt concrete softens more than Portland cement concrete at typical temperatures that roads attain.

*Cement concrete in new pavements.* Cement concrete consists of an inorganic binder or cement, which, after being mixed with water, can harden and hold together stony aggregate. The raw material of the cement contains lime (CaO), which is derived from limestone (calcium carbonate, $CaCO_3$) or oyster shells.

Portland cement contains clay, which has iron oxides, silica and alumina in it. The approximate composition (by weight percentage) of Portland cement is lime (60 per cent), silica (20 per cent), alumina (5 per cent), iron oxide (3 per cent), magnesia (2 per cent) and other (10 per cent) (Leighou, 1942). Depending upon the composition of the starting materials, a suitable mixture of them is ground together – for example, limestone contains 52 per cent lime and 3 per cent silica, but slag contains 42 per cent lime and 34 per cent silica, so the amount of clay (57 per cent silica) to be added would differ between limestone- and slag-based cements to get a final silica content of 20 per cent.

Concrete paving is the choice for very heavy traffic loads because the material does not deform as much as asphalt. In dry climates, for example, concrete is chosen when the traffic exceeds 70,000 cars per day. In wet climates, where the softer under-surface requires a stiffer road, concrete is preferred for traffic of 40,000 per day (Smart, 1994). However, the higher initial cost of concrete and the difficulty of modifying the surface favours the application of asphalt to roads that carry traffic in low volume and low weight, such as in residential areas and parking lots.

Cement is darkened by the presence of iron oxide, which can be reduced to get a whiter cement by using kaolin. Adding titanium dioxide makes cement whiter; but manganese oxide, present in slag, makes it browner. Measurements and literature searches (Pomerantz et al, 1997), give fresh cement concrete a solar reflectance of 0.35 to 0.40. As cement concrete ages it tends to get darker because of dirt, and the solar reflectance tends towards 0.25 to 0.30. Contrarily, asphalt concrete tends to get lighter as it ages because the black asphalt wears away to reveal the lighter aggregate.

It is possible to produce concrete with visible reflectivity approaching 68 per cent by using whiter cements and aggregates (Lehigh Cement, 1994).[4] The cement is white because the starting materials are selected to have low concentrations of coloured minerals, such as iron oxides. White aggregates, such as white sand, and some limestones are available at a cost premium of 10 to 20 per cent.

*Porous and grass pavers for new pavements.* Porous pavements are defined as pavements that deliberately allow water to pass through them. Permeability has the advantages of permitting rainwater to be stored in the earth, thus

*Source:* Akbari's personal files

**Figure 2.21** *Grasscrete lattices and a picture of a grasscrete application*

reducing the problems of flooding. A road surface made of grass has the added desirable qualities that the grass evapotranspires and thus cools the air above it, as well as being decorative. However, a grassy field as a parking lot or access road is soft when it is wet and is easily rutted permanently. These defects can be alleviated by enclosing the soil in a lattice structure that provides lateral containment. The lattice structure thus serves as a binder for the soil or gravel. We refer to such porous pavements as 'grass pavement'. All grass pavements must have sufficient water year round, which makes it ill-suited for dry climates.

Grass pavers are best suited for occasional use where perhaps one or two cars a day traverse it (e.g. parking for employees, sports facilities or overflow), or as fire lanes, because grass cannot survive frequent traffic. The lattices supporting the grass pavers are made either of concrete or plastic. Figure 2.21 shows two plastic grasscrete lattices and a picture of a grasscrete surface designed for car parking.

Another type of porous pavement is formed of asphalt or cement concrete (Brown, 1996) or is loosely packed so that water can percolate through it. To construct a permeable pavement entirely of asphalt or cement concrete, the aggregate is chosen to be a single size, usually about 9.5mm (0.37 inches) (so-called 'open-graded' aggregates). In the absence of fine aggregates and sand, the stones pack so loosely that there are channels through which moderate flows of water can filter (Asphalt Institute, 1974). This porous pavement is

*Source:* Akbari's personal files

**Figure 2.22** *Porous pavements*

usually placed over a solid pavement for strength and is domed so that the water leaks out the sides of the roadway. Blockage of the pores by dirt and fractures by freeze-thaw cycles may be problems. The porous surface has a safety advantage of avoiding standing water that can lead to aquaplaning by fast-moving vehicles. Another appealing benefit is that these surfaces tend to suppress tyre noise (Hugues and Heritier, 1995; Lefebvre and Marzin, 1995). Runoff of rainwater is reduced if it can percolate into the ground, relieving demand on a city's street drainage system. Figure 2.22 shows examples of asphalt concrete and cement concrete porous pavements.

*Tree-resin modified emulsions.* 'RoadOyl', a relatively new binder, is tan coloured because it is derived from pine tree pitch and resin (Loustalot et al, 1995). When it is mixed with stone or sand, it produces a light-coloured pavement. In the emulsified form it is water soluble, applied without heating and thus is particularly convenient to apply where access to large equipment is limited. After drying and setting it is water insoluble. It is comparable in strength to asphalt concrete in laboratory tests, but has not yet been extensively tested on city streets. RoadOyl comprises about 6 per cent by weight of the finished pavement. It is manufactured by Road Products Corp of Knoxville, Tennessee.

*Coal-tar resins.* In the south-eastern US, near coal-mining regions, coal-tar resins are used in a manner similar to asphalt binder. Because it is not applied much nationwide and is black, we will not discuss it any further here.

### Resurfacing pavements

*Asphaltic coatings.* Asphalt and asphalt-based materials are the most common for repair and resurfacing of roads (Raza, 1995). Asphalt adheres well to both older asphalt and to cement concrete. For large jobs, conventional hot-mix asphalt concrete at least an inch thick is commonly used.

Keeping asphalt in a fluid state is accomplished by having oil-fired heaters onboard the spreaders. For small repair jobs, room temperature bituminous binders have been developed. One such binder is asphalt dissolved in kerosene or creosote. This is called a 'cutback' asphalt. The solvent evaporates over a 'curing' time, after which the asphalt is hard. The emission of the organic solvents, however, has adverse effects on the environment, so the cutback asphalts have been superseded by water-soluble asphalt emulsions (AEMA, 1995). Here, the bitumen is ground to small particles and chemically treated with an emulsifier so that it remains in suspension in water. The emulsifier is chosen to be anionic or cationic to facilitate the wetting of the particular mineral aggregates that are mixed with the emulsion. After the spreading of the emulsion and aggregate, the water separates ('breaks') and evaporates harmlessly. The asphalt coats and binds the aggregate to form an asphalt concrete. Asphalt emulsions cost from 15 to 100 per cent more than bulk asphalt (Raza, 1995; Means, 2006). Emulsions have drying times of as little as a few hours, resulting in minimal disruption of traffic. A newer type of binder is formed by adding polymers to asphalt emulsions – this is called 'micro-surfacing',

*Source:* M. Pomerantz, personal files

**Figure 2.23** *Photo of a chip seal application (lower part of the picture) and asphalt pavement (upper part) in San José, California; note the lighter colour (higher solar reflectance) of the chip seal surface*

There are two general approaches to the resurfacing of existing pavements (Hunter, 1994). In both cases the new surface is a composite of binder and aggregate; the difference is whether these components are mixed after or before the binder is spread on the old surface. In a 'chip' seal application, the binder is spread first; the aggregate is then dropped on top of it and pressed into the binder. Otherwise, the aggregate and binder are premixed and then spread. The mixing is often done onboard the spreader vehicle just before the mixture is applied to the pavement. The premixed pavements are known as 'overlays', 'slurry coats', 'micro-surfaces', 'seal coats' or 'fog coats', depending upon the binder and the size of aggregate.

*Chip seal.* The binder in a chip seal is usually a fast-drying emulsified asphalt. As soon as possible after the binder is spread, the aggregate is dropped and rolled into the binder. The typical surface is about 6mm (0.25 inches) thick, which is determined by the diameter of the largest aggregate. When the chip seal is used to resurface an existing pavement it is sometimes referred to as a 'seal coat', which may be confused with the same word applied to a slurry coat containing fine aggregate (AEMA, 1995). When the chip seal is applied to a stony or soil surface, it may be referred to as a 'surface treatment' (AEMA, 1995).

Chip seals are usually applied to low-use roads, such as in rural areas. The rough aggregate on the surface is problematic in residential areas where children play, and loose aggregate thrown by car tires may pose another danger. The colour of the surface is strongly influenced by the colour of the

*Source:* M. Pomerantz, personal files

**Figure 2.24** *Machinery for slurry seal application and finished surface*

aggregate. When white limestone is used, where it is abundant, a quite white surface results. Figure 2.23 shows a picture of a chip seal application and contrasts its solar reflectance to that of an asphalt pavement.

*Hot-mix overlays.* For roads needing considerable repair or roads that must support large stresses, such as near stop signs where acceleration and turning are frequent, a sturdy repair can be done with a hot mix containing aggregate from 9.5mm to 12.5mm (0.37 to 0.5 inches) in maximum diameters.

*Slurries.* For surfaces with medium need of repair and that carry considerable traffic, resurfacing may be done with a mixture of asphalt emulsion and aggregate. The size of the aggregate and the formulation of the emulsion are determined by the expected traffic and the climate. The typical aggregate is about 6mm (0.25 inches) maximum diameter (ISSA, 1991). Figure 2.24 shows the machinery for slurry seal coating and the finished surface.

*Micro-surfacing.* When polymers are added to slurry binders the product is called 'micro-surfacing' (Raza, 1994a, 1994b). The polymer confers greater resistance to wear. In addition, it becomes possible to apply a layer in multi-stone thicknesses – it can be more than 1.5 times thicker than the largest aggregate and can be used for layers down to 7.5mm (0.3 inches).

*Seal coat or sand coat.* This consists of a mixture of emulsified asphalt and sand. Sometimes cement and other materials are added to the mix, but the aggregate particles must be smaller in diameter than about 0.04 inches. The preliminary preparation of the surface is relatively simple. Deposits of grease and oil must be removed or sealed over. Otherwise, the surface must be thoroughly cleaned of loose dirt or paving particles. The surface is then dampened with water, and the slurry is applied in a smooth coat. It is recommended that two coats be applied.

The colour of the material is essentially grey and is normally made darker by the addition of carbon black. Even when the carbon black is omitted, the grey surface has an albedo of 0.05. To lighten the colour, rutile (titanium dioxide, $TiO_2$) powder can be added. This increases the albedo to 0.10 with no loss of structural quality. An emulsion designed to rejuvenate asphalt, Reclamite (Erickson, 1989) is coated with sand. Thus, a lighter colour is achievable if white sand is used.

*Fog coat.* A thin layer of diluted asphalt emulsion is spread on existing pavement. It can be used as a protective layer, but also to change colour. The typical amount of asphalt applied is about 0.06 litres per square metre (0.03 gallons per square yard) (AEMA, 1995). This results in a coating of about 0.13mm (0.005 inches) thick. The cost of the labour would dominate the total cost because the amount of material is so small.

*Petroleum resin coatings.* A petroleum product that is not an asphalt is manufactured by Neville Chemical Co, Pittsburgh, Pennsylvania, and sold as Pavebrite® (Willockl, 1995). Similar products are distributed in Europe by French Shell Oil as Mexphalte C and by Total as LSC (Liants Synthetiques Clairs). These are synthetic resins derived from lighter fractions of petroleum and are chemically modified. The pure material is tan in colour; but colouring additives can achieve bright colours. The colour of the aggregates must be chosen to not interfere with the desired colour, as well as to provide the required mechanical strength. The aggregates are fine graded, meaning that they all pass through a number 8 mesh (about 2.5mm, or 0.1 inch) screen. This is necessary in order to prevent the colour of the aggregate from becoming significant as the pavement wears, if one desires that the pavement stay the colour of the binder. For the purposes of a whiter road, a white binder could be mixed with white rocks of any desired sizes. The mechanical properties of the paving are reported to be at least as good as comparable asphaltic pavings.

The typical use in the US has been for pavements at least 12.5mm (0.5 inches) thick. In Holland, there is some experience in using the binder in slurries. One tonne occupies about 0.42 cubic metres (15 cubic feet). When used for a 12.5mm (0.5 inch) pavement, it requires 1 tonne to make 34 square metres.

*Tree resin coatings.* A resinous material derived from pine trees, known as RoadOyl, is used for roads and dust suppression. In Marshall stability tests, it

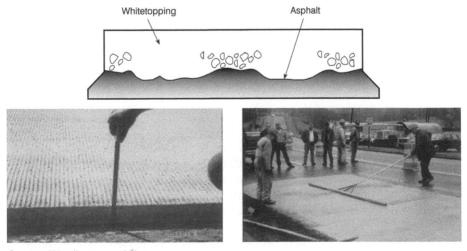

*Source:* Akbari's personal files

**Figure 2.25** *Schematic of a 'white topping' and application pictures*

is reported to perform at least as well as asphalt (SSC, 1995). It has not yet been completely evaluated as a slurry binder.

*Cement concrete coatings ('white topping').* Layers of concrete as thin as 5cm (2 inches) have been used for resurfacing roads. The procedure is still somewhat experimental and the long-term behaviour and proper practice are still under study. Figure 2.25 shows the schematic of a 'white topping' and its application to cover a portion of a street.

*Acrylics.* These are synthetic polymers that can be highly coloured. They are expensive, and thus far have been used mostly for special applications such as tennis courts. Reed and Graham Inc, San José, California, has produced experimental materials based on acrylics mixed with pigments, which proved to have acceptable structural strength as a roadway (Lungren and Goldman, 1996) and solar reflectivities of about 50 per cent (Berdahl and Wang, 1996).

## Shade trees and urban vegetation

Urban shade trees offer significant benefits by both reducing building air conditioning and lowering air temperature, thus improving urban air quality by reducing smog. Shade trees intercept sunlight before it warms a building. Trees also decrease the wind speed under their canopy and shield buildings from cold winter breezes. Akbari (2002) provides an overview of benefits and costs associated with planting urban trees. In a comprehensive study for Chicago, Illinois, McPherson et al (1994) provide a good review of the impact of an urban forest on the urban ecosystem.

In addition to their obvious aesthetic value, urban trees can modify the climate of a city and provide better urban thermal comfort in hot climates. A significant increase in the number of trees can moderate the intensity of the urban heat island by altering the heat balance of the entire city (see Figure 2.11).

Trees affect energy use in buildings through both direct and indirect processes. The direct effects are:

- reducing solar heat gain through windows, walls and roofs by shading; and
- reducing the radiant heat gain from the surroundings by shading.

The indirect effects include:

- reducing the outside air infiltration rate by lowering ambient wind speeds;
- reducing the heat gain into the buildings by lowering ambient temperatures through evapotranspiration in summer; and
- in hot and humid climates, increasing the latent air-conditioning load by adding moisture to the air through evapotranspiration (Huang et al, 1987).

*Shading* During the summer, properly placed and scaled trees around a building can block unwanted solar radiation from striking the building, reducing its cooling-energy use. In cold climates, shading of buildings can also increase the wintertime heating-energy use. Deciduous trees are particularly beneficial since they allow solar gain in buildings during the winter, while blocking it during the summer. The shade cast by trees also reduces glare and blocks the diffuse light reflected from the sky and surrounding surfaces (thereby altering the heat exchange between the building and its surroundings), providing natural insulation during both hot and cold weather. During the day, tree shading also reduces heat gain in buildings by reducing the surface temperatures of the surroundings. At night, trees block the heat flow from the building to the cooler sky and surroundings.

*Wind shielding (shelterbelts)* Trees act as windbreaks that lower the ambient wind speed, which can lower a building's cooling-energy use depending upon its physical characteristics. In certain climates, tree shelterbelts are used to block hot and dust-laden winds. In addition to energy-saving potentials, this will improve comfort conditions outdoors within the city.

*Evaporative cooling* The term evapotranspiration refers to the evaporation of water from vegetation and surrounding soils. On hot summer days, a tree can act as a natural 'evaporative cooler', using up to 100 gallons of water a day and thus lowering the ambient temperature (Kramer and Kozlowski, 1960). Evapotranspiration is most effective in the summer because of the presence of leaves on deciduous trees and the higher ambient temperatures.

Increased evapotranspiration during the summer from a significant increase in urban trees can produce an 'oasis effect' in which the urban ambient temperatures are significantly lowered. Although, in some cases, the amount of latent cooling (i.e. humidity removal) might be slightly increased on the whole, buildings in such cooler environments will consume less cooling power and energy.

## Energy and smog benefits of shade trees

*Direct energy savings* Data on measured energy savings from urban trees are scarce. Case studies (Laechelt and Williams, 1976; Buffington, 1979; Parker, 1981; Akbari et al, 1997b) have documented dramatic differences in cooling-energy use between houses on landscaped and un-landscaped sites. Akbari et al (1997b) conducted a 'flip-flop' experiment to measure the impact of shade trees on two houses in Sacramento. The experiment was carried out in three segments:

1  monitoring the cooling-energy use of both houses to characterize a base case energy use of the houses;
2  installing eight large and eight small shade trees at one of the sites for a period of four weeks; and then
3  moving the trees from one site to the other.

The experiment documented seasonal cooling-energy savings of about 30 per cent (about 4 kilowatt hour per day, kWh/day). The estimated peak electricity saving was about 0.7kW. In Florida, Parker (1981) measured the cooling-energy savings from well-planned landscaping and found that properly located trees and shrubs around a mobile trailer reduced the daily air-conditioning electricity use by as much as 50 per cent.

In computer simulation studies, Konopacki and Akbari (2000a, 2000b, 2002) investigated the energy-saving potential of urban trees in five US cities: Baton Rouge, Los Angeles; Chicago, Illinois; Houston, Texas; Sacramento, California; and Salt Lake City, Utah. The analysis included both direct (shading) and indirect (evapotranspiration) effects. The study considered planting an average of four shade trees per house, each with a top view cross-section of $50m^2$, and estimated net annual dollar savings in energy expenditure of US$5.2 million, $13.5 million, $27.8 million, $9.8 million and $1.1 million for Baton Rouge, Chicago, Houston, Sacramento, and Salt Lake City, respectively.

In another computer study, Taha et al (1996) analysed the impact of large-scale tree-planting programmes in ten US metropolitan areas: Atlanta, Georgia; Chicago, Illinois; Dallas, Texas; Houston, Texas; Los Angeles, California; Miami, Florida; New York City, New York; Philadelphia, Pennsylvania; Phoenix, Arizona; and Washington, DC. Both direct and indirect effects on air-conditioning energy use were addressed, using the DOE-2 building simulation program for energy calculations and a meso-scale simulation model for meteorological calculations. The energy analysis focused on residential and small commercial (small office) buildings (see Table 2.9). For most hot cities, the estimated total (direct and indirect) annual energy savings were US$10 to $35 per 100 square metres of single-storey residential and commercial buildings.

DeWalle et al (1983), Heisler (1989) and Huang et al (1990) have focused on measuring and simulating the wind-shielding effects of trees on heating-

**Table 2.9** DOE-2 *simulated heating, ventilating and air-conditioning (HVAC) annual energy savings from trees*

| Location | Old residence | | New residence | | Old office | | New office | |
|---|---|---|---|---|---|---|---|---|
| | Direct | Indirect | Direct | Indirect | Direct | Indirect | Direct | Indirect |
| Atlanta | 5 | 2 | 3 | 1 | 3 | 2 | 2 | 2 |
| Chicago | 3 | 2 | 1 | 0.5 | 1 | 1 | 2 | 1 |
| Los Angeles | 12 | 8 | 7 | 5 | 6 | 12 | 4 | 10 |
| Fort Worth | 6 | 6 | 5 | 4 | 4 | 5 | 2 | 4 |
| Houston | 10 | 6 | 6 | 4 | 3 | 5 | 3 | 3 |
| Miami | 9 | 3 | 6 | 3 | 3 | 2 | 2 | 2 |
| New York City | 3 | 2 | 2 | 1 | 3 | 3 | 2 | 2 |
| Philadelphia | –5 | 0 | –7 | 0 | 2 | 1 | 1 | 1 |
| Phoenix | 27 | 8 | 16 | 5 | 9 | 5 | 6 | 4 |
| Washington, DC | 3 | 2 | 1 | 1 | 3 | 1 | 2 | 1 |

*Note:* Three trees per house and per office are assumed. All savings are US$ per 100 square metres.
*Source:* Taha et al (1996)

and cooling-energy use. Their analyses indicated that a reduction in infiltration because of trees would save heating-energy use. However, in climates with cooling-energy demand, the impact of a windbreak on cooling is fairly small compared to the shading effects of trees. In cold climates, the wind-shielding effect of trees can reduce heating-energy use in buildings. Akbari and Taha (1992) simulated the wind-shielding impact of trees on heating-energy use in four Canadian cities. For several prototypical residential buildings, they estimated heating-energy savings in the range of 10 to 15 per cent.

Heisler (1990a) has measured the impact of trees in reducing ambient wind. Akbari and Taha (1992) used Heisler's data and analysed the impact of wind reduction on heating- and cooling-energy use of typical houses in cold climates. Simulations indicated that in cold climates, a 30 per cent uniform increase in urban tree cover can reduce winter heating bills in urban areas by about 10 per cent and in rural areas by 20 per cent. Savings in urban areas can almost be doubled if evergreen trees are planted strategically on the north side of buildings so that the buildings can be better protected from the cold north winter wind.

Heisler (1986, 1990b) has investigated the effect of tree placement around a house on heating- and cooling-energy use. Trees planted on the east and west sides of a building shade the walls and windows from sunlight in the morning and afternoon. Depending upon wall construction, the impact of morning heating may be seen in the late morning and early afternoon hours. Similarly, the impact of afternoon heating of the west walls may be seen in evening hours. Akbari et al (1993) performed parametric simulations on the impact of tree locations on heating- and cooling-energy use and found that savings can vary from 2 per cent to over 7 per cent; cooling-energy savings were higher for trees shading the west walls and windows.

**Table 2.10** *Number of additional trees planted in each metropolitan area and their simulated effects in reducing the ambient temperature*

| Location | Number of additional trees in the simulation domain (million) | Number of additional trees in the metropolitan area (million) | Maximum air temperature reduction in the hottest simulation cell (K) |
|---|---|---|---|
| Atlanta | 3.0 | 1.5 | 1.7 |
| Chicago | 12 | 5.0 | 1.4 |
| Los Angeles | 11 | 5.0 | 3.0 |
| Fort Worth | 5.6 | 2.8 | 1.6 |
| Houston | 5.7 | 2.7 | 1.4 |
| Miami | 3.3 | 1.3 | 1.0 |
| New York City | 20 | 4.0 | 2.0 |
| Philadelphia | 18 | 3.8 | 1.8 |
| Phoenix | 2.8 | 1.4 | 1.4 |
| Washington, DC | 11 | 3.0 | 1.9 |

*Note:* The simulated area is much larger than the metropolitan area.
*Source:* Taha et al (1996)

**Table 2.11** *Energy savings, ozone reduction and avoided peak power resulting from the addition of 11 million urban trees in the Los Angeles Basin*

| Benefits | Direct | Indirect | Smog | Total |
|---|---|---|---|---|
| 1  Cost savings from trees | | | | |
| (US$ million per year) | 58 | 35 | 180 | 273 |
| 2  Δ Peak power (GW) | 0.6 | 0.3 | | 0.9 |
| 3  Present value per tree (US$) | 68 | 24 | 123 | 211 |

*Source:* Rosenfeld et al (1998)

*Indirect energy and smog benefits*
Taha et al (1996) estimated the impact on ambient temperature resulting from a large-scale tree-planting programme in the selected ten cities. They used a three-dimensional meteorological model to simulate the potential impact of trees on ambient temperature for each region. The meso-scale simulations showed that, on average, trees can cool cities by about 0.3K to 1K at 2 pm; in some simulated cells the temperature was decreased by as much as 3K (see Table 2.10). The estimated air-conditioning savings resulting from ambient cooling by trees in hot climates ranges from US$5 to $10 per year per square metre of roof area of residential and commercial buildings. Indirect effects are smaller than the direct effects of shading and, moreover, require that the entire city be planted.

Rosenfeld et al (1998) studied the potential benefits of planting 11 million trees in the Los Angeles Basin. They estimate an annual total savings of US$270 million from direct and indirect energy savings and smog benefit; about two-thirds of the savings were from the reduction in smog concentration resulting from meteorological changes due to the evapotranspiration of trees (see Table 2.11). Peak demand savings was estimated to be 0.9GW.

The present value (PV) of savings is calculated to find out how much a homeowner can afford to pay for shade trees. Rosenfeld et al (1998) assumed the planting of small shade trees that would take about 10 to 15 years to reach maturity. Savings from trees before they reach maturity was neglected and the PV of all future savings was calculated to be US$7.5 for each $1 saved annually. On this basis, the direct savings to a homeowner who plants three shade trees would have a PV of about US$200 per home (US$68 per tree). The PV of indirect savings was smaller: about US$72 per home (US$24 per tree). The PV of smog savings was about US$120 per tree. Total PV of all benefits from trees was thus US$210 per tree.

Urban trees affect air pollution through two major processes:

1   cooling the ambient temperature and, hence, slowing the process of smog formation; and
2   dry deposition by which the airborne pollutants (both gaseous and particles) can be removed from the air.

Trees directly remove pollutant gases – carbon monoxide (CO), nitrogen oxide ($NO_x$), ozone ($O_3$) and sulphur dioxide ($SO_2$) – predominantly through leaf stomata (Smith, 1984; Fowler, 1985). Nowak (1994a) performed an analysis of pollutant removal by the urban forest in Chicago and concluded that through dry deposition, trees, on the average, remove about 0.002 per cent (0.34 grams per square metre per year, $g/m^2/yr$) of CO, 0.8 per cent ($1.24g/m^2/yr$) of $NO_2$, 0.3 per cent ($1.09g/m^2/yr$) of $SO_2$, 0.3 per cent ($3.07g/m^2/yr$) of $O_3$ and 0.4 per cent ($2.83g/m^2/yr$) of particulate matter with a diameter less than or equal to 10 microns (PM10) from the air.

Simulations performed by Taha et al (1997) for Los Angeles indicated that on a daily basis 1 per cent of the mass of ozone in the mixed layer would be scavenged by planting an additional 11 million trees (dry deposited). In addition to this amount of ozone being scavenged directly from the atmosphere, there is 0.6 per cent less ozone formation in the mixed layer due to the fact that vegetation also scavenges $NO_2$, an ozone precursor. The total effect of increased deposition by the additional vegetation is thus to decrease atmospheric ozone in the mixed layer by 1.6 per cent.

Taha et al (2000) refined their analysis and studied the effects of urban vegetation (and other heat-island reduction technologies, such as reflective roofs and pavements) on ozone air quality for Baton Rouge, Salt Lake City and Sacramento. The meteorological simulations indicated a reduction in daytime ambient temperature of the order of 1K to 2K. In Baton Rouge, the simulated reduction of 0.8K in the afternoon ambient temperature leads to a 4 to 5 parts per billion (ppb) reduction in ozone concentration. For Salt Lake City, the afternoon temperature and ozone reductions were 2K and 3 ppb to 4 ppb. And in Sacramento the reductions were 1.2K and 10 ppb (about 7 per cent of the peak ozone concentration of 139 ppb). Note that the reported reductions in ambient and ozone concentration have resulted from the combined effect of urban vegetation and reflective roofs and pavements. Preliminary simulations indicated that in dry climates, such as Sacramento and Salt Lake City, the contribution of urban vegetation and reflective surfaces to ambient air temperature and ozone reduction is about the same. In humid climates, such as Baton Rouge, increasing the reflectivity of surfaces is more effective in reducing ambient temperature and ozone than adding to the urban vegetation.

It is also suggested that trees improve air quality by dry depositing NOx, $O_3$ and PM10. Rosenfeld et al (1998) estimated that 11 million trees in Los Angeles will reduce PM10 by less than 0.1 per cent through dry deposition, worth about US$7 million per year.

Shade trees, by reducing peak power by 0.9GW, save about 0.5g of NOx per kilowatt hour avoided from power plants in the basin. Simulations have found that 4 tonnes of NOx per day are avoided – about 0.3 per cent of the base case.

## Other benefits of shade trees
There are other benefits associated with urban shade trees. Some of these include improvement in the quality of life, increased value of properties and

decreased rain runoff water – hence, protection against floods (McPherson et al, 1994). Trees also directly sequester atmospheric carbon dioxide.

Data for the rate of carbon sequestration by urban trees are scarce; most data are given in the units of tonnes per year of carbon per hectare of forested land. However, Nowak (1994b) has performed an analysis of carbon sequestration by individual trees as a function of tree diameter measured at breast height (dbh). He estimates that an average tree with a dbh of 31cm to 46cm (about 50 square metres in crown area) sequesters carbon at a rate of 19kg per year. The rate of carbon sequestration for several species of trees can be estimated using data by Frelich (1992) on the age, dbh, crown area and height for 12 species of trees around Twin Cities, Minnesota. Using this data, the volume of the wet biomass of the trunk can be estimated by assuming a cone-shape tree with a base area with the given diameter and height. The total volume of the tree accounting for main branches and roots is approximately 1.5 the volume of the tree trunk. The weight of the biomass can be estimated by multiplying the volume by a density of 900kg per cubic metre. The weight of the dry mass is estimated at 50 per cent of the wet mass and the amount of carbon is estimated to be 50 per cent of the dry mass. The calculation yields an average of about 4.5kg per year over the life of a tree until its crown has

**Table 2.12** *Annual carbon sequestration by individual trees*

| Tree type | Age | Diameter at breast height (dbh) (cm) | Height (m) | Average carbon sequestered (kg/yr) | Carbon sequestrated at maturity* (kg/yr) |
|---|---|---|---|---|---|
| Norway maple | 30 | 33.0 | 10.1 | 3.2 | 9.9 |
| Sugar maple | 29 | 29.5 | 11.2 | 2.9 | 7.8 |
| Hackberry | 25 | 27.4 | 10.3 | 2.7 | 8.5 |
| American and little-leaved linden | 33 | 41.4 | 11.5 | 5.3 | 13.8 |
| Black walnut | 32 | 31.0 | 11.2 | 3.0 | 8.0 |
| Green ash | 26 | 30.2 | 11.7 | 3.6 | 10.8 |
| Robusta and Siouxland hybrid | 33 | 52.1 | 20.5 | 14.9 | 29.6 |
| Kentucky coffee tree | 40 | 31.0 | 9.9 | 2.1 | 3.6 |
| Red maple | 24 | 27.4 | 10.2 | 2.8 | 8.9 |
| White pine | 34 | 34.5 | 13.6 | 4.2 | 15.2 |
| Blackhills (white) spruce | 60 | 37.6 | 15.9 | 3.3 | 7.7 |
| Blue spruce | 60 | 49.3 | 18.9 | 6.7 | 12.8 |
| Average | | | | 4.6 | 11.4 |
| Average excluding Robusta/Siouxland | | | | 3.6 | 9.7 |

*Notes:* Each tree is assumed to have a crown area of 50m². 
* Maturity is defined as when the tree has a crown area of 50m². 
*Source:* Frelich (1992)

grown to about 50 square metres (see Table 2.12). Data indicate that as trees grow, the rate of sequestration increases. The average sequestration rate for a tree with a crown area of 50 square metres was estimated at about 11kg per year.

This calculation suggests that urban trees play a major role in sequestering $CO_2$, thereby delaying global warming. Rosenfeld et al (1998) estimated that a tree planted in Los Angeles avoids the combustion of 18kg of carbon annually, and according to our calculations an average shade tree sequesters about 4.5kg to 11kg per year (as it would if growing in a forest). In that sense, one shade tree in Los Angeles is equivalent to three to five forest trees.

## Potential problems with shade trees

There are some potential problems associated with trees. Trees can contribute to smog problems by emitting volatile organic compounds (VOCs) that exacerbate the smog. The photochemical reaction of VOCs and NOx produces smog ($O_3$). Obviously, selection of low-emitting trees should be considered in a large-scale tree-planting programme. Benjamin et al (1996) have prepared a list of several hundred tree species with their average emission rate.

In dry climates and areas with a serious water shortage, drought-resistant trees are recommended. Unfortunately, this results in very little evapo-transpiration and, thus, very little ambient cooling. Some trees need significant maintenance that may entail high costs over the lifespan of the trees. Tree roots can damage underground pipes, pavements and foundations. Proper design is needed to minimize these effects. In addition, trees are a fuel source for fire; selection of appropriate tree species and planting them strategically to minimize the fire hazard should be an integral component of a tree-planting programme.

## Cost of trees

The cost of a citywide 'tree-planting' programme depends upon the type of programme offered and the types of trees recommended. At the low end, a promotional planting of trees with a height of 1.5m to 3m costs about US$10 per tree, whereas a professional tree-planting programme using fairly large trees could amount to US$150 to $470 a tree (McPherson et al, 1994). McPherson has collected data on the cost of tree planting and maintenance from several cities. The cost elements include planting, pruning, removal of dead trees, stump removal, waste disposal, infrastructure repair, litigation and liability, inspection and programme administration. The data provide details of the cost for trees located in or near parks, gardens, streets, highways and houses. The present value of all these life-cycle costs (including planting) is US$300 to $500 per tree. Over 90 per cent of the cost is associated with professional planting, pruning, and tree and stump removal. On the other hand, a tree-planting programme administered by the Sacramento Municipal Utility District (SMUD) and Sacramento Tree Foundation during 1992 to 1996 planted trees 6m in height at an average (low) cost of SU$45 per tree. This figure includes only the cost of a tree and its planting; it does not include pruning, removal of dead trees and stump removal. Tree costs can also be justified by other amenities that they provide beyond air conditioning and smog reduction. The low-cost programmes are then probably the information programmes that provide data on the energy and smog savings that trees offer to the communities and homeowners who have decided to plant trees for other reasons.

Two primary factors to be considered in designing a large-scale urban tree programme is the potential room (space available) for planting trees, and the types of programmes that utilize and employ the wide participation of the population. We recently studied the fabric (fraction of different land uses) of Sacramento by statistically analysing high-resolution aerial colour photographs of the city, taken at a 0.3m resolution (Akbari et al, 1999b; see Figure 2.8). On average, tree cover comprises about 13 per cent of the entire Sacramento metropolitan area. Assuming that trees can be planted in areas to cover barren land (8 per cent) and grass (15 per cent), tree cover in Sacramento would increase to 36 per cent. The design of a large-scale urban tree programme should take advantage of these types of data to plan the programme accurately for each neighbourhood.

## ANALYSIS TOOLS

Figure 2.11 depicts the overall methodology used in analysing the impact of heat-island mitigation measures on energy use and urban air pollution. Hourly building energy simulation models (such as DOE-2) are used to calculate the energy use and energy savings in buildings.[5] To calculate the direct effects, prototypical buildings are simulated with dark- and light-coloured roofs, and with and without shade trees. Typical weather data for each climate region of interest are used in these calculations. To calculate the indirect effects, the typical weather data input to the hourly simulation model are first modified to account for changes in the urban climate. The prototypical buildings are then simulated with the modified weather data to estimate savings in heating- and cooling-energy consumption.

Factors affecting the energy balance in urban areas include urban geometry, surface properties and release of anthropogenic heat. The extent and intensities of urban heat islands depend strongly upon temporal aspects (diurnal and seasonal) of the weather and synoptic conditions. They also depend upon other factors such as the location, topography, size of the city and its population density (Oke, 1987, 1988).

To understand the impacts of large-scale increases in albedo and vegetation on urban climate and ozone air quality, meso-scale meteorological and photochemical models are used. For example, Taha et al (1995) and Taha (1996, 1997, Taha et al, 2001) used the Colorado State University Mesoscale Model (CSUMM) and the Pennsylvania State University/National Center for Atmospheric Research (PSU/NCAR) (known as MM5) to simulate the meteorology of several urban areas and its sensitivity to changes in surface properties. The Urban Air Shed Model (UAM) and the California Institute of Technology (CIT) air shed model were used to simulate the impact of the changes in meteorology and emissions on ozone air quality. The CSUMM, MM5, CIT and UAM essentially solve a set of coupled governing equations representing the conservation of mass (continuity), potential temperature (heat), momentum, water vapour and chemical species continuity to obtain prognostic meteorological fields and pollutant species concentrations. The governing equations are summarized below:

$$\frac{\delta \rho}{\delta t} = -(\nabla \cdot \rho \mathbf{V}) \qquad \text{conservation of mass [2]}$$

$$\frac{\delta \theta}{\delta t} = -\mathbf{V} \cdot \nabla \theta + S_\theta \qquad \text{conservation of energy [3]}$$

$$\frac{\delta \mathbf{V}}{\delta t} = -\mathbf{V} \cdot \nabla \mathbf{V} - \frac{1}{\rho} \nabla p - g\mathbf{k} - 2\Omega \times \mathbf{V} \quad \text{conservation of momentum [4]}$$

$$\frac{\delta q}{\delta t} = -\mathbf{V} \cdot \nabla q + S_q \qquad \text{conservation of moisture [5]}$$

$$\frac{\delta C_i}{\delta t} + \nabla \cdot \mathbf{V} C_i = \nabla \cdot (K \nabla C_i) + R_i + S_i + D_i \quad \text{conservation of species [6]}$$

where:
- $\rho$ = density of the air;
- $\mathbf{V}$ = wind velocity vector;
- $\theta$ = potential temperature;
- $S_\theta$ = source or sink term for potential temperature;
- $p$ = pressure;
- $g$ = gravitational acceleration;
- $\mathbf{k}$ = unit vector in vertical direction;
- $\Omega$ = Earth angular velocity;
- $q$ = specific humidity;
- $S_q$ = source or sink term for humidity;
- $C_i$ = concentration of species $i$;
- $K$ = turbulent diffusion coefficient;
- $R_i$ = reaction rate for species $i$;
- $S_i$ = source rate for species $i$;
- $D_i$ = sink (or deposition) rate for species $i$.

The CSUMM is a hydrostatic, primitive equation, three-dimensional Eulerian model originally developed by Pielke (1974). The model is incompressible (uses the incompressibility assumption to simplify the equation for conservation of mass), and employs a terrain-following coordinate system. It uses a first-order closure scheme in treating sub-grid scale terms of the governing differential equations. The model's domain is about 10km high with an underlying soil layer about 50cm deep. The CSUMM generates three-dimensional fields of prognostic variables, as well as a boundary layer height profile that can be input to the air quality models.

The MM5 is a state-of-science, non-hydrostatic, three-dimensional (Eulerian) primitive equation model that is gaining wide acceptance in the US scientific and regulatory communities. The MM5 has been used by researchers, meteorologists and scientists in numerous application, including weather forecasting; air pollution forecasting; frontogenesis; thunderstorms; hurricanes; urban-scale phenomena, such as urban heat islands and related convective circulations; land–sea breeze circulations; and topographically induced flows. Although utilized worldwide, the MM5 is mostly used in the US in both research and forecast/operational modes. The modelling system comprises several components collectively referred to as the MM5. The model has been under continuous development since the late 1970s and is based on an original formulation (Anthes and Warner, 1978; Anthes et al, 1987) that was developed and maintained by the Pennsylvania State University in collaboration with the National Center for Atmospheric Research. More recently, the model has undergone significant changes and improvements (Dudhia, 1993; Grell et al, 1994).

The UAM and CIT are three-dimensional, Eulerian photochemical models that are capable of simulating inert and chemically reactive atmospheric pollutants. These models are used in various urban air-shed areas to study the effects of air quality improvement technologies. The UAM and CIT simulate the advection, diffusion, transformation, emission and deposition of pollutants. They treat about 30 chemical species and use the carbon bond CB-IV mechanism (Gery et al, 1988). The models account for emissions from area and point sources, elevated stacks, mobile and stationary sources, and vegetation (biogenic emissions). For a detailed discussion of the use and adaptation of these models and the study of the impact of the heat island mitigation strategies in the Los Angeles Basin, see Taha (1996, 1997).

Examples of outputs from these simulations are shown in Figures 2.26 and 2.27. Figure 2.26 shows the predicted reduction in air temperature in Los Angeles at 2 pm on 27 August as a result of increasing the urban albedo and vegetation cover by moderate amounts (average increases of 7 per cent). Figure 2.27 shows corresponding changes in ozone concentrations. Because of the combined effects of local emissions, meteorology, surface properties and topography, ozone concentrations increase in some areas and decrease in others. The net effect, however, is a decrease in ozone concentrations. The simulations also predict a reduction in population-weighted exceedance exposure to ozone (above the California and National Ambient Air Quality Standards) of 10 to 20 per cent (Taha, 1996). This reduction, for some smog scenarios, is comparable to ozone reductions obtained by replacing all gasoline on-road motor vehicles with electric cars.

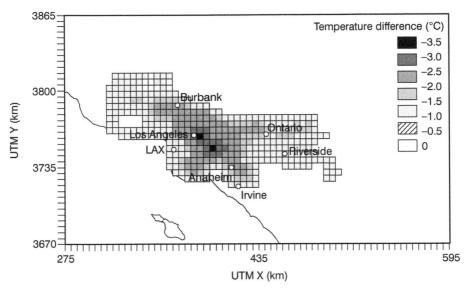

*Source:* Taha (1996, 1997)

**Figure 2.26** *Temperature difference (from the base case) for a case with increased surface albedo and urban forest; the temperature difference is at 2 pm on a late August day in Los Angeles*

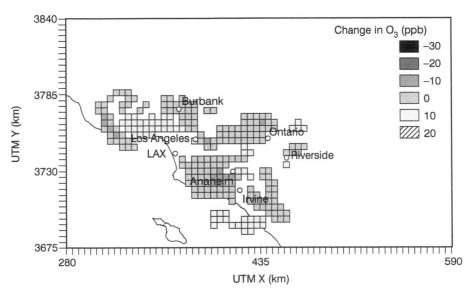

*Source:* Taha (1996, 1997)

**Figure 2.27** *Ozone concentrations difference (from the base case) for a case with increased surface albedo and urban forest; the difference is shown for 2 pm on a late August day in Los Angeles*

# CONCLUSION

Most urban areas are warmer than their surrounding rural areas. The temperature difference between urban and rural areas is commonly referred to as the urban heat-island effect. With the rapid expansion of cities in the last five decades, heat islands are growing and are affecting the world's ever increasing urban population. Increasing urban ambient temperatures raise building cooling-energy use, worsen the urban air quality and reduce citizens' comfort. Cool surfaces (cool roofs and cool pavements) and urban trees can have a substantial effect on urban air temperature and, hence, can reduce cooling-energy use and smog. In the US, it is estimated that about 20 per cent of the national cooling demand can be avoided through a large-scale implementation of heat-island mitigation measures. This amounts to 40TWh per year savings, worth over US\$4 billion per year by 2015 in cooling-electricity savings alone. Once the benefits of smog reduction are accounted for, the total savings could add up to over US\$10 billion per year.

Achieving these potential savings is conditional on receiving the necessary governmental and local community support. Scattered programmes for planting trees and increasing surface albedo already exist; but to start an effective and comprehensive campaign would require a more aggressive agenda. Much of the fundamental work to promote heat-island mitigation measures are already in place. The American Society for Testing of Materials (ASTM) has developed standards for measuring solar reflectance of roofing and pavement materials. The Cool Roof Rating Council (CRRC) has been organized to measure, rate and label the solar reflectance and thermal emittance of roofing materials. Many industrial leaders have introduced cool roofing materials on the market. In contrast, the development of cost-effective solutions for cool pavement has been very slow. The cool roofs criteria and standards are incorporated within the Building Energy Performance Standards of the American Society of Heating, Refrigerating and Air-Conditioning Engineers (ASHRAE), the California Title 24 Building Code and the California South Coast's Air Quality Management Plans. Many field projects have demonstrated the energy benefits of cool roofs and shade trees. The South Coast Air Quality Management District and the US Environmental Protection Agency (EPA) now recognize that air temperature is as much a cause of smog as NOx or volatile organic compounds. In 1992, the EPA published a milestone guideline for tree planting and light-coloured surfacing (Akbari et al, 1992). Many countries have joined efforts in developing heat-island reduction programmes to improve urban air quality. The efforts in Japan are of quite notable interest.

Trees can potentially reduce energy consumption in a city and improve air quality and comfort. These potential savings are clearly a function of climate: in hot climates, deciduous trees shading a building can save cooling-energy use; in cold climates, evergreen trees shielding the building from the cold winter wind can save heating-energy use. Trees also improve urban air quality by lowering the ambient temperature and, hence, reducing the formation of urban smog, and by dry deposition to directly absorb gaseous pollutants and PM10 from the air. Low-emitting trees should be considered in designing a tree-planting programme so that volatile organic compound-emitting trees would not undermine efforts. Finally, a major cost of a tree-planting programme is that associated with planting and maintaining by tree professionals. The cost of the water consumption of trees in most climates is small compared to planting and maintenance costs. It is quite possible to design a low-cost tree-planting programme that utilizes and employs the full voluntary participation of the population.

Pavements cover a surprisingly large fraction of a city's surface and typically are among the darkest and hottest surfaces. There are well-accepted methods of creating lighter-coloured pavements, such as chip seals using whiter aggregate. The difficulty in implementing cooler pavements is in taking a long-term and citywide view of the situation. Most often, the decision about pavements is made on the basis of initial cost, without regard for the shortened lifetime of hot pavements or the heat island effects. When these are taken into account, as in the study by Ting et al (2001), the lifetime costs of cooler pavements may be lower for many kinds of roads.

## ACKNOWLEDGEMENTS

The work summarized in this chapter has benefited from many years of research contributions by Paul Berdahl, Steven Konopacki, Ronnen Levinson, Melvin Pomerantz, Shea Rose, Arthur Rosenfeld and Haider Taha. Writing of this chapter has been supported by the California Energy Commission and the US Department of Energy under contract DE-AC02-05CH11231.

## ENDNOTES

1  When sunlight (including ultraviolet, visible and near-infrared light) hits an opaque surface, some of the sunlight is reflected (this fraction is called the albedo, or *a*), and the rest is absorbed (the absorbed fraction is $1 - a$). Low *a* surfaces, of course, become much hotter than high *a* surfaces.

2  The night-time heat island is typically greater than the daytime summer heat island. The night-time heat island is caused by the differential in cooling between the rural and urban areas during the early evening hours, and its magnitude is typically largest in winter.

3  In March 2005, a major manufacturer of roofing shingles in California announced the availability of cool-coloured shingles in three popular colours.

4  White cement is available, for example, from Lehigh Portland Cement Co, Allentown, Pennsylvania, 18195.

5  DOE-2 is an hourly simulation program that simulates the heating- and cooling-energy demand of the building. DOE-2 input includes building and heating, ventilation, and air-conditioning characteristic data, operating schedules, occupancy and hourly weather data (BESG, 1990).

# REFERENCES

AEMA (Asphalt Emulsion Manufacturers Association) (1995) *Recommended Performance Guidelines*, second edition, Asphalt Emulsion Manufacturers Association, Washington, DC

Akbari, H. (2002) 'Shade trees reduce building energy use and $CO_2$ emissions from power plants', *Environmental Pollution*, vol 116, ppS119–S126

Akbari, H. (2003) 'Measured energy savings from the application of reflective roofs in 2 small non-residential buildings', *Energy*, vol 28, pp953–967

Akbari, H., Berdahl, P., Levinson, R., Wiel, S., Desjarlais, A., Miller, W., Jenkins, N., Rosenfeld, A. and Scruton, C. (2004) 'Cool colored materials for roofs', in *Proceedings of the 2004 ACEEE Summer Study on Energy Efficiency in Buildings*, vol 1, Pacific Grove, CA, p1

Akbari, H., Berdahl, B., Levinson, R., Wiel, S., Miller, W. and Desjarlais, A. (2006) *Cool Color Roofing Materials*, LBNL-59886, Lawrence Berkeley National Laboratory, Berkeley, CA, March

Akbari, H., Bretz, S., Hanford, J., Kurn, D., Fishman, B., Taha, H. and Bos, W. (1993) *Monitoring Peak Power and Cooling Energy Savings of Shade Trees and White Surfaces in the Sacramento Municipal Utility District (SMUD) Service Area: Data Analysis, Simulations and Results*, LBL-34411, Lawrence Berkeley National Laboratory, Berkeley, CA, December

Akbari, H., Bretz, S., Kurn, D. and Hanford, J. (1997a) 'Peak power and cooling energy savings of high-albedo roofs', *Energy and Buildings*, vol 25, pp117–126

Akbari, H., Davis, S., Dorsano, S., Huang, J. and Winnett, S. (eds) (1992) *Cooling Our Communities: A Guidebook on Tree Planting and Light-Colored Surfacing*, US Environmental Protection Agency, Office of Policy Analysis, Climate Change Division, Washington, DC

Akbari, H., Levinson, R. M. and Rainer, L. (2005) 'Monitoring the energy-use effects of cool roofs on California commercial buildings', *Energy and Buildings*, vol 37, no 10, pp1007–1016

Akbari, H. and Konopacki, S. (2005) 'Calculating energy-saving potentials of heat-island reduction strategies', *Energy Policy*, vol 33, pp721–756

Akbari, H., Konopacki, S., Eley, C., Wilcox, B., Van Geem, M. and Parker, D. (1998) 'Calculations for reflective roofs in support of Standard 90.1', *ASHRAE Transactions*, vol 104, no 1, pp984–995

Akbari, H., Konopacki, S. and Pomerantz, M. (1999a) 'Cooling energy savings potential of reflective roofs for residential and commercial buildings in the United States', *Energy*, vol 24, pp391–407

Akbari, H., Kurn, D., Taha, H., Bretz, S. and Hanford, J. (1997b) 'Peak power and cooling energy savings of shade trees', *Energy and Buildings – Special Issue on Urban Heat Islands and Cool Communities*, vol 25, no 2, pp139–148

Akbari, H., Pomerantz, M. and Taha, H. (2001) 'Cool surfaces and shade trees to reduce energy use and improve air quality in urban areas', *Solar Energy*, vol 70, no 3, pp295–310

Akbari, H. and Rose, L. S. (2001a) *Characterizing the Fabric of the Urban Environment: A Case Study of Salt Lake City, Utah*, LBNL-47851, Lawrence Berkeley National Laboratory, Berkeley, CA, February

Akbari, H. and Rose, L. S. (2001b) *Characterizing the Fabric of the Urban Environment: A Case Study of Chicago, Illinois*, LBNL-49275, Lawrence Berkeley National Laboratory, Berkeley, CA, October

Akbari, H., Rose, L. S. and Taha, H. (1999b) *Characterizing the Fabric of the Urban Environment: A Case Study of Sacramento, California*, LBNL-44688, Lawrence Berkeley National Laboratory, Berkeley, CA, December

Akbari, H., Rosenfeld, A. and Taha, H. (1990) 'Summer heat islands, urban trees, and white surfaces', *ASHRAE Transactions*, vol 96, no 1, pp1381–1388

Akbari, H. and Taha, H. (1992) 'The impact of trees and white surfaces on residential heating and cooling energy use in four Canadian cities', *Energy: The International Journal*, vol 17, no 2, pp141–149

Akridge, J. (1998) 'High-albedo roof coatings – impact on energy consumption', *ASHRAE Technical Data Bulletin* vol 14, no 2

Anthes, R. A., Hsie, E. Y. and Kuo, Y. H. (1987) *Description of the Penn State/NCAR Mesoscale Model Version 4 (MM4)*, NCAR/TN-282+STR, National Center for Atmospheric Research, Boulder, CO

Anthes, R. A. and Warner, T. T. (1978) 'Development of hydrodynamic models suitable for air pollution and other meteorological studies', *Monthly Weather Review*, vol 106, pp1045–1047

Asphalt Institute (1974) *Open-Graded Asphalt Friction Courses*, Construction Leaflet no 10, Asphalt Institute, Lexington, KT

Asphalt Institute (1989) *The Asphalt Handbook*, Asphalt Institute, Lexington KT

Bally, M. (1998) Personal communication on the prices for superpave grades of asphalt charged by Koch Performance Asphalt Co. in Nevada

Benjamin, M. T., Sudol, M., Bloch, L. and Winer, A. M. (1996) 'Low-emitting urban forests: A taxonomic methodology for assigning isoprene and monoterpene emission rates', *Atmospheric Environment*, vol 30, no 9, pp1437–1452

Berdahl, P. and Bretz, S. (1997) 'Preliminary survey of the solar reflectance of cool roofing materials', *Energy and Buildings – Special Issue on Urban Heat Islands and Cool Communities*, vol 25, no 2, pp149–158

Berdahl, P. and Wang, F. (1996) Unpublished measurements at LBNL of solar reflectivities of seal coatings, pers comm

BESG (Building Energy Simulation Group) (1990) *Overview of the DOE-3 Building Energy Analysis Program*, version 2.1D, LBL-19735, Rev 1, Lawrence Berkeley National Laboratory, Berkeley, CA

Boutwell, C. and Salinas, Y. (1986) *Building for the Future – Phase I: An Energy Saving Materials Research Project*, Mississippi Power Co, Rohm and Haas Co and the University of Mississippi, Oxford, MS

Bretz, S. and Akbari, H. (1994) 'Durability of high-albedo roof coatings', in *Proceedings of the ACEEE 1994 Summer Study on Energy Efficiency in Buildings*, vol 9, p65

Bretz, S. and Akbari, H. (1997) 'Long-term performance of high-albedo roof coatings', *Energy and Buildings – Special Issue on Urban Heat Islands and Cool Communities*, vol 25, no 2, pp159–167

Bretz, S., Akbari, H. and Rosenfeld, A. (1997) 'Practical issues for using high-albedo materials to mitigate urban heat islands', *Atmospheric Environment*, vol 32, no 1, pp95–101

Brown, D. C. (1996) 'Porous asphalt pavements rescue parking lots', *Asphalt Contractor*, pp70–77

Brown, L. R. (ed) (1988) *State of the World: A World Watch Institute Report on Progress Toward a Sustainable Society*, W. W. Norton and Co, New York, Chapter 5, pp83–100

Buffington, D. E. (1979) 'Economics of landscaping features for conserving energy in residences', in *Proceedings of the Florida State Horticultural Society*, vol 92, pp216–220

CDC (US Centers for Disease Control and Prevention) (2006) www.cdc.gov

CDIAC (Carbon Dioxide Information Analysis Center) (2006) Oak Ridge National Laboratory, Oak Ridge, TN, http://cdiac.esd.ornl.gov

Cominsky, R. J., Huber, G. A., Kennedy, T. W. and Anderson, M. (1994) *The Superpave Mix Design Manual for New Construction and Overlays*, SHRP-A-407, National Research Council, Washington, DC

DeWalle D. R., Heisler, G. M. and Jacobs, R. E. (1983) 'Forest home sites influence heating and cooling energy', *Journal of Forestry*, vol 81, no 2, pp84–87

Doulos L., Santamouris, M. and Livada, I. (2004) 'Passive cooling of outdoor urban spaces: The role of materials', *Solar Energy*, vol 77, pp231–249

Dudhia, J. (1993) 'A non-hydrostatic version of the Penn State/NCAR mesoscale model: Validation tests and simulation of an Atlantic cyclone and cold front', *Monthly Weather Review*, vol 121, pp1493–1513

Dunn, B. H. (1996) 'What you need to know about slurry seal', *Better Roads*, March, pp21–25

EIA (Energy Information Administration) (1997) *DOE/EIA-0383(97): Annual Energy Outlook*, EIA, Washington, DC, Tables A8, A19

Erickson, B. (1989) 'Helena develops money-saving maintenance formula', *Public Works Magazine*

F. W. Dodge (2005) *Construction Outlook Forecast*, F. W. Dodge Market Analysis Group, 24 Hartwell Avenue, Lexington, MA, 02421

Fowler, D. (1985) 'Deposition of $SO_2$ onto plant canopies', in Winner, W. E., Mooney, H. A. and Goldstein, R. A. (eds) *Sulfur Dioxide and Vegetation*, Stanford University Press, Stanford, CA, pp389–402

Frelich, L. E. (1992) *Predicting Dimensional Relationship for Twin Cities Shade Trees*, Department of Forest Resources, University of Minnesota – Twin Cities, St Paul, MN

Gartland, L., Konopacki, S. and Akbari, H. (1996) 'Modeling the effects of reflective roofing', in *Proceedings of the ACEEE 1996 Summer Study on Energy Efficiency in Buildings*, vol 4, Pacific Grove, CA, pp117–124

Gery, M. W., Whitten, G. Z. and Kills, J. P. (1988) *Development and Testing of the CBM-IV for Urban and Regional Modeling*, Report EPA-600/3/88-012, US EPA, Research Triangle, North Carolina

Grell, G. A., Dudghia, J. and Stauffer, D. R. (1994) *A Description of the Fifth Generation of the Penn State/NCAR Mesoscale Model (MM5)*, NCAR Technical Note, NCAR TN-398-STR, Boulder, CO

Goodridge, J. (1987) 'Population and temperature trends in California', in *Proceedings of the Pacific Climate Workshop*, Pacific Grove, CA, 22–26 March

Goodridge, J. (1989) *Air Temperature Trends in California, 1916 to 1987*, Chico, CA 95928

Hall, J. V., Winer, A. M., Kleinman, M. T., Lurmann, F. M., Brajer, V. and Colome, S. D. (1992) 'Valuing the health benefits of clean air', *Science*, vol 255, pp812–817

Heisler, G. M. (1986) 'Effects of individual trees on the solar radiation climate of small buildings', *Urban Ecology*, vol 9, pp337–359

Heisler, G. M. (1989) *Effects of Trees on Wind and Solar Radiation in Residential Neighborhoods*, Final report on site design and microclimate research, ANL no 058719, Argonne National Laboratory, Argonne, IL

Heisler, G. M. (1990a) 'Mean wind speed below building height in residential neighborhoods with different tree densities', *ASHRAE Transactions*, vol 96, no 1, pp1389–1396

Heisler, G. M. (1990b) 'Tree plantings that save energy', in Rodbell, P. D. (ed) *Proceedings of the Fourth Urban Forestry Conference*, 15–19 October 1989, St Louis, MO, American Forestry Association, Washington, DC, pp58–62

HIG (Heat Island Group) (2006) Lawrence Berkeley National Laboratory, Berkeley, CA, http://HeatIsland.LBL.gov

Hildebrandt, E., Bos, W. and Moore, R. (1998) 'Assessing the impacts of white roofs on building energy loads', *ASHRAE Technical Data Bulletin*, vol 14, no 2

Huang, Y. J., Akbari, H. and Taha, H. (1990) 'The wind-shielding and shading effects of trees on residential heating and cooling requirements', *ASHRAE Transactions*, vol 96, no 1, pp1403–1411

Huang, Y. J., Akbari, H., Taha, H. and Rosenfeld, A. (1987) 'The potential of vegetation in reducing summer cooling loads in residential buildings', *Climate and Applied Meteorology*, vol 26, no 9, pp1103–1116

Hugues, J. and Heritier, B. (1995) 'Tapiphone: A technique for reducing noise in towns', *Revue Generale des Routes*, no 735, pp37–40

Hunter, R. N. (ed) (1994) *Bituminous Mixtures in Road Construction*, Thomas Telford Ltd, London

ISSA (International Slurry Seal Association) (1991) *Recommended Performance Guidelines for Emulsified Asphalt Slurry Seal*, A105 (revised), ISSA, Washington, DC

Konopacki, S. and Akbari, H. (1998) *Simulated Impact of Roof Surface Solar Absorptance, Attic, and Duct Insulation on Cooling and Heating Energy Use in Single-Family New Residential Buildings*, LBNL-41834, Lawrence Berkeley National Laboratory, Berkeley, CA

Konopacki, S. and Akbari, H. (2000a) 'Energy savings calculations for urban heat island mitigation strategies in Baton Rouge, Sacramento and Salt Lake City', in *Proceedings of the 2000 ACEEE Summer Study on Energy Efficiency in Buildings*, vol 9, Pacific Grove, CA, p215

Konopacki, S. and Akbari, H. (2000b) *Energy Savings Calculations for Heat Island Reduction Strategies in Baton Rouge, Sacramento and Salt Lake City*, LBNL-42890, Lawrence Berkeley National Laboratory, Berkeley, CA

Konopacki, S. and Akbari, H. (2001) *Measured Energy Savings and Demand Reduction from a Reflective Roof Membrane on a Large Retail Store in Austin*, LBNL-47149, Lawrence Berkeley National Laboratory, Berkeley, CA

Konopacki, S. and Akbari, H. (2002) *Energy Savings of Heat-Island-Reduction Strategies in Chicago and Houston (Including Updates for Baton Rouge, Sacramento and Salt Lake City*, LBL-49638, Lawrence Berkeley National Laboratory, Berkeley, CA

Konopacki, S., Akbari, H., Gabersek, S., Pomerantz, M. and Gartland, L. (1997) *Cooling Energy Saving Potentials of Light-Colored Roofs for Residential and Commercial Buildings in 11 US Metropolitan Areas*, Lawrence Berkeley National Laboratory Report LBNL-39433, Berkeley, CA

Konopacki, S., Akbari, H., Gartland, L. and Rainer, L. (1998) *Demonstration of Energy Savings of Cool Roofs*, LBNL-40673, Lawrence Berkeley National Laboratory, Berkeley, CA

Kramer, P. J. and Kozlowski, T. (1960) *Physiology of Trees*, McGraw Hill, New York

Laechelt, R. L. and Williams, B. M. (1976) *Value of Tree Shade to Homeowners*,

Alabama Forestry Commission, Montgomery, AL

Lefebvre, J. P. and Marzin, M. (1995) 'Pervious cement concrete wearing course offering less than 75dB(A) noise level', *Revue Generale des Routes*, vol 735, pp33–36

Lehigh Cement (1994) *White Concrete Median Barriers*, Lehigh Portland Cement Company, Allentown, PA

Leighou, R. B. (1942) *Chemistry of Engineering Materials*, McGraw-Hill, New York

Levinson, R., Berdahl, P., Akbari, H., Miller, W., Joedicke, I., Reilly, J., Suzuki, Y. and Vondran, M. (2007) 'Methods of creating solar-reflective nonwhite surfaces and their application to residential roofing materials', *Solar Energy Materials & Solar Cells*, vol 91, pp304–314

Levinson, R., Akbari, H., Konopacki, S. and Bretz, S. (2005c) 'Inclusion of cool roofs in nonresidential Title 24 prescriptive requirements', *Energy Policy*, vol 33, no 2, pp151–170

Levinson, R., Berdahl, P. and Akbari, H. (2005a) 'Spectral solar optical properties of pigments. Part I: Model for deriving scattering and absorption coefficients from transmittance and reflectance measurements', *Solar Energy Materials and Solar Cells*, vol 89, no 4, pp319–349

Levinson, R., Berdahl, P. and Akbari, H. (2005b) 'Spectral solar optical properties of pigments. Part II: Survey of common colorants', *Solar Energy Materials and Solar Cells*, vol 89, no 4, pp351–389

Loustalot, P., Cibray, J.-C., Genardini, C. and Janicot, L. (1995) 'Clear asphalt concrete on the Paris ring road', *Revue Generale des Routes et des Aerodromes*, vol 735, pp57–60

Lungren, B. and Goldman, C. (1996) 'High albedo seal coatings were formulated and tested at Reed and Graham, Inc, San José, CA, by Carol Goldman and Bart Lungren: The coloring was from Asphacolor of Toluca Lake, CA', pers comm

McPherson, E. G., Nowak, D. J. and Rowntree, R. A. (1994) *Chicago's Urban Forest Ecosystem: Results of the Chicago Urban Forest Climate Project*, Forest Service, US Department of Agriculture, NE

Means, R. S. (2006) *Site and Landscape Costs*, R. S. Means Company, Inc, Kingston, MA

Monismith, C. L., Coplantz, J. S., Deacon, J. A., Finn, F. N., Harvey, J. T. and Tayebali, A. A. (1994) *Fatigue Response of Asphalt-Aggregate Mixtures*, SHRP-A-404, Strategic Highway Research Program, National Research Council, Washington, DC

Myrup, L. O. and Morgan, D. L. (1972) *Numerical Model of the Urban Atmosphere, Volume I: The City–Surface Interface*, Department of Agricultural Engineering, Department of Water Science and Engineering, University of California, Davis, CA, October

Nowak, D. J. (1994a) 'Air pollution removal by Chicago's urban forest', in McPherson, E. G., Nowak, D. J. and Rowntree, R. A (eds) *Chicago's Urban Forest Ecosystem: Results of the Chicago Urban Forest Climate Project*, Forest Service, US Department of Agriculture, NE, pp63–81

Nowak, D. J. (1994b) 'Atmospheric carbon dioxide reduction by Chicago's urban forest', in McPherson, E. G., Nowak, D. J. and Rowntree, R. A (eds) *Chicago's Urban Forest Ecosystem: Results of the Chicago Urban Forest Climate Project*, Forest Service, US Department of Agriculture, NE, pp83–94

Oke, T. R. (1987) *Boundary Layer Climates*, second edition, Methuen, London

Oke, T. R. (1988) 'The urban energy balance', *Progress in Physical Geography*, vol 12, p471

Parker, J. H. (1981) *Use of Landscaping for Energy Conservation*, Department of Physical Sciences, Florida International University, Miami, FL

Parker, D., Huang, J., Konopacki, S., Gartland, L., Sherwin, J. and Gu, L. (1998a) 'Measured and simulated performance of reflective roofing systems in residential buildings', *ASHRAE Transactions*, vol 104, no 1, pp963–975

Parker, D. S., Sherwin, J. R. and Sonne, J. K. (1998b) 'Measured performance of reflective roofing systems in a Florida commercial buildings', *ASHRAE Transactions*, vol 104, no 1

Parker, D., Sonne, J. and Sherwin, J. (1997) *Demonstration of Cooling Savings of Light Colored Roof Surfacing in Florida Commercial Buildings: Retail Strip Mall*, FSEC-CR-964-97, Florida Solar Energy Center, Cocoa, FL

Parker, D. S., Sonne, J. K. and Sherwin, J. R. (2002) 'Comparative evaluation of the impact of roofing systems on residential cooling energy demand in Florida', in *Proceedings of the 2002 ACEEE Summer Study on Energy Efficiency in Buildings*, vol 1, Pacific Grove, CA, p219

Pielke, R. (1974) 'A three-dimensional numerical model of the sea breeze over South Florida', *Monthly Weather Review*, vol 102, pp115–139

Pomerantz, M., Akbari, H., Chen, A., Taha, H. and Rosenfeld, A. H. (1997) *Paving Materials for Heat Island Mitigation*, LBNL-38074, Lawrence Berkeley National Laboratory, Berkeley, CA

Raza, H. (1994a) *Design, Construction, and Performance of Microsurfacing*, FWHA-SA-94-072, FHWA, DoT, Washington DC

Raza, H. (1994b) *State-of-the-Practice Design, Construction, and Performance of Micro-Surfacing*, FWHA-SA-94-051, FHWA, US DoT, Washington DC

Raza, H. (1995) *An Overview of Surface Rehabilitation Techniques for Asphalt Pavement*, FWHA-SA-94-074, FWHA, US DoT, Washington DC

Rose, L. S., Akbari, H. and Taha, H. (2003) *Characterizing the Fabric of the Urban Environment: A Case Study of Greater Houston, Texas*, LBNL-51448, Lawrence Berkeley National Laboratory, Berkeley, CA

Rosenfeld A., Akbari, H., Taha, H. and Bretz, S. (1992) 'Implementation of light-colored surfaces: Profits for utilities and labels for paints', in *Proceedings of the ACEEE 1992 Summer Study on Energy Efficiency in Buildings*, vol 9, p141

Rosenfeld, A. H., Romm, J. J., Akbari, H. and Pomerantz, M. (1998) 'Cool communities: Strategies for heat islands mitigation and smog reduction', *Energy and Buildings*, vol 28, pp51–62

Santamouris, M. (2001) *Energy and Climate in the Urban Built Environment*, James and James, London

Santamouris, M. (2007) 'Heat island research in Europe – The state of the art', *Journal of Advances in Building Energy Research (ABER)*, vol 1, pp123–150

Sarrat, C., Lemonsu, A., Masson, V. and Guedali, D. (2006) 'Impact of urban heat island on regional atmospheric pollution', *Atmospheric Environment*, vol 40, no 10, pp1743–1758

Simpson, J. R. and McPherson, E. G. (1997) 'The effect of roof albedo modification on cooling loads of scale residences in Tucson, Arizona', *Energy and Buildings*, vol 25, pp127–137

Smart, M. T. (1994) Texas Department of Transportation Practice, personal communication

Smith, W. H. (1984) 'Pollutant uptake by plants', in Treshow M. (ed) *Air Pollution and Plant Life*, John Wiley and Sons, New York, NY

SSC (Soil Stabilization Corporation) (1995) *RoadOyl*, Product brochure, SSC, Merced, CA

Stathopoulou, E., Mihalakakou, P. and Santamouris, M. (2006) 'On the impact of temperature on tropospheric ozone concentration levels in urban environments', Paper submitted for publication

Synnefa, A. and Santamouris, M. (2006) 'Development and performance of cool colored coatings', in *Proceedings of the Conference EUROSUN*, Edinburgh, Scotland

Synnefa, A., Santamouris, M. and Apostolaki, K. (2006) 'On the development, optical properties and thermal performance of cool colored coatings for the urban environment', Paper submitted for publication

Taha, H. (1996) 'Modeling the impacts of increased urban vegetation on the ozone air quality in the South Coast Air Basin', *Atmospheric Environment*, vol 30, no 20, pp3423–3430

Taha, H. (1997) 'Modeling the impacts of large-scale albedo changes on ozone air quality in the South Coast Air Basin', *Atmospheric Environment*, vol 31, no 11, pp1667–1676

Taha, H. S., Chang, C. and Akbari, H. (2000) *Meteorological and Air-Quality Impacts of Heat Island Mitigation Measures in Three US Cities*, LBL-44222, Lawrence Berkeley National Laboratory Berkeley, CA

Taha, H., Chang, S. C. and Akbari, H. (2001) *Sensitivity of the Houston-Galveston Meteorology and Ozone Air Quality to Local Perturbations in Surface Albedo, Vegetation Fraction, and Soil Moisture: Initial Modeling Results*, Report prepared for the Global Environment and Technology Foundation, Center for Energy and Climate Solutions, LBNL-47663, Lawrence Berkeley National Laboratory, Berkeley, CA

Taha, H., Douglas, S. and Haney, J. (1997) 'Mesoscale meteorological and air quality impacts of increased urban albedo and vegetation', *Energy and Buildings – Special Issue on Urban Heat Islands and Cool Communities*, vol 25, no 2, pp169–177

Taha, H., Douglas, S., Haney, J., Winer, A., Benjamin, M., Hall, D., Hall, J., Liu, X. and Fishman, B. (1995) *Modeling the Ozone Air Quality Impacts of Increased Albedo and Urban Forest in the South Coast Air Basin*, LBL-37316, Lawrence Berkeley Laboratory, Berkeley, CA

Taha, H., Konopacki, S. and Gabersek, S. (1996) *Modeling the Meteorological and Energy Effects of Urban Heat Islands and their Mitigation: A 10-Region Study*, LBL-38667, Lawrence Berkeley Laboratory, Berkeley, CA

Ting, M., Koomey, J. and Pomerantz, M. (2001) *Preliminary Evaluation of the Lifecycle Costs and Market Barriers of Reflective Pavements*, LBNL-45864, Lawrence Berkeley Laboratory, Berkeley, CA

*Western Roofing, Insulation and Siding* (2002) http://WesternRoofing.net

Willockl, G. (1995) Pers comm about the properties of Pavebrite

Yoder, E. J. and Witzak, M. W. (1975) *Principles of Pavement Design*, Wiley and Sons, New York, NY

# 3

# Solar Control

---

*Karsten Voss, with Tilmann E. Kuhn, Peter Nitz,*
*Sebastian Herkel, Maria Wall and Bengt Hellström*

## INTRODUCTION

Transparent components are essential to the design and performance of a
building. They influence its indoor comfort and energy budget in many diverse
ways – for example, transmitted sunlight illuminates indoor rooms throughout
the year (Aschehoug et al, 2000), and solar energy is used passively to heat
buildings (primarily residential buildings) in winter (Voss and Wittwer, 2003).
If the room temperature exceeds the desired comfort range, solar heat sources
become solar loads, which have to be compensated for by adequate heat sinks,
such as ventilation, evaporation or cooling (see Figure 3.1). Taking the limited
heat capacity of typical building constructions and the similarly limited
capacity of natural heat sinks into account, effective control of solar loads is
thus a pre-condition if a passive cooling concept is to operate successfully.

The amount of heat which enters modern buildings by conduction via
opaque areas of the envelope is usually small due to the small temperature
differences in summer and the level of thermal insulation that is already
common in many countries. Conductive gains may still be relevant in warmer
climates with less need for thermal insulation. In older buildings, inadequately
insulated roof areas, for example, present a main cause of solar loads in upper
storeys: dark-coloured roofing materials are a main factor for increasing
cooling loads in air-conditioned buildings during sunny summer days (Simpson
and McPherson, 1997). Dark-coloured façade claddings may also increase
cooling loads, especially in buildings with façade-integrated openings for
continuous ventilation.

The amount of transmitted heat, or solar load, of transparent areas of a
building envelope is determined primarily by:

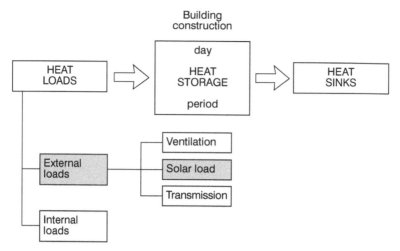

Note: The resulting increase in room temperature can be attenuated by effective heat storage in the building construction and heat removal by suitable sinks with a certain time delay. The time delay is given by the thermal inertia of a building and the potential to activate it for short (a few hours, up to a day) and medium time (period of a few days up, to a week) heat storage. This chapter concentrates on reducing solar loads.

**Figure 3.1** *The external loads, together with the internal loads, constitute the total amount of heat introduced into a building*

- the size of the glazed areas;
- the orientation of the glazed areas with respect to the sun;
- external obstructions by surrounding buildings, trees, etc.;
- glazing properties;
- the properties of sun-shading devices; and
- how they are operated.

In order to discuss the performance of different strategies and components for effective solar control, this chapter begins by highlighting the importance of solar load control compared to internal loads, thereby focusing on office buildings. The relevant fundamentals of meteorology, physics and standardization are then presented. Next, the chapter provides examples of advanced solutions from:

- glazing technology;
- moveable sun-shading devices; and
- fixed sun-shading components.

Automatic and user control of moveable or switchable shading systems is discussed on the basis of experimental investigations. Advanced calculation methods for the performance of combinations of windows and shading systems are then introduced. Finally, the chapter summarizes the findings with criteria for the selection of appropriate solar-control systems and system combinations for buildings.

## THE IMPORTANCE OF SOLAR LOAD CONTROL

Office buildings represent the main application for passive cooling today (Voss et al, 2005). In addition to the objective of maintaining acceptable room temperatures in summer and limited temperature fluctuations, the prevalence of computer work means that high visual comfort must be guaranteed. In particular, glare due to inappropriately high or uneven luminance levels within the field of view should be avoided. Where the first task is solar control, the second is defined as glare protection. The high luminance levels encountered in the direction of windows with direct incident sunlight reduce the effective contrast on computer monitors and can cause glare. However, unobstructed visual contact with the outdoors is simultaneously an important criterion for worker satisfaction. For economic reasons, solar control and glare protection are often both provided by a single device. Only a few products – for example, Venetian blinds – succeed in providing a convincing compromise for both requirements (see the section on 'Components for effective solar control'). This chapter concentrates on solar load control only; an overview of current work on glare protection for daylight environments is given in Wienold and Christoffersen (2006).

Effective solar control is imperative in order to reduce solar heating loads to the point that usual expectations on indoor thermal comfort in summer can be met without active air conditioning under Central European climatic conditions. It is much harder to meet this goal in the Earth's warmer climatic zones, especially those without sufficient temperature drop during the night or with long-lasting hot periods where the damping effect of the thermal inertia of a building is exhausted. Without effective solar control, the solar heat load is usually significantly larger than typical internal heating loads in offices of today (Wilkings and Hosni, 2000).

Figure 3.2 illustrates the solar load as a function of the weighted solar aperture of a room for different levels of irradiation. The solar aperture increases with the glass area (per floor area) and the total energy transmittance of the glass used. Sun shading devices can effectively reduce the total energy transmittance. As the internal loads are determined by building usage, an upper limit for the allowable solar loads can be established directly from the average heat extraction power of accessible heat sinks (see Figure 3.3) and the heat storage capacity of the building construction. This limit is a crucial criterion for the design and planning of building façades.

On a sunny day – 3000 watt hours per square metre per day (Wh/m$^2$/d) – the solar gain with these parameters is about three times higher than a typical internal gain in an office of about 200Wh/m$^2$/d).

Solar load control is a main topic of architecture, but is often not considered during the early design phase of a building. In buildings with large

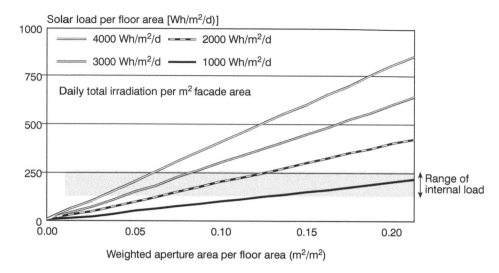

Solar load per floor area [Wh/m²/d)]

Weighted aperture area per floor area (m²/m²)

*Note:* The aperture area (glass area) is weighted with the effective total solar energy transmittance (g$_{eff}$, also known as the solar heat gain coefficient or solar factor) and refers to the floor area of the adjacent office. Example:
- net floor area = 10 square metres;
- glass area = 4.5 square metres;
- g$_{eff}$ value = 0.5;
- weighted aperture = 0.23 square metres.

**Figure 3.2** *Total daily solar load per floor area as a function of the weighted specific aperture area and the total daily irradiation on a building façade*

glazed areas and external sun-shading devices, the latter dominate the external appearance during hot periods. The desired transparency from inside and outside is thus absent over several months of the year (see Figure 3.4). Solar control must be considered not only by the architect in the initial stages of determining basic parameters and design concepts, but also by all specialists involved in consistent implementation as part of an integrated planning approach.

Within six hours of operation at an average temperature difference of 5K, about 120 watt hours per square metre (Wh/m²) can be removed. The comparison to the loads (see Figure 3.2) clearly underlines the importance of effective solar control and thermal storage in the building structure.

## SOLAR RADIATION

The solar radiation incident on the Earth's surface consists of electromagnetic waves within the ultraviolet, visible and near-infrared part of the spectrum. On average, 48 per cent of the energy is attributed to the visible spectrum of between 380 and 780 nanometres (nm); near-infrared accounts for 46 per cent and the remaining 6 per cent comprises ultraviolet radiation (Duffie and

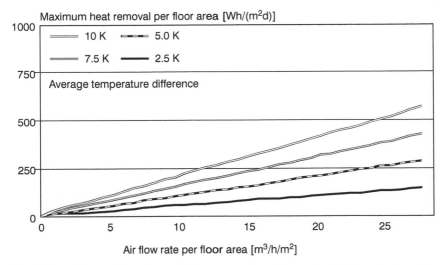

**Figure 3.3** *Maximum heat removal via night ventilation as a function of the airflow rate per floor area and the average temperature difference between indoor and outdoor air*

Beckmann, 1991). This energy distribution changes slightly for different degrees of cloud cover. The main change, however, is in the ratio of diffuse to beam radiation. Whereas only 10 to 20 per cent of the incident radiation is diffuse on clear days, this component increases to 90 to 100 per cent with a completely overcast sky.

The intensity and beam-to-diffuse ratio of solar radiation incident on building envelope surfaces varies according to the location, orientation and tilt. As shown later, this results in important criteria for designing and selecting solar-control systems. A decisive feature is that the solar radiation load, in contrast to the air temperature, differs among the variously orientated walls of a building. This is a fundamentally different situation from the commonly adopted approach of designing all façades to be identical (symmetry).

The maximum solar altitude decreases as the latitude increases. The solar azimuth, the horizontal projection of the angle between the normal to the surface and the sun, is determined by the sky direction from which the sun shines on the surface during the course of its apparent motion from sunrise to sunset. Different situations are found depending upon the latitude and the orientation of the surface (see Figure 3.5).

The orientation and tilt of a surface determine the amount of incident energy, its distribution over time and the splitting into a direct and diffuse

**Figure 3.4** *Widely used external Venetian blinds in front of an entirely glazed office building façade*

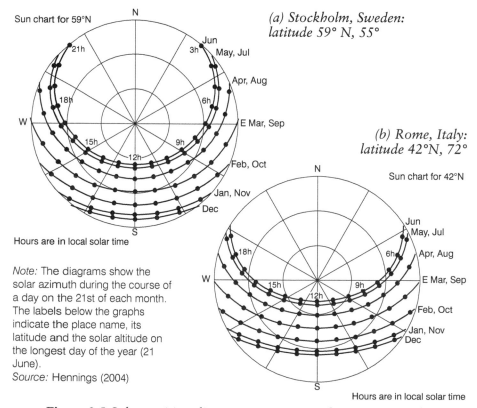

*(a) Stockholm, Sweden: latitude 59° N, 55°*

*(b) Rome, Italy: latitude 42°N, 72°*

Sun chart for 59°N

Hours are in local solar time

*Note:* The diagrams show the solar azimuth during the course of a day on the 21st of each month. The labels below the graphs indicate the place name, its latitude and the solar altitude on the longest day of the year (21 June).
*Source:* Hennings (2004)

Sun chart for 42°N

Hours are in local solar time

**Figure 3.5** *Solar position diagrams as stereographic projections for two selected locations in Northern and Southern Europe*

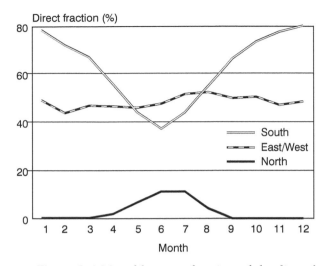

Note: The example is based on a climate data analysis for the location Rome, Italy; but a similar behaviour occurs for the other location analysed in Figure 3.5. The high solar altitude during the summer period causes the decreasing direct fraction for the southern façade. There are almost constant fractions for eastwards- and westwards-oriented façades.
Source: MeteoNorm (2003)

**Figure 3.6** *Monthly mean fraction of the direct beam irradiation to the total irradiation for differently oriented façades*

component. As the tilt becomes less steep, the portion of sky 'seen' by the surface increases and, thus, the amount of diffuse radiation received. Direct sunlight reaches a north-oriented façade, in the Northern Hemisphere, only early in the morning and late in the day during summer. The amount of radiated energy on eastern or western façades in summer is similar, or even higher due to the lower solar altitude, than on a southern façade. However, direct radiation is incident during only half of the day (see Figures 3.6 and 3.7). The fact that the maximum radiative power on a west façade coincides with the daily maximum outdoor temperature is particularly critical. In contrast to a southern façade, the maximum daily radiation totals also correlate with the outdoor temperature (see Figure 3.8). High radiative intensity often occurs on eastern to north-eastern façades before daily work in a building has started. This means that automatic operation of mechanical sun-shading devices is needed to prevent overheating of the rooms in the morning (see the section on 'Automatic and manual control').

## Data sources

Solar radiation data are included, for example, in the German (DWD, 2004) and European (Lund, 1985) test reference years, the US Department of Energy's Weather file database (DOE, undated), as well as in the MeteoNorm data bank (MeteoNorm, 2003). These sources provide hourly average values of the irradiance and other meteorological parameters, in some cases using synthesized data. MeteoNorm is a worldwide data bank of all important meteorological parameters. It allows interpolation for any choice of location and the calculation of radiation data for tilted surfaces (the database also includes the former test reference years for Switzerland). The data and models for Europe

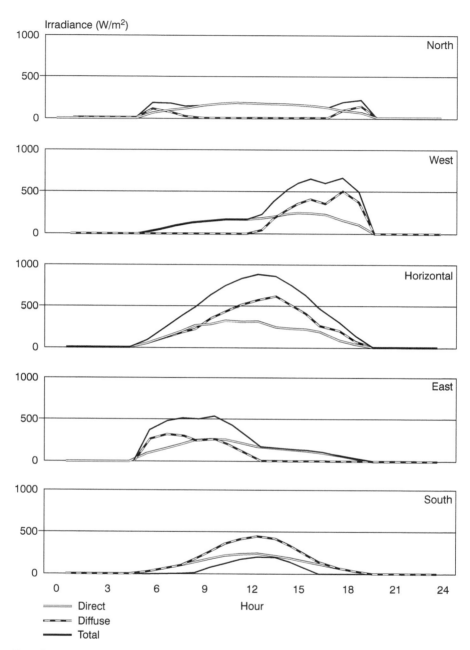

*Note:* Direct beam radiation hits a north-oriented façade only in the early morning and evening. The maximum irradiance on eastern or western façades in summer is similar, or even higher due to the lower solar altitude, than on a southern façade; but direct radiation is incident during only half the day.
*Source:* MeteoNorm (2003)

**Figure 3.7** *Examples of daily profiles of the irradiance on a horizontal plane (centre) and differently oriented façades on a sunny day in Rome*

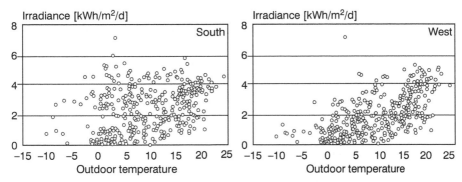

*Note:* Whereas high irradiance totals on the southern façade are found for practically all outdoor temperatures, high values on the western façade correlate strongly with high outdoor temperatures. This also applies during the course of a single day (see Figure 3.7). This emphasizes the importance of effective solar control on west-oriented surfaces.
*Source:* Reise (2005)

**Figure 3.8** *Daily solar irradiance totals on a southern (left) and a western façade (right) located in Würzburg, Germany*

are reliable and validated. Irradiance data are also available from a satellite server (Fontoynont et al, 1998). The data include monthly and hourly averages for any location in Europe from 1996 to 2000, which were calculated from images acquired by the Meteosat meteorological satellite. A further worldwide source of data is the electronic climatic atlas (Müller and Hennings, undated).

# PHYSICAL FUNDAMENTALS FOR SOLAR CONTROL

## Tilmann E. Kuhn

Two fundamental principles are applied to achieve solar control with respect to thermal comfort:

1  *Limitation of solar loads:* this indirectly reduces the heating of a room. The solar loads do not act immediately, but only after a certain time delay determined by the heat storage capacity of the building construction. If the solar loads are not limited until room overheating has already become perceptible, further heating cannot be prevented. The delayed effect represents a fundamental difference to many other protective functions (e.g. glare protection).
2  *Protection against direct irradiation by the sun:* if a person is irradiated directly by the sun, the body surface absorbs the solar energy immediately and perceptibly. Under summer conditions, this is generally not desirable for the work environment. For this reason, there is a requirement on all solar-control systems to prevent direct solar radiation from entering a

room, or at least to reduce it very strongly. This specification also applies where computer monitors are used in order to ensure glare protection.

Many other criteria beyond the primary objective of solar control play a role in practice and affect the system selection (see the section on 'Criteria of solar-control system selection').

## Limitation of solar loads

Solar gains are determined by the prevailing solar radiation intensity; the total solar energy transmittance, or so-called 'g value' of the façade element; and the area of the aperture. The total solar energy transmittance (also known as the solar heat gain coefficient or solar factor) is a physically measurable parameter corresponding to the total fraction of the incident solar energy that is transferred through a building component at normal incidence irradiation (see Equation 1) (DIN EN 410, undated). The g value consists of two parts – namely, the solar transmittance $\tau_e$ and the secondary internal heat transfer factor $q_i$. A g value of 0.3 means that 30 per cent of the incident solar energy reaches the room (see Figure 3.9):

$$g = total \; solar \; energy \; transmittance = \left[ \frac{total \; solar \; gain}{total \; incident \; radiance} \right]_{normal \; incidence}$$

[1]

$$g = \left[ \tau_e + q_i \right]_{normal \; incidence}$$

[2]

with:
- $\tau_e$ = solar transmittance; and
- $q_i$ = secondary heat gain factor.

Whereas the position of the aperture is significant for the spatial distribution of daylight in a room, this aspect is of little consequence for the solar heat gain. Accordingly, if the area of a window is doubled but the g value is halved, the solar heat gain through the window remains practically unchanged. Since transparent components with low g values are often the more expensive alternative, economic considerations are decisive, in practice.

It is important to recognize that, unlike laboratory measurements under standardized conditions, the total solar energy transmittance in practical applications is not a constant value. The g value depends upon several factors, such as the following parameters:

- the direction of the incident radiation; and
- wind conditions: the wind at the external surface determines how well the absorbed energy is transported back to the surroundings.

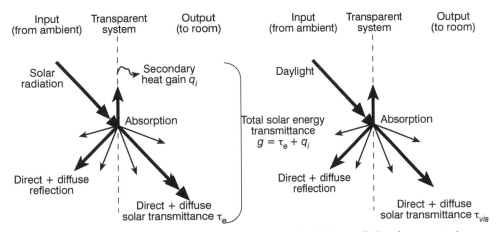

Note: All quantities defined in standards are defined for normally incident radiation. A non-normal angle of incidence has been used in the diagrams only for reasons of clarity.

**Figure 3.9** *Solar (left) and light transmittance (right) for transparent systems*

In switchable systems, the g value depends upon the setting of the system (e.g. the tilt angle of the slats or the switching state of the glazing).

If a sun-shading device is combined with a glazing unit, the g value depends upon both the glazing unit and the sun-shading system. This means that the solar-control effect differs, depending upon whether the shading device is combined with solar-control glazing or with heat-mirror glazing. A detailed discussion concerning the influencing factors and evaluation methods is given in Kuhn et al (2000) and Kuhn (2006a). The g value can be measured directly calorimetrically. There are also different approaches to calculate the g value (e.g. ISO 15099, undated; prEN 13363, undated). The choice of the most appropriate method or combination of different methods for a specific question must be decided individually for each particular case.

The g value as defined above is a parameter that only applies to specific boundary conditions. In particular, the g value is defined in the standards (DIN EN 410, undated) only for solar irradiation at normal incidence, which is an unrealistic special case. It is therefore difficult to use these standard conditions as a basis for a realistic and reliable comparison of the efficiency of façade variants with differing aperture areas and shading devices. One realistic approach is the use of the so-called 'effective g value $g_{eff}$', defined as an hourly or monthly average value under project-specific boundary conditions (Kuhn et al, 2000; Kuhn, 2006a, b). This means that a calculation is performed to determine how much solar energy impinges directly and diffusely on the outer surface of the façade during a certain period (e.g. in one hour or one month) and how much solar energy enters the room. The ratio of the hourly or monthly totals of the transmitted to the incident solar energy is defined as the effective g value for the relevant time period and location:

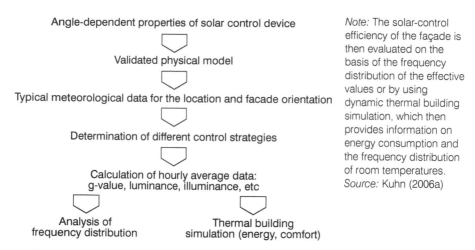

Angle-dependent properties of solar control device

Validated physical model

Typical meteorological data for the location and facade orientation

Determination of different control strategies

Calculation of hourly average data:
g-value, luminance, illuminance, etc

Analysis of
frequency distribution

Thermal building
simulation (energy, comfort)

*Note:* The solar-control efficiency of the façade is then evaluated on the basis of the frequency distribution of the effective values or by using dynamic thermal building simulation, which then provides information on energy consumption and the frequency distribution of room temperatures.
*Source:* Kuhn (2006a)

**Figure 3.10** *Average hourly values (effective values) are calculated for a specific location and a specific façade orientation, based on validated model calculations for solar control, glare protection and daylight supply*

$$g_{eff} = \frac{\text{monthly (or hourly) total solar gains}}{\text{monthly (or hourly) total incident irradiance}} \qquad [3]$$

The frequency distribution of the average hourly values for different types of systems and usage then makes it possible to compare the efficiency of the different systems realistically. In particular, it is possible to state how robustly a system reacts in the case of real or suspected incorrect operation. A comparison of average hourly values is not only limited to the thermal protection effect in summer and the g value. The method can also be applied very generally to daylight supply and glare protection. The general steps of the evaluation are shown in Figure 3.10.

Two examples are given to illustrate a typical calculation procedure (e.g. for a window with a Venetian blind system). A certain type of control strategy for the slats of the blind is the basis for calculating these average values. Evaluation of the 'best case' and the 'worst case without incorrect operation' provided valuable information.

### Strategy 'closed' ('best case')

With this strategy, the blind is fully extended and the slats are always completely closed whenever the façade is irradiated directly by the sun. The blind is fully retracted when the façade is in the shade or when there is no direct illumination. For a given combination of blind and glazing, this control strategy maximizes the overheating protection and glare protection when the sun shines directly on the façade. This control strategy ignores the need for visual contact to the exterior. The dimensions of the room and the windows will determine whether the supply of daylight is sufficient or not.

**Strategy 'cut-off' ('worst case')**

As for the first strategy, the blind is fully retracted when the façade is in the shade or when there is no direct illumination. When the sun is shining directly on the façade, the slats are tilted into the cut-off position. The 'cut-off' slat position depends upon the actual position of the sun. In this position, the slats are opened as far as possible without letting the sun shine directly through the blind. For the cut-off strategy, the tilt angle of Venetian blinds is determined by the profile angle of the sun. The profile angle is the projection of the solar altitude angle on a vertical plane perpendicular to the surface of the façade. This control strategy ensures at least some overheating protection. The visual contact to the exterior between the opened slats is an advantage, at least for higher positions of the sun. Because the room is protected against direct irradiation, the strategy ensures that there are not any disturbingly bright, directly lit stripes on desks or workbenches. The dimensions of the room and the windows will determine whether the supply of daylight is sufficient or not. This does not imply at all that an automatic adjustment of the slats is necessary. The user is free to close the slats more than with the cut-off position; but overheating protection is not guaranteed if the slats are opened further than the cut-off position.

## Protection against direct irradiation

The ability of a solar protection device to protect persons and surroundings against direct irradiation is determined by the direct–direct solar transmittance $\tau_e$, dir–dir of the device in combination with a glazing unit. $\tau_e$, dir–dir is the ratio of the directly transmitted flux of solar radiation to the incident flux of solar radiation. $\tau_e$, dir–dir is only the directly transmitted component of the solar transmittance: $\tau_e = \tau_e$, dir–dif + $\tau_e$, dir–dir (see Figure 3.11). For the sake of simplicity, it is often assumed that the radiation is normally incident to the surface. This means that the normal–normal solar transmittance $\tau_e$, n–n is used instead of $\tau_e$, dir–dir. $\tau_e$, n–n should be measured according to prEN 14500 (2006). In prEN 14501 (2005), different performance classes are given for $\tau_e$, n–n (see Table 3.1).

## COMPONENTS FOR EFFECTIVE SOLAR CONTROL

## Glazing technology

### Solar-control glazing

Architectural glazing today is usually produced as functional insulating glazing. The type of glass, the number of panes, and the tinting, coatings and gas fill are optimized for the required application (Hutchins, 2003). Protection against overheating in summer is provided primarily by so-called solar-control glazing. The main characteristic is a low $g$ value. In principle, this can be achieved by tinting, printing or coating one or more panes, or by introducing components into the cavity between the panes.

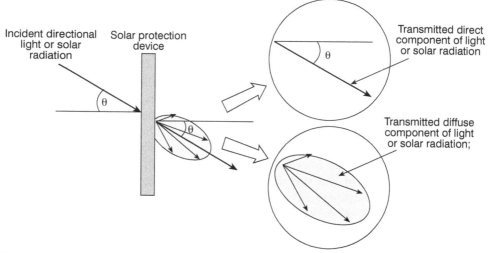

Notes: θ = angle of incidence.

**Figure 3.11** *Direct and diffuse components of transmitted radiation*

Tinting of glass mainly decreases the transmittance by increasing the absorptance. For a sufficiently low $g$ value, it is important to limit the secondary heat gain, resulting from the increased absorptance. Therefore, tinted glass is used for the outer glazing only. A variable $g$ value might be achieved in some application by seasonally turning the inside glass to the outside and/or seasonally venting the glazing gap (Erell, 2002): winter mode – tinted glass facing inside; summer mode – tinted glass facing outside. Printing on glass is generally based on the same physical principles. Light colours are preferred to increase the reflection to the outside.

Coated solar-control glazing units dominate the market today. With a thin coating of noble metals on the inside surface of the outer pane, they are optimized as far as possible in order to only transmit the visible portion of the

**Table 3.1** *Performance classes of solar-control systems for the normal–normal solar transmittance* $\tau_e$, *n–n*

| Protection of individuals and surroundings against direct irradiation | 0 (very little effect) | 1 (little effect) | 2 (moderate effect) | 3 (good effect) | 4 (very good effect) |
|---|---|---|---|---|---|
| $\tau_e$, n–n | $\tau_e$, n–n $\leq 0.20$ | $0.15 \leq$ $\tau_e$, n–n $< 0.20$ | $0.10 \leq$ $\tau_e$, n–n $< 0.15$ | $0.05 \leq$ $\tau_e$, n–n $< 0.10$ | $\tau_e$, n–n $< 0.05$ |

Note: Shading devices with non-perforated slats are class 4 when the slats are tilted in such a way that there is no direct penetration of the sun.
Source: prEN 14500 (2006)

solar radiation, but to reflect the thermal radiation (selectivity). Selectivity is not achieved with tinted or printed glass. In the case of the physical optimum, the light transmittance $\tau_{vis}$ of a solar-control glazing unit is almost twice as high as its g value, so the selectivity coefficient S is just under two (see Figure 3.12):

$$ S = \frac{\tau_{vis}}{g} [\ -\ ] \qquad\qquad [4] $$

Recent progress in coating technology has resulted in various products that almost reach this physical limit today. Still greater differences between the g value and the light transmittance can only be achieved by sacrificing colour neutrality and indoor colour rendering, so that this type of glazing is not suitable for rooms which are regularly occupied for working. Tinting or printing on glass reduces the light transmittance significantly. These approaches are thus less suitable for the combination of effective solar control and high luminous efficacy (see Equation 4) needed in office buildings.

Solar-control glazing alone usually does not provide sufficient protection against solar gains and certainly cannot protect adequately against glare due

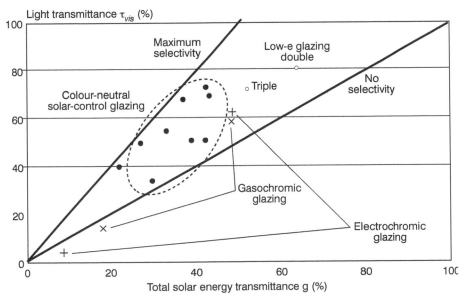

*Note:* The graph shows the light transmittance versus the *g* value for normal incidence according to DIN EN 410 (undated).

Two examples of prototype glazing with variable properties are also included (see Table 3.3).

*Source:* manufacturers' specifications

**Figure 3.12** *Selectivity of typical solar-control glazing based on noble-metal coatings, compared to low-e glazing with low thermal transmittance but high g values*

**Table 3.2** *Typical luminance values for the sky in candela (cd)*
*per square metre*

| view direction | Dark overcast | Light overcast | Hazy | Clear |
|---|---|---|---|---|
| | | Sky Conditions | | |
| Sun | 3000 | 12,000 | $10^9$ | $10^9$ |
| Sky near sun | 3000 | 5000–7000 | 15,000–30,000 | 10,000–15,000 |
| Sky away from the sun | 1500 | 3000–4000 | 1700–25,000 | 600–8000 |

*Source:* VBG (2002)

to direct sunlight. The luminance in the direction of the window is still clearly too high and direct visual contact to the sun is retained (see Table 3.2). Only if the light transmittance becomes unacceptably low (with respect to daylighting under overcast sky conditions) will the luminance be reduced to values that are acceptable for computer work. Thus, additional fixed or moveable devices are needed. However, not all of these are suitable as glare protection either (see the section on 'Moveable sun-shading devices').

The decisive advantage of solar-control glazing is that it achieves a basic level of solar control: regardless of user behaviour with respect to mechanical sun-shading devices, extreme indoor temperatures in summer are already prevented so that the robustness of a purely passive cooling concept is increased (see the section on 'Criteria of solar-control system selection'). A disadvantage is the increasing space heating demand due to less passive solar gains in winter. This effect is in the order of magnitude of 10 per cent for a well-insulated office building in a mid-European climate. The selectivity of the glazing increases the luminous efficacy, η, which is defined as the ratio of illuminance, L, to solar irradiance, I:

$$\eta = \frac{L}{I} \left[ \frac{\text{lm}}{\text{W}} \right]$$
[5]

Whereas solar radiation has a luminous efficacy of 90 to 110 lumen per watt (lm/W), depending upon the proportion of diffuse radiation, the radiation filtered by solar-control glazing has a luminous efficacy of up to 200lm/W (see Figure 3.13). If it is dosed and distributed appropriately, daylight can thus illuminate a room three times more efficiently than the T5 fluorescent lamps (up to 90lm/W) mounted in efficient luminaires (efficiency > 80 per cent) that have been introduced in offices today. The summer heat load in the room thus decreases (Reinhart and Wambsganß, 2005).

In addition to coated glazing units, coated and adhesive polymer films are manufactured specially for the renovation of existing buildings glazing. They

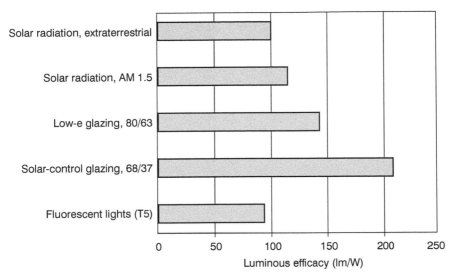

*Note:* The numbers after the glazing types specify the light transmittance and *g* value according to DIN EN 410 (undated).
*Source:* manufacturers' specifications

**Figure 3.13** *Luminous efficacy for solar radiation outside buildings (for extraterrestrial and terrestrial AM 1.5 spectra), behind low-e and solar-control glazing, and for a highly efficient T5 fluorescent lamp*

consist of polyester films with evaporated or sputtered reflective coatings, similar to the materials for roller blinds (see the section on 'Sun-shading devices between glass panes'). The films are further treated with transparent protective and adhesive layers in order to form a self-adhesive, weather-resistant laminate that can be applied to the outer surface of a glazing unit. Due to the permanent weather exposure, the durability of the films is not comparable to coatings on glass as part of sealed double-glazing units.

**Special glazing**

There are many types of insulating glazing units which combine coated panes with special elements in the hermetically sealed cavity between the panes. These elements modify the incident light and reduce the *g* value (Aschehoug et al, 2000). In most cases, they are designed to block direct solar radiation from a selected angular range by ordinary reflection or total internal reflection (angular selectivity). Specially formed reflective slats, prismatic structures (façades) or reflective two-dimensionally periodic structures (roofs) are suitable for this purpose. In the blocking range, radiation is not transmitted directly so that only secondary heat transfer remains (see Figures 3.14 and 3.15). Due to absorption of some solar radiation by the profiles or prisms, the reduction in total solar energy transmission is usually less than for the light transmission, so it is the opposite of the effect of selectively coated solar-

*Note:* Apart from the obstruction of certain angular ranges of the direct radiation, high reflectance (> 85 per cent) of the surfaces is important in order to achieve low *g* values. A coating of very pure aluminium is most commonly used for this purpose. This is essential to limit the secondary heat input.
*Source:* Klein (2000)

**Figure 3.14** *Examples of rays through double-glazing units with specially formed reflective slats (above, 'Okasolar') and reflective two-dimensionally periodic structures (below, 'Sieteco')*

control glazing. For this reason, the elements are usually combined as compensating components with selectively coated glass. Compared to solar-control glazing without additional inserts, the light and solar radiation transmission depends more strongly upon the direction; above all, the transmission behaviour is no longer rotationally symmetric. This latter property means that the energy-relevant effect of this type of glazing is not reproduced correctly in the building energy simulation programs that are in general use (see the section on 'Combinations of glazing units and solar control: Calculation of properties and performance').

All of these fixed inlays in glazing cavities more or less disturb the view outdoors. Therefore, they are not suitable as the only glazing in offices. The main applications are as glass façades and roofs of circulation areas, staircases or atria. In these kinds of applications, the disadvantage of only moderate solar load reduction due to the secondary heat gain is largely acceptable.

More recent developments have the aim of replacing the macroscopic elements by micro-structured polymer or glass surfaces and combining this with a spatially selective coating of certain structure surfaces (Gombert et al, 2004; Nitz et al, 2004). The structures developed for this purpose have

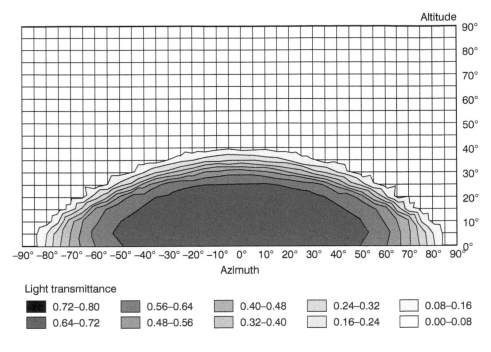

**Light transmittance**

| | | | | |
|---|---|---|---|---|
| ■ 0.72–0.80 | ■ 0.56–0.64 | ▨ 0.40–0.48 | ▢ 0.24–0.32 | ▢ 0.08–0.16 |
| ■ 0.64–0.72 | ▨ 0.48–0.56 | ▨ 0.32–0.40 | ▢ 0.16–0.24 | ▢ 0.00–0.08 |

*Note:* The system is designed so that no rays originating from a solar altitude greater than 30° are transmitted directly into the room, but only via reflections (cut-off). Rays originating from a low sun are deflected by the system. It should be noted that the additional secondary heat transfer is not represented here.
*Source:* Klein (2000)

**Figure 3.15** *Transmission behaviour (direct hemispherical) of a double-glazing unit, including reflective profile elements of the 'Okasolar' type, obtained by ray-tracing simulation*

dimensions in the range between 10μm and 100μm. They can be produced, for example, by interference lithographic processes. In this process, a split laser beam interferes on the surface of a photoresist plate. After developing the photoresist exposed to the interference pattern, the resulting surface profile is copied to a metallic master plate by electroforming. Alternatively, metallic master structures can be generated by ultra-precision diamond turning. The generated master structures are used for large-area replication by ultraviolet (UV) curing or embossing in polymer substrates or films and coatings on glass. One advantage is that the tiny structures cannot be resolved by the eye. This means that their appearance is homogeneous over the entire surface, unlike macroscopic structures, and they can be tilted for façades that are not oriented due south. Further advantages include savings in materials and weight, simpler integration into façade elements and less absorption in the thin coatings or films.

## Switchable glazing

*Peter Nitz*

In contrast to many moveable systems, some types of so-called switchable glazing can combine solar-control functionality with provision of sufficient daylight, while retaining visual contact to the outdoor surroundings. There is no need for moving parts that are exposed to the weather. Incompatible requirements such as visual contact and privacy can be fulfilled at different times, as required, with glazing that can be switched between transparent and light-scattering states or darkened by pressing a button. Solar control in summer and large solar gains in winter are no longer mutually exclusive demands, but can be met as needed with switchable glazing. In office buildings, glare protection is a particularly important control criterion for the user, in addition to solar control. For this reason, special attention should be paid to the visual comfort that can be offered by switchable glazing.

Although most of the 'products' reported are in a non-commercial stage of development, an overview will be given to the variety of system approaches under investigation.

Switchable glazing can be classified according to the optical switching mechanism or the configuration of active layers on or between glass panes:

- *electrochromic:* switching by darkening (blue coloration) induced by an electric current;
- *gasochromic or hydrochromic:* switching by darkening due to contact with a gas;
- *thermochromic or thermotropic:* switching as a change in colour or light-scattering properties upon crossing a certain temperature threshold;
- *suspended particle devices (SPDs):* switching to a more transmissive state due to the orientation of optically anisotropic absorbing particles by application of a voltage;
- *photochromic:* switching by darkening induced by illumination;
- *photoelectrochromic:* electrochromic switching activated by solar radiation;
- *polymer-dispersed liquid crystal (PDLC):* switching to a transparent state due to the orientation of light-scattering particles by application of a voltage (also feasible in the reverse mode – that is, light scattering on application of a voltage);
- *polymer network cholesteric liquid crystal (PNLC):* switching of the reflection or absorption colour and light scattering by application of a voltage;
- *twisted nematic liquid crystal (TNLC):* switching due to polarization rotation on orientation of nematic liquid crystals by application of a voltage; and
- *switchable mirrors based on metal hydrides:* transition from a metallic mirror to a transparent semiconductor via an absorbing intermediate state due to contact with a gas.

Some types are still in the laboratory phase and are available only as small-area operational samples, whereas other systems have already been manufactured in test or pilot production series and tested in demonstration projects. In the following sub-sections, greater detail will be presented on some systems that appear to be suited for architectural glazing and are either already available or have reached a development phase that could allow market introduction within the foreseeable future. Electrochromic, gasochromic, thermotropic and suspended particle device glazing fall into this category.

*Electrochromic glazing* If a voltage is applied to an electrochromic glazing unit and an electric current flows, the optically active layer changes its properties. If this effect is switched by the user, such windows might be suitable for daylight applications in offices. The transmittance can be changed continuously and good visual contact to the surroundings is always retained. The switching is visible as intensive blue coloration for the most commonly used electrochromic material: tungsten oxide. Solar-control specifications are met entirely; but complete glare protection against directly incident sunlight is not achieved. The solar transmittance is reduced by increased absorption of longer-wavelength visible and near-infrared radiation. This can lead to high temperatures in the glass panes or electrochromic layers, which must be taken into account in the glazing unit construction. In one type of electrochromic glazing, the active layer is sandwiched in a laminate between two glass panes that are each coated with transparent electrodes. In other types, the active layer together with a counter-electrode, a solid-state ion conductor and two transparent conducting layers are deposited as a thin-film coating onto a glass pane. Glazing units in architectural dimensions have been manufactured and tested in buildings (Platzer, 2003; Lee et al, 2006). Small panes are available commercially in the automotive sector, large-area glazing for architectural applications is available both in the US and in Europe. Other electrochromic glazing systems are being developed by several different parties across the world.

*Gasochromic glazing* Gasochromic glazing also displays the characteristic deep blue colour of partly reduced tungsten oxide in the dark state. Comparable to the electrochromic glass, the switching might also be user controlled. Intermediate states are possible and viewing retention is good (see Figure 3.16). The active layers are coloured when a gas with a small concentration of hydrogen flows over them and are bleached by contact with oxygen. The hydrogen and oxygen can be supplied within a closed-loop system from an electrically powered gas supply unit that is integrated within the window frame. Gasochromic glazing requires fewer coating layers than electrochromic glazing. Specifically, there is no need for the ionic conductor/electrolyte or the two transparent electrodes, which results in higher transmittance values in the bleached state. To date, glazing units with areas up to 1.5 by 1.8 metres have been produced and tested in the first pilot façades

*Note:* For comparison, a conventional glazing unit is shown to the right and a completely coloured gasochromic glazing unit is shown to the upper left of each photo: (a) bleached; (b) partly bleached; (c) completely coloured.
*Source:* Fraunhofer ISE/Interpane Entwicklungs- und Beratungsgesellschaft

**Figure 3.16** *Gasochromic glazing in different switching states*

(Platzer, 2003). Gasochromic glazing for architectural applications is available on request in limited quantities (ISE, 2005).

*Thermotropic glazing* Thermotropic layers switch automatically (i.e. without activation by a user) when a critical temperature is exceeded, from a clear to a strongly light-scattering, opaque white state. Since they do not allow active user control they might be applied in skylights, etc. The light scattering in the opaque state is caused by small particles that have a different refractive index from the surrounding matrix, similar to the finely distributed cream droplets in milk. Incident radiation is diffusely reflected due to the strong light scattering, and solar control is guaranteed. Many types of materials show this type of behaviour. Apart from lyotropic liquid crystals (see Seeboth et al, 2000 for an overview of many types of materials) and other phase-change materials (Byker et al, 2000), hydrogels, polymer blends and cast resin systems are primarily suitable for applications in architectural glazing. In hydrogels, the light-scattering particles re-dissolve on cooling in an aqueous gel (see Figure 3.17); in polymer blends two polymers separate and mix reversibly (Raicu et al, 2002; Nitz and Hartwig, 2005). In thermotropic cast resin systems, the permanently existing scattering particles consist of a material with a temperature-dependent refractive index, which typically suddenly changes at the melting point (Wirth and Horn, 2002). As the refractive index of the particles does not match that of the cast resin perfectly even at low temperatures, these panes also display a minimal haziness in the 'clear' state.

Since the view is completely obstructed in the opaque state, thermotropic glazing is primarily suited for clerestory glazing or glazing in rooms that are not continuously occupied. Applications in overhangs or in combination with printed panes have also been discussed and tested (Hartwig, 2003). Several groups have investigated active switching by electrically heating the panes (Fischer et al, 2000; Inoue and Harimoto, 2005). Due to the light scattering, thermotropic glazing in the opaque state is perceived as a source of illumination, which can lead to (reflective) glare under unfavourable

*Note:* The clerestory and side windows are fitted out with thermotropic glazing from Affinity Co Ltd, Japan: (a) clear state; (b) diffusing state.
*Source:* Hartwig (2003)

**Figure 3.17** *View of a 1:2.5 scale model of an office façade equipped with laminated hydrogel glazing*

conditions. On the other hand, glare due to directly illuminated areas is avoided and the light distribution within a room becomes more even. While European developments have been suspended, thermotropic glazing is still being produced on a pilot line in Japan and is marketed in limited quantities (Watanabe, 1998).

*Suspended particle devices (SPDs)* Microscopic particles with a high absorptance, which are suspended but still mobile within a polymer matrix, are responsible for the darkening of SPD glazing. The panes become lighter when a voltage is applied because the flakes or flat rod-shaped particles are orientated in the electric field so that viewing between the particles becomes possible (see Figure 3.18). However, even in the lighter state, there is still some slight haziness and colour, which can be disturbing. The clear state requires continuous application of an electric field, which is associated with a power consumption of < 5W per square metre. SPD glazing is available in dimensions up to 1.2 by 2.8 metres (Cricursa Cristales Curvados SA, 2005).

*Characteristic data* The characteristics of the different types of switchable glazing with respect to their application as solar-control systems are most usefully presented for double- or triple-insulating glazing units (see Table 3.3). In a comparison, the system-specific advantages and disadvantages, as discussed above, should also be taken into account.

Other types of switchable glazing are still being developed in the laboratory or are only of limited suitability for solar-control purposes. Polymer-dispersed liquid crystal (PDLC) glazing is commercially available in dimensions up to 1 by 3 metres, but is primarily used for switchable privacy or projection screens. This glazing is not suitable for solar control since the total solar energy transmittance hardly changes between the clear and scattering

*Source:* Research Frontiers, Inc, New York

**Figure 3.18** *Suspended particle device (SPD) glazing in the clear and coloured states*

states (Saint Gobain Glass, 2005). Developments of other PDLC variants are still in the laboratory phase. Thermochromic coatings based on vanadium oxide (e.g. Burkhardt et al, 2002) switch primarily at wavelengths exceeding 500nm. At shorter wavelengths, the coatings absorb strongly, which greatly limits their utility as switchable solar-control systems. Switchable mirrors based on metal hydrides – also known as hydrogen switchable mirrors (HYSWIMs) – (Huiberts et al, 1996; Wittwer et al, 2004) have been produced in the laboratory, and materials research is currently in progress. At present, an application in glazing products is hindered by the unsatisfactory reversibility and durability of the switching films, among other aspects. Nematic liquid crystals are used throughout the world in displays. Applications in architectural glazing have so far been prevented by the costs of the liquid crystals and polarizers, and the manufacturing processes for large areas. Independent of the switching state, the crossed polarizers absorb at least 50 per cent of the incident radiation, which results in a correspondingly low transmittance even in the lighter state and fundamentally limits the switching range that can be achieved. The components that are currently available are not optimized for solar-control purposes. Thus, the effects that have been observed in investigations (Haase, 2004) are restricted essentially to the visible spectral range. Prototypes of a polymer network cholesteric liquid crystal (PNLC) glazing unit with an area of 0.68 by 1.2 metres were produced within a European Union (EU) project (SMARTWIN, 2005). In addition to the clear state, this system displayed both reflective and light-scattering states. Unfortunately, technical data for this system have not been published. The optically active component of electrochromic and gasochromic glazing, tungsten oxide, is also the switchable component in so-called

**Table 3.3** *Specification data for switchable double- or triple-insulating glazing units, as available at the end of 2005*

| | System thickness (mm) | u value (W/m²/K) | Light transmittance | | Solar transmittance | | g value | | Source |
|---|---|---|---|---|---|---|---|---|---|
| | | | Light | Dark | Light | Dark | Light | Dark | |
| Gasochromic | 39 | 0.9 | 0.60 | 0.15 | 0.40 | 0.08 | 0.48 | 0.18 | Platzer (2003) |
| Electrochromic* | 27 | 1.6 | 0.62 | 0.04 | 0.40 | 0.02 | 0.48 | 0.09 | Sage Glass (2005) |
| Thermotropic hydrogel | 29 | 1.3 | 0.74 | 0.16 | 0.44 | 0.09 | 0.55 | 0.14 | Georg et al (1998) |
| Cast resin | 36 | 0.9 | 0.63 | 0.28 | 0.41 | 0.20 | 0.49 | 0.27 | Wirth (2003) |
| Suspended particle device (SPD)** | 25 | 1.7 | 0.33 | 0.03 | 0.30 | 0.14 | 0.43 | 0.24 | Cricursa Cristales Curvados SA, 2005) |

*Notes:* The table includes only those types that have been produced and tested in large dimensions and as glazing prototypes. The only product which is currently commercially available in Europe is suspended particle device (SPD) glazing. Glazing containing thermotropic hydrogels is distributed in Japan and electrochromic glazing in the US.
\* = available in different versions; the data listed here are for type 'Classic' in combination with a laminated glass.
\*\* = available in different versions (A, B, C, D) with different degrees of maximum coloration; the data listed here are for type C.

photoelectrochromic films. The charge carriers needed for coloration are generated in the element itself by irradiation. This is achieved by using the processes of a dye-sensitized solar cell (Georg and Georg, 2006). The charge carriers generated during irradiation lead to coloration. Opening and closing the electric circuit allows active switching (e.g. bleaching under irradiation). The principle has already been implemented successfully as a laboratory sample.

A promising new development consists of photochromic films based on tungsten oxide (ISE, 2005). Again, the electrochromic coloration of tungsten oxide is combined with generation of charge carriers. The film stack is much simpler than the photoelectrochromic variant, which results in higher transmittance in the bleached state and reduces colour effects. The temperature dependence of this system is much weaker than the systems used in photochromic sunglasses, which are not suitable for architectural applications for various reasons. The new photochromic systems exist as laboratory samples. Further development to produce architectural glazing still lies ahead.

User surveys and studies have demonstrated a high degree of acceptance for switchable glazing (Platzer, 2003). These studies have also confirmed the high potential for saving electricity needed for cooling and artificial lighting, as had already been predicted in a number of preliminary investigations.

Whether switchable glazing becomes installed in large areas will also be decided by its price, which, in turn, depends upon the potential for using inexpensive materials and industrial manufacturing processes. Fundamental criteria for successful market penetration, however, include an acceptable lifetime for a building component and a flawless appearance, also in large dimensions, as desired by architects, planners and building owners.

## Moveable sun-shading devices

*Tilmann E. Kuhn and Karsten Voss*

Moveable sun-shading devices in various forms are regarded as standard fittings today. They are usually required to guarantee solar control and to protect against glare. In the following discussion, the devices are classified according to their position with reference to the glazing units of the window.

### Externally mounted sun-shading devices

The greatest degree of solar control is generally achieved with an externally mounted sun-shading device. If identical sun-shading devices are mounted externally, internally or in the space between glass panes, the total solar energy transmittance is reduced most strongly for the external position. In many countries, casement windows open into the room. Thus, only external sun-shading devices ensure that the solar-control effect is retained even when the window is opened. A disadvantage is their exposure to wind loads – particularly for awnings – and the higher production and maintenance costs, both related to the effects of weathering. With regard to the passive use of solar energy for heating in winter, the reduction of solar gains by an external sun-shading device is a disadvantage. However, experience has shown that this has little effect on the energy balance of low-energy office buildings in a Central European climate (Voss et al, 2005).

### External Venetian blinds

Regardless of the slat form or colour, if the slats are completely closed, $g$ values of 10 per cent and less can almost always be achieved if the blinds are mounted in front of glazing units with U values of $1.1 Wm^2/K$ or better. When the blind is closed, the lowest $g$ values are obtained with highly reflective slats (e.g. white slats). In the cut-off position, highly reflective slats in externally mounted blinds lead to a higher $g$ value than less reflective slats because more radiation is re-directed into the room via multiple reflections. Thus, with this blind position, the situation is reversed: better daylighting is achieved, but less solar control with the more reflective slats. The $g$ value for externally mounted Venetian blinds in the cut-off position can be higher than 15 per cent.

*Note:* In the upper section, with dimensions that can be specified as required, the slats can be turned independently of the lower section. In this way, a room can still benefit from daylight while effective solar control and glare protection are achieved. It is also possible to close the entire blind completely if the irradiation level is very high.
*Source:* University Karlsruhe, fbta

**Figure 3.19** *Venetian blind divided into two independently functional sections*

Perforated slats offer a certain degree of visual contact even when the blind is almost closed. However, only part of each slat should be perforated, so that when the irradiation level is extremely high and the blind is closed completely, the perforated section can be blocked by the opaque section of the adjacent slat.

Blinds that allow divided adjustment of the slats in the lower and upper part of the window improve glare protection and daylighting (see Figure 3.19).

**External roller blinds**

External roller blinds are made of either water-repellent, weather-resistant textiles or stainless steel. Textile roller blinds expose a large continuous area to the wind and are therefore more susceptible to wind loads than Venetian blinds. In contrast to Venetian blinds, roller blinds, which can largely be moved vertically only, do not allow angle-selective blocking of direct solar radiation to achieve solar control. Thus, the solar-control effect of a textile blind is mainly determined by its transmittance. To achieve sufficient glare protection, only woven materials with a very low total transmittance (below 0.8) and practically no direct–direct transmittance can be used in front of windows that

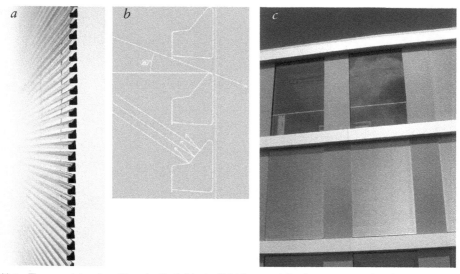

*Note:* The special rod profile selectively blocks light from certain sky segments. A rod cross-section typically has dimensions of 4mm × 5mm. Combined with low-e double glazing, $g_{tot}$ values below 0.1 are possible.
*Source:* Clauss Markisen

**Figure 3.20** *Stainless steel roller blind as a weather-resistant external sun-shading device*

are exposed to direct sunlight. A stainless steel blind, comprising a flexible band of thin horizontal rods, is more wind resistant due to its greater mass and can offer some angular selectivity, depending upon the rod profiles (see Figure 3.20). In connection with solar control, it is important to note that the users often do not pull down the blinds completely, so that only part of the window is shaded. The daylight supply toward the back of the room is significantly lower than with sectioned Venetian blinds.

**Indoor sun-shading devices**
The advantages of indoor devices are the ease of installation (also as a renovation measure) and their comparatively favourable purchasing and maintenance costs since they are not exposed to weathering. They can also provide solar control in tall or exposed buildings with high wind loads. In order to achieve adequate solar control for a building with purely passive cooling, indoor sun-shading devices are almost always combined with solar-control glazing. Its higher price reduces the cost advantage. With special Venetian or roller blinds, similarly low g values to externally mounted devices can then be achieved (0.15). Figure 3.21 shows two examples. As transparent polymer films do not scatter light, in contrast to woven materials, the view is retained even for very low transmittance values. A disadvantage of all internally mounted devices is the doubled path length for radiation through the glazing unit due to reflection by the sun-shading device, with the

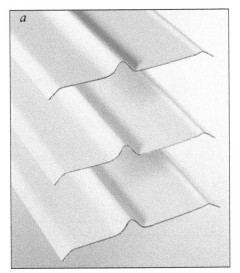 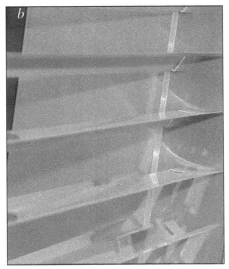

*Note:* (a) The slat profile has a special form, which, in combination with a highly reflective ($\rho$ = 85 per cent) white coating, ensures that most of the incident radiation is reflected outwards without multiple reflections when the slats are in the cut-off position; (b) the special profile, together with the reflective coating on the upper side of the slats, causes incident direct light to be redirected towards the room ceiling for low solar altitudes and to be retro-reflected outwards for high solar altitudes.
*Source:* (a) Hüppelux; (b) Retrolux

**Figure 3.21** *Internal Venetian blinds with special slat profiles and coatings can achieve low $g_{tot}$ values of about 0.15, when combined with highly selective solar-control glazing*

corresponding absorption of some of the radiation in the glass. The solar-control effect is poor for devices behind light-scattering glazing (frosted glass) or radar-screening glass. Depending upon the type of fitting, the solar-control effect is negligible or limited if the windows are opened.

### Sun-shading devices between glass panes

The construction types are the same as those for indoor devices. The mechanical protection for the devices is better than for indoor devices. For identical sun-shading elements, the solar-control performance of the system is between that of externally mounted and indoor devices.

Glazed double-skin façades have frequently been equipped with these systems in recent years. The selection and position of the sun-shading device must be harmonized with the window and ventilation system. The aim is that when the sun-shading device is activated in summer and a window is opened, the air entering the room should be as cool as possible, not additionally heated by the warm air rising past the activated sun-shading device (see Figure 3.22). The colour of the entire curtain wall and the sun-shading device is chosen to achieve particularly high reflectance values so that heating of the space between the external and internal glazing is minimized.

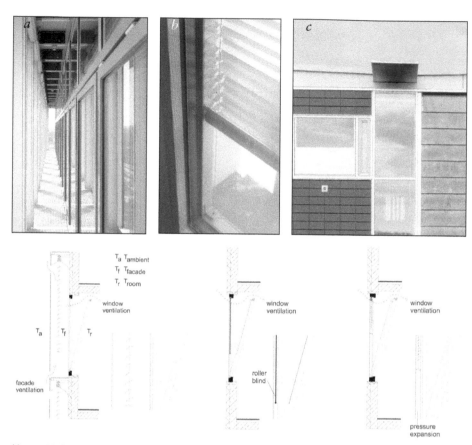

*Notes:* (a) Double-skin glazed façade with a Venetian blind in the space between the outer and inner glazing layers. There is free convection through this space to remove a large proportion of the absorbed heat in summer. This can create the specific problem of warm air entering the room if windows are opened. The overall sun-shading performance of a double-skin façade considerably depends upon the design of the façade space (segmentation, colour, blind position, window type, etc.; see Pasquay, 2004). (b) Multiple-glazed window with an integrated blind. The window is glazed with a single outdoor pane and an indoor insulating glazing unit. The Venetian blind is located in the space in between, which is at the same pressure as outdoors (not sealed to the ambient). The outdoor section of the frame can be opened to allow maintenance of the blind. (c) Roller blind in the gas-filled cavity of a sealed double-glazing unit. The blind is operated by an electric motor integrated within the glazing spacer.
The figure shows an example with the roller blind closing upward. This allows daylight to enter through the top section of the window, while complete glare protection is retained. Coated polyester films are applied to minimize the blind's space demand.
*Source:* (a) Fraunhofer-Gesellschaft; (b) University Wuppertal btga; (c) Banz & Riecks Architects

**Figure 3.22** *Examples of sun-shading devices between glass panes*

Within a multiply glazed window, the sun-shading device can be mounted in part of the window frame, which can be opened separately. Usually the device is positioned between a single outdoor pane and an indoor insulating glazing unit. The insulating glazing unit improves both the u value and the *g* value. In contrast to the double-skin façade, air does not pass through the space between the single and double glazing; there are only small openings to allow pressure equalization and dehumidification. Warm air from this space does not enter the room when the window is opened.

Only thin sun-shading devices can be mounted within the hermetically sealed gap of an insulating glazing unit (maximum gap width is about 40mm). Temperature changes of the gas in the gap result in pressure differences and mechanical stress for the spacer sealant. Roller or small Venetian blinds can be installed. As the glazing unit cannot be opened for maintenance purposes, the specifications concerning long-term functionality are very stringent. For this reason, some manufacturers of Venetian blinds only allow the slats to be turned, but prevent vertical motion of the blind. Motion might be transferred by magnetic force to prevent wiring through the spacer sealing. The panes are usually coated with so-called 'hard coatings' based on doped oxide semiconductors – for example, tin oxide ($SnO$) and zinc oxide ($ZnO$) – as they are not easily scratched by contact with the sun-shading device.

## Solar control with fixed building components

Solar control can also be achieved without moving parts – to a limited extent – by large external structures on the building envelope. They are designed individually as part of the architecture. Like solar-control glazing, they improve the robustness of the indoor climatic concept as the solar-control effect is independent of user behaviour (see the final section on 'Criteria of solar-control system selection'). Furthermore, they allow an 'undisturbed' view of the surroundings for a longer period of the year since the supplementary moveable systems that are generally needed can be used less frequently. However, the view is already restricted to a greater or lesser extent by the external structure itself.

Like reflective profile elements or Venetian blinds, the transmission behaviour of external structures is not rotationally symmetric with respect to the incident direction of the radiation. The fundamental concept in dimensioning the sun-shading structures is to block the largest possible angular range of direct radiation in summer (Olgyay, 1963). Highly reflective surfaces are not necessary here to achieve solar control as the heat generated by absorption in the structures is transferred completely to the ambient air. However, care should be taken that the heat released locally at the façade does not cause an additional thermal load to the building via the ventilation system.

The solar-control effect that can be achieved with these fixed components fluctuates with the seasonally varying angle of incidence of the direct radiation. Horizontally projecting components on south-facing façades are particularly suitable in climatic zones with a high proportion of direct radiation, and are most effective in geographic regions where the solar altitude in summer is high. It is only under these conditions that large areas of a façade can be shaded by an overhang with dimensions small enough to be acceptable to an architect. In order to completely shade a window that is 1.5m high at the summer solstice in Rome, an overhang of 0.5m is needed; in Stockholm, the projection would be twice as wide. The reduction in incident diffuse radiation is independent of the shading effect for direct radiation, is constant throughout the year and is determined by the reduction in the sky sector area that is visible from the transparent aperture (view factor).

The seasonal change in solar transmittance might be increased by using a thermotropic glazing as fixed external-shading structures. The seasonal change of the outdoor temperature might induce switching from the transparent to a translucent stage (see the section on 'Switchable glazing'; Hartwig, 2003).

Light-reflecting projections that divide a window horizontally are called 'light shelves'. Due to weathering, it has been found that it is not possible to prepare surfaces that retain a very high reflectance and gloss factor permanently. Light shelves are installed with the aim of positively influencing the light distribution in a room (Aschehoug et al, 2000). Part of the direct light that has been reflected, usually diffusely, by the upper side of the light shelf passes through the glazed area above it and is reflected by the room ceiling towards the back of the room. The available daylight is distributed more evenly over the depth of the room, but is not increased as the aperture area remains unchanged (see Figure 3.23). The solar-control effect of this variant is slight. At the same time, it is particularly significant for effective daylighting under overcast sky conditions to note that the luminance at the zenith is a factor 3 higher than at the horizon. As a result, a fixed light shelf above a window reduces the available daylight from an overcast sky appreciably so that an application in climatic zones with predominantly diffuse daylight does not really come into question. Nevertheless, if rooms are not intended to be permanently occupied, fixed sun-shading elements also present a robust variant with scope for distinctive design.

Fixed sun-shading components can be dimensioned with the usual tools for opaque, completely diffusely reflecting elements in most dynamic building simulation programs. In addition, programs are now available that can take account of special grid patterns (Wall and Bülow-Hübe, 2001, 2003) or apply optimization procedures to generate geometric configurations that shade particularly effectively (Kaftan and March, 2005).

# AUTOMATIC AND MANUAL CONTROL

## *Sebastian Herkel*

Switchable or moveable solar-control systems are operated by automatic controls and/or manually. The control simultaneously affects the visual contact to the outdoors, the daylight availability (light transmittance and luminance) and the thermal comfort (solar transmittance and direct irradiation). Whereas energy-intensive air-conditioning systems compensate for 'faults' in actively cooled buildings, buildings with passive cooling only rely on applied control for their thermal comfort.

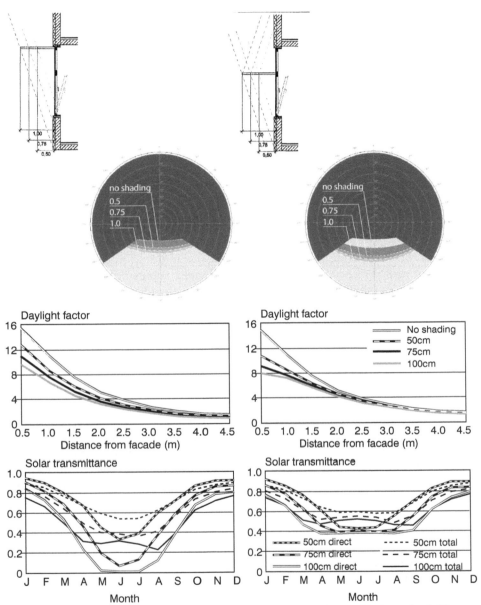

Note: Calculations are based on the weather data of Rome, Italy (maximum altitude angle of 72°). The material of the light shelf was assumed to be reflecting with solar and light reflectance values of 50 per cent at a secularity of 50 per cent. The figure shows from top to bottom: schematic representation; shading mask with a viewpoint 1m behind the façade at working desk height; daylight factor as a function of the room depth (overcast sky conditions, centre axis); average monthly radiation shading factor $F_c$ for the total exposed glass area given for direct and total radiation. Calculations were performed with Ecotect v5.5 and Desktop Radiance v3.7.
Source: Hans (2006)

**Figure 3.23** *Energy-relevant and lighting characteristics of an opaque overhang (left) and a light shelf (right) in front of a glazed façade (window height 1.25m, skylight 0.75m)*

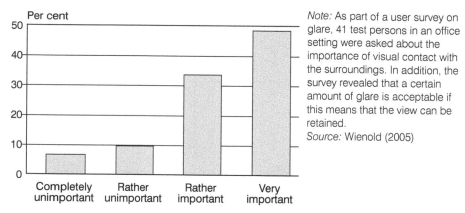

Note: As part of a user survey on glare, 41 test persons in an office setting were asked about the importance of visual contact with the surroundings. In addition, the survey revealed that a certain amount of glare is acceptable if this means that the view can be retained.
Source: Wienold (2005)

**Figure 3.24** *Importance of a view outdoors*

As the daily irradiance profiles shown in Figure 3.7 indicate, automatic solar control is of primary importance to protect rooms oriented towards the east from overheating in the early morning in advance of the building utilization time. As there is no direct irradiance at the end of the daily utilization time, manual control would only result in a lack of solar control in the morning. Thermal comfort in the whole building profits from automatic solar control when the building is unoccupied, such as at the weekends. Automatic control ensures that the solar load in single unoccupied rooms is minimized. Post-occupancy evaluations have highlighted, in many buildings, that fully automatic control does not appeal to the buildings' users. A mixed form of automatic and manual operation is preferable. Experience shows that users are generally more willing to accept automatic opening than closing of sun-shading devices (Rubin et al, 1978). One of the main reasons for this result is the importance of an undisturbed visual contact with the outdoors (see Figure 3.24). An exception to fully automatic control is the situation in openplan offices, with the need to control daylight and thermal comfort according to the requirements of a large group of users.

User behaviour concerning the operation of sun-shading devices has been investigated in several field tests (Reinhart and Voss, 2003). These reveal strong variation from person to person, but consistent behaviour from each individual, which correlates, for example, with the irradiation. The same user always adjusts his or her sun-shading device in the same way.

If solar gains are to be kept low, it is particularly important that the users employ the available potential to reduce the $g$ value to its full extent. The effect as overheating protection in summer is only perceived immediately if the person is subjected to direct irradiation by the sun. This behaviour is well described by the depth of solar penetration into the room. The lower the solar altitude, the greater the number of closed blinds (see Figures 3.25 and 3.26). Thus, the sun-shading device is activated more frequently for eastern and western orientations than for a southern orientation. Visual contact to the

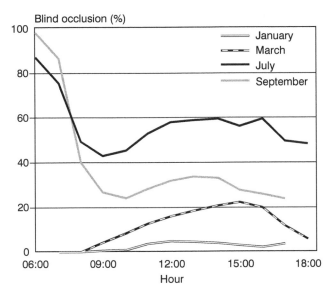

*Note:* The blinds were only operated manually except during the months of July and September, when the blinds were closed automatically at 4 am. The observations revealed that most of the blinds were opened as soon as the user arrived at the office. If the blind is open on arrival – the default state in winter – it is most frequently closed when direct radiation is incident on the façade. *Source:* Herkel and Löhnert (2005)

**Figure 3.25** *Observed average daily occlusion status of external Venetian blinds in 16 south-facing offices in Freiburg, Germany, on working days between September 2002 and July 2003 (100 per cent occlusion = completely closed)*

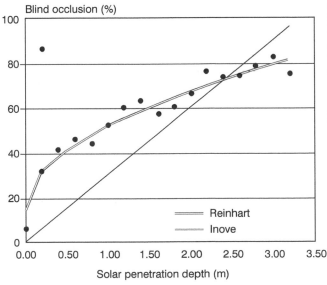

*Note:* Only data from occupied offices have been included in this analysis. The two functions shown were derived from field tests by Inoue et al (1988) and Reinhart and Voss (2003). *Source:* Reinhart and Voss (2003)

**Figure 3.26** *Average occlusion status of the sun-shading device as a function of the solar penetration depth for direct irradiation of the façade > 50W/m²* $$(100 \text{ per cent occlusion = blind completely closed})$$

surroundings is obstructed or at least reduced by most sun-shading devices. As a result, the user makes a decision between glare protection, solar control and view retention. How frequently a sun-shading device is operated also depends upon cultural and climatic boundary conditions. Investigations in North America and Japan demonstrated that a sun-shading device, once it has been closed, is seldom opened, whereas in Europe this happens more frequently (Inoue et al, 1988; Lindsay and Littlefair, 1993).

Models based on systematic field tests have been introduced into simulation programs to describe user behaviour. Such models have already been implemented in the dynamic lighting simulation programs DAYSIM and LIGHTSWITCH 2002 (Reinhart, 2004). Preliminary steps for wider application in thermal simulation programs have also been presented (Bourgeois, 2006).

## COMBINATIONS OF GLAZING UNITS AND SOLAR CONTROL: CALCULATION OF PROPERTIES AND PERFORMANCE

It is common practice for most building energy simulation software (e.g. TRNSYS, ESP-r, TAS, IDA, DOE-2 and Ecotect) to model solar control by globally reducing the irradiation on a window. The only exceptions are fixed external-shading structures. The reduction might be time controlled or controlled by a set point of the direct solar irradiance on the glazing or the indoor temperature. Applying a global reduction factor means that the angle-dependent properties of a solar-control system are treated as if they were rotationally symmetric compared to usual glazing. Detailed investigations have shown that the assumption of rotationally symmetric properties is very inaccurate in the case of Venetian blinds or blinds with vertical slats, although it might be sufficiently accurate for roller blinds (Kuhn, 2006b). If an advanced, rotationally asymmetric, calculation model is applied, the effect of the different symmetry assumptions on the total solar gain of a window system can be quantified at stationary conditions with the calculation of a combined $g$ value or for a time period with the $g_{eff}$ values as introduced early in this chapter. The combined $g$ or $g_{eff}$ value then includes the correctly modelled properties of the solar-control systems, as well as the glazing. Such an advanced calculation has been described and validated in detail as a stand-alone tool or future part of a building simulation program in Kuhn (2006a).

The following section describes two advanced, already publicly available, software tools. Whereas the first one is specially suited for detailed calculation of window properties at stationary conditions (WIS), the second one is a fully integrated single-zone building-simulation software program (ParaSol). Both tackle solar-control systems with a calculation that is rotationally asymmetric.

## Window property calculation tool:
## The Window Information System (WIS)

The Window Information System (WIS) is a software program that determines the thermal and optical characteristics of windows in detail. It is a result of European research work, headed by TNO Building and Construction Research (TNO Bouw) within the Joule programme. In order to extend commonly used window performance calculation tools, such as WINDOW 5.2 (WIN, 2005), it includes modelling of solar-control systems and vented glazing cavities. WIS contains databases with component properties and routines for calculating the thermal/optical interactions of components in a window (WIS, 2005). WIS enables calculations of the overall window properties according to the European standards at stationary, user-specified design conditions (irradiation, temperature, wind speed, etc.). It has been developed with an open object-oriented structure to enable it to respond to future developments.

Ray-tracing calculations are possible for both slatted type blinds (such as Venetian blinds and louvres) and for pleated blinds. At present WIS includes four different types of scattering layers:

1  slatted-type blinds (Venetian and louvres);
2  pleated blinds;
3  roller blinds and screens; and
4  generic diffusing devices (such as diffusing panes) and other scattering devices that do not belong to one of the other categories.

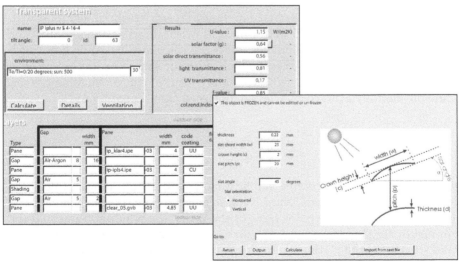

*Note:* The output is presented as a summary table.
*Source:* http://windat.ucd.ie

**Figure 3.27** *Input screens of the Window Information Systems (WIS) software; the example shows a calculation for a double-glazing window with internal Venetian blinds with curved slats*

All optical layers can be described in detail or taken from databases. The output of the calculation is presented in the form of a data report for the combined window and solar-control system as a whole, and for the absorption, reflection and transmittance fractions of the radiation at each optical layer.

## Single-zone building simulation tool: ParaSol

*Bengt Hellström and Maria Wall*

ParaSol is a graphical interface for calculating solar transmittance values and cooling/heating demands for a single (reference) room with different kinds of glazing and sunshades (ParaSol, 2001). The sunshade can be external, internal or inter-pane, or a combination. The software is based on DEROB-LTH a monolithic, detailed hour-by-hour building energy simulation program. The special features are as follows.

*Direct calculation of* $\tau$ *, g and U values of a window or a window/shading system when a window or an internal/inter-pane solar shading is selected.* The calculation is made for normal incidence beam irradiation, using standardized boundary conditions.

*Calculation of monthly average values of solar transmittance and total solar transmittance for a window with a sunshade.* Values for both the window and the combined system with the sunshade are obtained. Values for the sunshade, taken as the ratio between the values of the system and of the window, are also calculated. The results are based on four different yearly simulations: with and without solar shading, and with and without solar irradiation. The monthly average solar transmittance values are calculated based on how much of the solar irradiation on the window (without solar shading) is directly transmitted to the room through the window or the window/shading system, while the monthly average g values are obtained from Equation 6:

$$g_{eff} = \frac{(C - H)_{solar} - (C - H)_{no\ solar}}{I} \qquad [6]$$

where $C$ and $H$ are the cooling and heating demands for a month, calculated with or without solar loads, and $I$ is the solar irradiation on the window. The solar absorption for the outer faces of the walls, the window frame and the embrasure are (for this case) set to zero, so that the resulting $g_{eff}$ values are valid strictly for the glazing or for the glazing/sunshade system. The influence from the heat capacity of the walls in the room is low since the heating and cooling set points are both set to 20°C. Some absorbed solar irradiation will temporarily be stored in the walls; but since the integration time is as long as one month, the impact on the results is negligible.

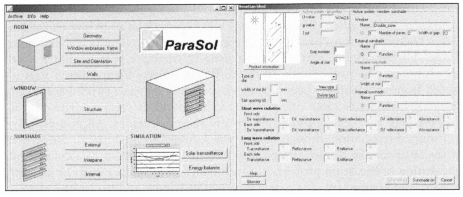

*Source:* www.parasol.se

**Figure 3.28** *Input screens of the ParaSol software programme; the example shows a calculation for a Venetian blind in the gap of a double-glazing unit*

*Calculation of heating and cooling demands for a room.* Annual peak and integrated heating and cooling demands are obtained and compared for the cases with and without the solar shading. The internal load and different ventilation parameters can be varied. In addition, the set-point room temperatures for heating and cooling and the set-point irradiation level for shading control can be chosen. A post-processor for thermal and visual comfort is also available.

*The geometry of awnings, blinds and projecting solar screens (horizontal slatted baffles) is defined by parametric input values.* There are libraries for the materials that the solar shadings are made of – for example, fabrics for awnings, Italian awnings, screens and intermediate/interior curtains, and slats for Venetian blinds. New materials can also be added to the libraries. In the simulations, the external convection heat-transfer coefficient is set to a fixed value ($15 \text{W/m}^2/\text{K}$), while the internal heat-transfer coefficient is calculated from the temperature difference to the ambient air and the inclination of the surface. Currently, all external shadings are assumed to be ventilated, while all inter-pane screens are assumed to be non-ventilated. Internal shadings are (since version 3) assumed to have an open air gap between the shading and the window, allowing ventilation through natural convection; but there is also an option for a closed air gap. The internal and inter-pane solar shadings can be specularly and diffusely reflecting, as well as directly and diffusely transmitting. For external screens, the reflectance is currently limited to being diffuse. For external awnings, the transmittance is also assumed to be diffuse. Currently, all Venetian-blind models in ParaSol assume that the slats are non-curved.

*For internal and inter-pane Venetian blinds, the beam irradiation is absorbed or reflected/transmitted diffusely or specularly/directly at each impact on the slats.* The propagation of the beam irradiation on the slats is tracked nine times. The remaining part, which is usually small, is then added to the absorptance. The propagation of the diffusely reflected/transmitted part is solved from an equation system, using view factors between two adjacent slat surfaces, each divided into five parts, and the boundary surfaces. The optical properties of the Venetian-blind 'layer' are then calculated. All non-directly transmitted irradiation through the layer is assumed to be diffused. Properties of the layer for sky diffuse and ground-reflected irradiation are calculated by integration of the beam properties, respectively, over the sky and the ground parts of the hemisphere. For external Venetian blinds (and horizontal slatted baffles), the same model as for internal and inter-pane Venetian blinds is used, although transmission or specular reflection of the slats is not yet an option.

*Calculation of shading of beam irradiation.* All surfaces are divided into nine by nine parts. Investigations are made as to whether the vector towards the sun from the centre of each part is intersected by any other surface, opaque or transparent. If so, the beam irradiation on the part is reduced by the beam transmittance of the intersecting surface. The irradiation on the surface is taken as the sum of the irradiation on the parts. Shading of diffuse irradiation is calculated from the projections of the surfaces on each other for different solid angles (Källblad, 1998). Multiple reflections between the outer surfaces are taken into account through an iteration process, using view factors between the surfaces. For the distribution of light within the room, an equation system based on view factors between the inner surfaces is solved.

The description of ParaSol above refers to the program's version 3. Results of calculations with ParaSol, version 2, have been compared with measurements (Wall and Bülow-Hübe, 2003). Some modifications of the models have since then been made in version 3. A comparison with measurements is also planned to be made for this version.

## CRITERIA OF SOLAR-CONTROL SYSTEM SELECTION

For buildings with passive cooling, only highly effective solar-control concepts are suitable. This can be achieved either with a single component or the combination of several components. Based on experience, Zimmermann (2003) recommends a $g_{tot}$ value of 0.15 as the upper limit for the case of fully activated solar control when used in office buildings in a Central European climate. This corresponds to a performance classification of 3 ($0.1 < g < 0.15$) to 4 ($g < 0.1$) according to prEN 14501 (2005) (see Table 3.4). As discussed previously, the criterion is to be applied to the complete system of glazing and sun-shading device(s) and may need to be further differentiated in the case of angle-selective sun-shading elements, taking the local meteorological conditions and façade orientation into account ($g_{eff}$). The shading effect of the

**Table 3.4** *Performance classification of combined systems (glazing + solar-control system) concerning the total solar energy transmittance*

| Class | 0 (very little effect) | 1 (little effect) | 2 (moderate effect) | 3 (good effect) | 4 (very good effect) |
|---|---|---|---|---|---|
| $g_{tot}$ | $g_{tot} \geq 0.5$ | $0.35 \leq g_{tot} < 0.50$ | $0.15 \leq g_{tot} < 0.35$ | $0.10 \leq g_{tot} < 0.15$ | $g_{tot} < 0.1$ |

*Source:* prEN 14501 (2005)

shading device only can be expressed by calculating the shading factor $F_C$, the ratio of the solar factor of the glazing + solar-control system ($g_{tot}$) to the solar factor of the glazing only ($g$):

$$F_c = \frac{g_{tot}}{g} \; [ \; - \; ] \tag{7}$$

where:

- $g_{tot}$ = total solar energy transmittance of the combined system (glazing + shading device);
- $g$ = total solar energy transmittance of the glazing (no shading device);
- $F_c$ = shading factor.

Many of the moveable, externally installed sun-shading devices meet this criterion very well, but cannot be installed in all buildings, such as tall buildings where the wind load is too high. In addition, such devices may not be favoured from an aesthetic point of view. The significant progress made recently in developing colour-neutral, selective solar-control glazing has extended the scope for alternative concepts considerably. Combinations of highly selective solar-control glazing and sun-shading devices mounted indoors or between the panes are increasing in importance. With careful planning, they can attain similarly low $g_{tot}$ values.

Apart from low $g_{tot}$ or $g_{eff}$ values, there are numerous other criteria (see the previous section on 'Components for effective solar control'). Table 3.5 gives an overview.

If the solar-control effect depends essentially upon the users operating moveable or switchable systems appropriately, buildings with passive cooling can thus be defined as sensitive. Since inappropriate behaviour cannot be compensated for by higher energy consumption, the thermal comfort zone may deteriorate. Due to thermal interaction within a building, rooms are also affected indirectly when users have operated their systems correctly. Since the effect of solar loads is not immediately evident, but only after a certain time delay, many users do not understand the process of cause and effect, so that inappropriate behaviour is not recognized as such.

Table 3.5 Selected performance criteria for common and future (switchable glazing) solar-control systems for office rooms

| | | Glazing technologies | | | Mechanical shading devices | | | Fixed shading devices | |
| | | Fixed properties | | Switchable properties | | | | | | |
| Group | Criterion | Solar-control glazing | Special glazing* | Electrochromic, gasochromic, SPD | Thermotropic* | Venetian blind | Roller blind with opaque slats | Transparent polymer roller blind | Overhang | Light shelf |
|---|---|---|---|---|---|---|---|---|---|---|
| Solar control | Load reduction** | Moderate | Moderate, seasonal variation | High | High/moderate | High to moderate, depending on the position and slat angle | | | Moderate, seasonal variation | Moderate, seasonal variation |
| | User or control dependent | No | No | Yes | No | Yes | Yes | Yes | No | No |
| Daylighting | Visual contact | Excellent | Disturbed | Variable, moderate | Variable, poor | Variable | Variable | Variable, excellent | Good | Good |
| | Glare protection | No | Yes | No | No | Variable | Variable | Variable | Good | Good |
| | Main daylight effects | Slightly reducing | Direct light cut-off | Variable, not colour neutral | Light diffusing | Variable, Light-guiding | Variable, diffusing | Variable, reducing | Seasonal direct light cut-off | Seasonal direct light cut-off, light-guiding |
| | Single system sufficient to tackle both issues? | No | No | No | No | Generally yes, but not all products | | | No | No |
| Economy | Passive solar gain reduction (winter) | Moderate | Moderate | Slightly | Slightly | No | No | No | Slightly | Slightly |
| | Investment costs | Moderate | High | Not yet clear | | Low to moderate, depending on position | Moderate to high, depending on position | No | High | High |
| | Maintenance | No | No | Not yet clear | | Moderate to high, depending on position | No | Yes | Low | Low |
| | Site-dependent solution? | No | Yes | No | | No | No | No | Yes | Yes |
| | Identical system for all orientations? | Yes | No | Yes | | Yes | Yes | Yes | No | No |
| | Potential synergetic effects | No | No | No | | No | No | No | Yes (security path, façade cleaning) | No |
| | Design dominance | Low | Moderate | Moderate | | Moderate to high, depending on mounting position | | | High | High |
| Mounting position | Exterior | | | | | X | X | X | X | |
| | Interior | | | | | X | X | X | X (no load reduction) | |
| | Inter-pane | | | | | X | X | X | X (second skin façade) | |

*Selected architectural design consideration*

*Notes:* * not suitable for office glazing due to obstructed visual contact.
** high: g < 20 per cent; moderate: 20 < g < 40 per cent; poor: g > 40 per cent.

# CONCLUSION

This chapter concludes by highlighting the need for a permanent basic level of solar control in combination with further solar-control elements that can be activated by users when thermal loads are very high or glare protection is required. Since visual contact with surroundings has a high priority for users (see Figure 3.24), the basic solar-control component can reduce the duration when this visual contact is impaired by a moveable sun-shading element. Solar-control glazing or (depending upon the orientation and local climate) fixed shading devices, such as overhangs, are suitable components for this basic solar control. It can also be expected that moveable or switchable systems offering good visibility (e.g. gasochromic or electrochromic glazing and transparent polymer blinds) will be activated frequently, although the transmitted light may be somewhat coloured.

# REFERENCES

Aschehoug, O., Christoffersen, J., Jakobiak, R., Johnsen, K., Lee, E., Ruck, N. and Selkowitz, S. (eds) (2000) *Daylight in Buildings*, IEA SHC Task 21/ECBCS, Annex 29, Lawrence Berkeley National Laboratory with support from Energy Design Resources

Bourgeois, D., MacDonald, I. and Reinhart, C. F. (2006) 'Adding advanced behavioral models in whole building energy simulation: A study on the total energy impact of manual and automated lighting control', *Energy and Buildings*, vol 38, no 7, pp814–823

Burkhardt, W., Chjristmann, T., Franke, S., Kriegseis, W., Meister, D., Meyer, B. K., Niessner, W., Schalch, D. and Scharmann, A. (2002) 'Tungsten and fluorine co-doping of $VO_2$ films', *Thin Solid Films*, vol 402, pp226–231

Byker, H., Byker, J., Millett, F. A. and Ogburn, P. H. (2000) *Thermoscattering Materials and Devices*, Patent US 6,362,303 B1 (filed 2000), ww.pleotint.com

Cricursa Cristales Curvados SA (2005) *Cri-Regulite Product Information*, www.cricursa.com

DIN EN 410 (undated) *Bestimmung der lichttechnischen und strahlungsphysikalischen Kenngrößen von Verglasungen*, Beuth-Verlag, Düsseldorf

DOE (US Department of Energy) (undated) Weather file database, www.eere.energy.gov/buildings/energyplus/cfm/weather_data.cfm

Duffie, J. A. and Beckmann, W. A (1991) *Solar Engineering of Thermal Processes*, second edition, John Wiley and Sons, New York

DWD (Deutscher Wetterdienst) (2004) *Testreferenzjahre für Deutschland für mittlere und extreme Klimaverhältnisse (TRY)*, Deutscher Wetterdienst, Offenbach, www.dwd.de/TRY

Erell, E. (2002) *SOLVENT: Development of a Ventilated, Solar-Screen Glazing System*, Enerbuild project leaflet, http://erg.ucd.ie/enerbuild/pdfs/SOLVENT.pdf

Fischer, T., Lange, R. and Seeboth, A. (2000) 'Hybrid solar and electrically controlled transmission of light filters', *Solar Energy Materials and Solar Cells*, vol 64, pp321–331

Fontoynont, M. et al (1998) 'Satellite: A WWW server which provides high quality daylight and solar radiation data for Western and Central Europe', in *Proceedings of the 9th Conference on Satellite Meteorology and Oceanography*, Paris, www.satel-light.com

Georg, A. and Georg, U. (2006) 'Opara Krašovec: Photoelectrochromic window with Pt catalyst', *Thin Solid Films*, vol 502, no 1–2, pp246–251

Georg, A., Graf, W., Schweiger, D., Wittwer, V., Nitz, P. and Wilson, H. R. (1998) 'Switchable glazing with a large dynamic range in total solar energy transmittance (TSET)', *Solar Energy*, vol 62, no 3, pp215–228

Gombert, A., Bläsi, B., Bühler, C., Nitz, P., Mick, J., Hoßfeld, W. and Niggemann, M. (2004) 'Some application cases and related manufacturing techniques for optically functional micro structures on large areas', *Optical Engineering*, vol 43, no 11, pp2525–2533

Haase, W. (2004) *Adaptive Strahlungstransmission von Verglasungen mit Flüssigkristallen [Adaptive Radiation Transmission of Glazing with Liquid Crystals]*, PhD thesis, University of Stuttgart, Germany

Hans, O. (2006) *Shading by Fixed External Elements – A Typology*, MSc thesis, University Wuppertal, Germany

Hartwig, H. (2003) *Konzepte für die Integration selbstregelnder, thermotroper Schichten in moderne Gebäudehüllen zur passiven Nutzung der Sonnenenergie [Concepts for the Integration of Automatically Controlled, Thermotropic Layers in Modern Building Envelopes for the Passive Use of Solar Energy]*, PhD thesis, University of Munich, Germany

Hennings, D. (2004) *Sunpath*, www.eclim.de

Herkel, S. and Löhnert G. (2005) 'Nutzerverhalten und Betriebsführung', in Voss, K., Löhnert, G., Herkel, S., Wagner, A. and Wambsganß, A. (eds) *Bürogebäude mit Zukunft, 2, Auflage*, Solarpraxis Verlag, Berlin

Huiberts, J. N., Griessen, R., Rector, J. H., Wijmgaarden, R. J., Dekker, J. P., de Groot, D. G. and Koeman, N. J. (1996) 'Yttrium and lanthanum hydride films with switchable optical properties', *Nature*, vol 380, pp231–234

Hutchins, M. G. (2003) 'Spectrally selective materials for efficient visible, solar and thermal radiation control', in Santamouris, M. (ed) *Solar Thermal Technologies for Buildings: The State of the Art*, James and James, London

Inoue, T. and Harimoto, K. (2005) 'Effects of combination use of thermotropic glass and transparent heat-generating layer for solar shading and daylighting', in *Proceedings of the CISBAT 2005 Conference*, Lausanne, http://cisbat.epfl.ch

Inoue, T., Kawase, T., Ibamoto, T., Takakusa, S. and Matsuo, Y. (1988) 'The development of an optimal control system for window shading devices based on investigations in office buildings', *ASHRAE Transactions*, vol 104, pp1034–1049

ISE (2005) Fraunhofer ISE, Freiburg, Germany, pers comm

ISO 15099 (International Organization for Standardization 15099) (undated) *Thermal Performance of Windows, Doors and Shading Devices – Detailed Calculations*, Beuth-Verlag, Düsseldorf

Kaftan, E. and March, A. (2005) 'Integrating the cellular method for shading design with thermal simulation', in *Proceedings of the International Conference on Passive and Low Energy Cooling in the Built Environment*, Sanourini, Greece, pp965–970

Källblad, K. (1998) Thermal Models of Buildings: Report TABK 98/1015, PhD thesis, Lund University, Lund Institute of Technology, Department of Building Science, Lund, Sweden

Klein, M. (2000) *Tageslichtsysteme – Ein Vergleich*, Diploma thesis, University Stuttgart, Fraunhofer ISE, Freiburg

Kuhn, T. E. (2006a) 'Solar control: A general evaluation method for façades with Venetian blinds or other solar control systems to be used "stand-alone" or within building simulation programs', *Energy and Buildings*, vol 38, no 6, pp648–660

Kuhn, T. E. (2006b) 'Solar control: Comparison of two new systems with the state of the art on the basis of a new general evaluation method for facades with Venetian blinds or other solar control systems', *Energy and Buildings*, vol 38, no 6, pp661–672

Kuhn, T. E., Bühler, C. and Platzer, W. J. (2000) 'Evaluation of overheating protection with sun-shading systems', *Solar Energy*, vol 69 (supplement), pp59–74

Lee, E. S., DiBartolomeo, D. L. and Selkowitz, S. E. (2006) 'Daylighting control performance of a thin-film ceramic electrochromic window: Field study results', *Energy and Buildings*, vol 38, no 1, pp30–44

Lindsay, C. R. T. and Littlefair, P. J. (1993) *Occupant Use of Venetian Blinds in Offices*, PD 233/92, Building Research Establishment, Watford

Lund, H. (1985) *Test Reference Years (TRY): Report*, Commission of the European Communities, DG XII, Brussels

MeteoNorm (2003) *MeteoNorm Version 5.0*, Bezug, MeteoTest, Fabrikstrasse 14, CH-3012 Bern, www.meteotest.ch

Müller, J. and Hennings, D. (undated) *Global Climate Data Atlas*, www.climate1.com

Nitz, P., Gombert, A., Bläsi, B., Georg, A., Walze, G. and Hossfeld, W. (2004) *Lichtlenkende Mikrostrukturen mit optisch-funktionalen Beschichtungen*, Tagungsband 10, Symposium Innovative Lichttechnik in Gebäuden, Staffelstein, 29–30 January, OTTI Energie-Kolleg, Germany, pp30–35

Nitz, P. and Hartwig, H. (2005) 'Solar control with thermotropic layers', *Solar Energy*, vol 79, no 6, pp573–582

Olgyay, V. (1963) *Design with Climate*, Princeton University Press, Princeton, NJ

ParaSol (2001) www.parasol.se

Pasquay, T. (2004) 'Energy performance of double skin façades – three examples', *Energy and Buildings*, vol 36, no 4, pp381–389

Platzer, W. J. (ed) (2003) *Architectural and Technical Guidelines: Handbook for the Use of Switchable Facade Technology*, www.eu-swift.de or www.ise.fraunhofer.de

prEN 13363 (undated) *Solar Protection Devices Combined with Glazing: Calculation of Solar and Light Transmittance. Part 1*: Simplified Method (prEN 1336-1, 2003-10). Part 2: Reference Calculation Method (prEN 13363-2:2002, Corrections EN 13363-2:200/AC:2006)

prEN 14500 (2006) *Blinds and Shutters – Thermal and Visual Comfort: Test and Calculation Methods*, draft

prEN 14501 (2005) *Blinds and Shutters – Thermal and Visual Comfort: Performance Characteristics and Classification*, draft

Raicu, A., Wilson, H. R., Nitz, P., Platzer, W., Wittwer, V. and Jahns, E. (2002) 'Façade systems with variable solar control using thermotropic polymer blends', *Solar Energy*, vol 72, no 1, pp31–42

Reinhart, C. F. (2004) 'Lightswitch: A model for manual control of electric lighting and blinds', *Solar Energy*, vol 77, no 1, pp15–28

Reinhart, C. F. and Voss, K. (2003) 'Monitoring manual control of electric lighting and blinds', *Lighting Research and Technology*, vol 35, no 3, pp243–260

Reinhart, C. and Wambsganß, M. (2005) 'Zusammenspiel von Tageslicht und elektrischer Beleuchtung', in: Voss, K., Löhnert, L., Herkel, S., Wagner, A. and

Wambsganß, A. (eds) *Bürogebäude mit Zukunft, 2. Auflage*, Solarpraxis Verlag, Berlin

Reise, C. (2005) 'Klima und Mikroklima', in Voss, K., Löhnert, L., Herkel, S., Wagner, A. and Wambsganß, A. (eds) *Bürogebäude mit Zukunft, 2. Auflage*, Solarpraxis Verlag, Berlin

Rubin, A. I., Collins, B. L. and Tibbott, R. L. (1978) *Window Blinds as a Potential Energy Saver – Case Study*, NSB Building Science Series 112, National Bureau of Standards, Washington, DC

Sage Glass (2005) *Sage Glass – The Power to Change*, Product brochure, SAGE Electrochromics, Inc, Faribault, MN, www.sage-ec.com

Saint Gobain Glass (2005) *Product Information SGG PRIVA-LITE*, www.sggprivalite.com

Seeboth, A., Schneider, J. and Patzak, A. (2000) 'Materials for intelligent sun protecting glazing', *Solar Energy Materials and Solar Cells*, vol 60, pp263–277

Simpson, J. R. and McPherson, E. G. (1997) 'The effects of roof albedo modification on cooling loads of scale model residences in Tucson, Arizona', *Energy and Buildings*, vol 25, pp127–137

SMARTWIN (2005) *European Research Project SMARTWIN II: New Liquid Crystal Smart Window and its Production Process*, Contract No ENK6-CT-2001-00549, Project No NNE5-2001-00146, Publishable Final Technical Report, Technical University of Denmark (DTU), Lyngby

VBG (Verwaltungsberufsgenossenschaft) (2002) *Sonnenschutz im Büro*, Design guide, http:www.vbg.de

Voss, K., Löhnert, L., Herkel, S., Wagner, A. and Wambsganß, A. (2005) (eds) *Bürogebäude mit Zukunft – Konzepte, Analysen, Erfahrungen, 2. Auflage*, Solarpraxis Verlag, Berlin

Voss, K. and Wittwer, V. (2003) 'Passive solar heating of buildings', in Santamouris, M. (ed) *Solar Thermal Technologies for Buildings – The State of the Art*, James and James, London

Wall, M. and Bülow-Hübe H. (eds) (2001) *Solar Protection in Buildings, Report TABK—01/3060*, Lund University, Lund Institute of Technology, Department of Construction and Architecture, Lund, Sweden

Wall, M. and Bülow-Hübe, H. (eds) (2003) *Solar Protection in Buildings. Part 2: 2000–2002*, Report EBD-R—03/1, Lund University, Lund Institute of Technology, Department of Construction and Architecture, Lund, Sweden

Watanabe, H. (1998) 'Intelligent window using a hydrogel layer for energy efficiency', *Solar Energy Materials and Solar Cells*, vol 54, pp203–211

Wienold, J. (2005) 'Blendschutz', in Voss, K. , Löhnert, L., Herkel, S., Wagner, A. and Wambsganß, A. (eds) *Bürogebäude mit Zukunft – Konzepte, Analysen, Erfahrungen, 2. Auflage*, Solarpraxis Verlag, Berlin

Wienold, J. and Christoffersen, J. (2006) 'Evaluation methods and development of a new glare protection model for daylight environments with the use of CCD cameras', *Energy and Buildings*, vol 38, pp743–757

Wilkings, C. and Hosni, M. H. (2000) 'Heat gain from office equipment', *ASHRAE Journal*, June, pp33–39

WIN (2005) http://windows.lbl.gov/software/window/window.html

Wirth, H. (2003) Okalux GmbH, pers comm

Wirth, H. and Horn, R. (2002) 'Entwicklung von selbstregulierenden Sonnenschutzgläsern' ['Development of automatically controlled solar-control glazing'], in *Proceedings of the Eighth Conference, Symposium Innovative*

*Lichttechnik in Gebäuden*, Staffelstein, 24–25 January, OTTI Energie-Kolleg, Germany, pp124–128

WIS (Window Information System) (2005) http://windat.ucd.ie/wis/html/index.html

Wittwer, V., Datz, M., Ell, J., Georg, A., Graf, W. and Walze, G. (2004) 'Gaschromic windows', *Solar Energy Materials and Solar Cells*, vol 84, pp305–314

Zimmermann, M. (ed) (2003) *Handbuch der Passiven Kühlung*, Fraunhofer IRB Verlag, Stuttgart, Germany

# 4

# Ventilation for Cooling

*Maria Kolokotroni and Mat Santamouris*

## FUNCTIONS OF VENTILATION

Ventilation is an essential part of building design and operation in order to deliver a comfortable thermal environment and adequate indoor air quality, but it brings with it a significant energy load. For this reason, ventilation strategies have been developed and applied in buildings to minimize energy demand and to utilize the cooling potential of outdoor air for certain conditions.

Traditionally, ventilation has been provided in buildings in order to maintain acceptable indoor air quality (IAQ) by fulfilling the following functions (BRE, 1994):

- providing sufficient oxygen;
- diluting body odours; and
- diluting to acceptable levels the concentration of carbon dioxide produced by occupants and combustion, as well as other internally and externally generated pollutants.

The ventilation rates to be provided for each of these functions are quite different. The requirements for oxygen are far less than those for metabolic carbon dioxide dilution. These, in turn, are less than the air needed for diluting body odours and much less than the ventilation needed to dilute other pollutants (including tobacco smoke) to acceptable levels.

Ventilation is increasingly used to provide internal thermal comfort in buildings (i.e. controlling temperature inside buildings and thus avoiding overheating). In this case, ventilation rates required are quite different (usually much higher) than ventilation rates required for IAQ functions. In many cases, ventilation is provided by natural means, although hybrid strategies and coupling with internal thermal mass are increasingly popular. In some cases, mechanical ventilation is used for providing thermal comfort, although this option is usually energy consuming because of the energy required by the fans.

This chapter focuses on the second function of ventilation – the provision of thermal comfort. Thermal comfort principles and new developments for naturally ventilated buildings have been covered in Chapter 1. Ventilation for controlling internal temperatures is location and building-operation dependent. In very broad terms, the following climatic classifications might apply:

- *Climatic regions with a high cooling load.* In such climates, ventilation strategies are usually designed to provide some cooling to reduce reliance on active air-conditioning systems. Such ventilation strategies would most probably be implemented in climates with a hot summer and where buildings require no or very little heating during winter. Ventilation strategies for thermal comfort (cooling) would, in most cases, be combined with other passive and/or active cooling methods in addition to passive and/or active heating methods.
- *Climatic regions with a high heating load.* In such climates, ventilation strategies are mainly designed to provide indoor air quality. Such ventilation strategies would most likely be implemented in climates with a cold winter and where buildings require no or very little cooling during summer. Ventilation strategies for IAQ (heating) would, in most cases, be efficient mechanical ventilation strategies, perhaps combined with passive cooling strategies for the summer.
- *Climatic regions with moderate heating and cooling loads.* In such climates, ventilation strategies are designed to provide thermal comfort during the summer. Such ventilation strategies might be also implemented in climates where, in addition, high moisture levels impose a further load. Natural ventilation strategies may be able to satisfy cooling load requirements for a range of buildings with moderate internal heat gains.

## BUILDING DESIGN STRATEGIES, SYSTEMS AND COMPONENTS

During recent years, a number of research reports, collaborative projects and information papers on natural ventilation have been published. Although the cooling function of ventilation is discussed in most of these publications, it is treated together with other functions of ventilation. This section draws out and summarizes their findings, with a specific focus on cooling potential. As a result, the section outlines some of the available passive design strategies for cooling buildings using ventilation. Design strategies can be divided into categories according to:

- *Building system/component used.* In broad terms, these are openings in the façade of the building (windows) or specific components to enhance passive stack and (in some cases) wind-driven ventilation. Such components can be termed 'ventilation towers'. The ventilation and cooling function of windows has been addressed in detail in Santamouris and Asimakopoulos (1996), where methods of determining ventilation rates have also been outlined. The function of ventilation towers is discussed below. In addition, components for ambient air pre-conditioning are used in buildings. These could be placed outside the building and are classified as ground-cooling (air) components (see Chapter 5) or can be integrated within the structure of the building using natural driving forces or by incorporating fans. The later components are described in this chapter. Finally, the building façades can be used for ventilation cooling and double façades have been used in many buildings. These are also briefly described below.
- *Operational principles.* Ventilation can provide cooling during the day in some climates, but in many buildings is also used during the night to enhance the cooling potential. This strategy has been termed 'night cooling' and has been used in many built examples. The operational principles of night cooling are discussed below.
- *Cooling effectiveness.* In many cases, due to building type, internal heat load and/or external climatic conditions, natural ventilation is combined with mechanical ventilation for maximum cooling effectiveness and is termed hybrid or 'mixed mode' ventilation. The operational principles of hybrid ventilation for cooling are discussed below. Urban buildings are a special case for which important results are available – these are outlined in the section on 'Ventilation of urban buildings'.

## Ventilation towers

Ventilation towers, in the form of stacks or chimneys, are currently popular in modern low-energy ventilation designs. Ventilation towers work by amplifying natural driving forces and serve to extend the depth of space over which a cross-ventilation regime can be applied. There are two main types, wind and thermal chimneys, but several variants exist as classified by NiRiain and Kolokotroni (2000).

Thermal chimneys (see Figure 4.1a) utilize buoyancy forces generated by the vertical density (temperature) differences in a space and across the building envelope in order to drive a vertical circulation across the envelope. Temperature differences are caused by heat gains generated within the spaces themselves from occupants, and electrical and heating equipment. The general purpose of thermal stacks is to remove warm stale air at a high level.

A variation of this is the solar chimney (see Figure 4.1b), which is usually positioned at the south façade of the building and acts as a concentrator of substantial solar heat gains. This further enhances thermal buoyancy. Solar stacks are designed to increase ventilation during the summer months under

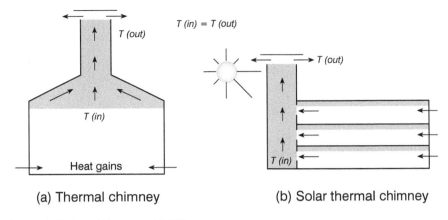

*Source:* NiRiain and Kolokotroni (2000)

**Figure 4.1** *Thermal and solar chimneys utilize buoyancy forces to enhance ventilation flow*

hot still conditions. As the driving force is upwards, inlet location is lower down on the façade of the building.

Wind chimneys or towers are aerodynamically designed to enhance the wind pressure differences that occur when air flows around obstacles. There are two main design types: pressurization (inlet) and depressurization (outlet) types for either supply or extraction of the air, respectively – the inlet variety are usually referred to as wind catchers or scoops (see Figure 4.2a). Both are roof-mounted devices to capitalize on increased wind velocity with height and may be combined in a single strategy – indeed, in a single element – for supply and extraction. The outlet type (see Figure 4.2b) draws air from the building using negative (suction) pressures created at the outlet, usually assisted by existing buoyancy forces in the building. The suction pressure is either due to a lee pressure or Venturi effect operating at the outlet. Both can be created through either roof-profiling, terminal design or both. For wind catchers, the inlet is windward to 'catch' the air at high level and push it down into the building. The outlet in this case is often also a roof-mounted wind chimney, as described above. This is sometimes co-located on the leeward side of the inlet using a split-duct system (see Figure 4.2c).

In modern low-energy buildings, ventilation stacks are rarely employed as the primary ventilation driver, but to enhance a ventilation strategy – either a natural cross-ventilation strategy or, latterly, low-energy mechanical. Stacks have also been employed to assist atrium or glazed-façade ventilation strategies. Solar stacks have been used as a back-up system to increase cross-ventilation rates during the summer on hot still days. Some applications of these systems in the UK are outlined as follows.

Thermal chimneys have been used as the primary ventilation driver for the auditoria in the Queens Building at De Montfort University (DETR, 1999b), and this application has been proved successful. More recent examples have

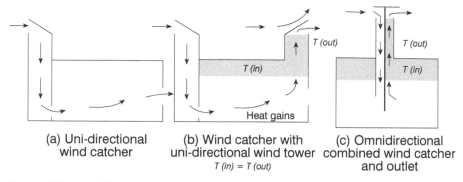

(a) Uni-directional wind catcher

(b) Wind catcher with uni-directional wind tower
T (in) = T (out)

(c) Omnidirectional combined wind catcher and outlet

*Source:* NiRiain and Kolokotroni (2000)

**Figure 4.2** *Wind chimneys or towers utilize aerodynamic designs to enhance wind pressure differences and thus increase air flow*

been constructed, as outlined in Cook and Short (2005). Two large non-domestic buildings – Frederick Lanchester Library, Coventry University (see Figure 4.3) and Lichfield Garrick Performing and Static Arts Centre, Staffordshire, UK (see Figure 4.4) – are described, in which some form of thermal chimney was used. Computational fluid dynamics (CFD) modelling was used during the design of the buildings in order to increase confidence, and post-occupancy evaluation in both buildings has indicated occupant satisfaction with environmental conditions. Another application is modern school buildings incorporating thermal chimneys to enhance ventilation rates, as well as for cooling purposes (Kolokotroni et al, 2002). Solar chimneys were employed in the office block of the BRE Environmental Building (NatVent,

*Source:* Cook and Short (2005)

**Figure 4.3** *Frederick Lanchester Library, Coventry University*

*Source:* Cook and Short (2005)

**Figure 4.4** *Lichfield Garrick Performing and Static Arts Centre*

1998) to assist with cross-ventilation strategies. Since stack pressures are very low, mechanical assistance is provided, in many cases, in the form of low-power fans incorporated at the top of the ventilation stacks in order to assist ventilation when drivers are low.

The use of wind chimneys and wind catchers (or scoops) is increasingly becoming popular in the UK. A wide variety of design options exist for terminal design and roof profiling (Parker and Teekaram, 2005), and the design procedure is complex, usually involving physical or numerical fluid flow simulation. Assessment of their performance is included in Parker and Teekaram (2005), and simplified calculation tools have been developed (Wind Towers, 1996). The wind towers of the Ionica Headquarters Building (DETR, 1999a) in Cambridge are situated on the glazed pitched-roof of an atrium and are capped by aerofoil-type 'rain hats'. Since the building is comparatively exposed, wind pressure at the façade inlets and the buoyancy generated by the atrium will be significant.

The application of wind catchers requires a relatively constant wind direction. Additionally, the wind pressure at the inlet must exceed any naturally occurring (thermal) stack pressure to be effective, or the inflow pathway down into the building must be cooled in order to reduce stack pressure. Design solutions adopted to overcome these restrictions include:

- air distribution via displacement ventilation strategies, using the internal buoyancy of the space to provide much needed additional buoyancy force to drive this type of circuit; and

**Figure 4.5** *Jubilee Campus, University of Nottingham*

• systems such as passive downdraught evaporative cooling (PDEC) (Ford, 2002, 2003), which cool the intake air to overcome the temperature gradient conflicts.

In inland or urban areas, omni-directional or variable-direction wind catcher designs have been employed. Split-duct wind catchers (inlet and outlet combined in a single terminal) have been applied to assist mechanical displacement ventilation strategies in urban areas where wind access and air quality at lower levels are poor. Rotating split-duct wind catchers are used with displacement ventilation to reduce the energy consumption of the mechanical strategy in the Jubilee Campus at the University of Nottingham, UK (see Figure 4.5) (Palmer, 1999; CIBSE, 2005). The orientation and opening of outlets are controlled such that the inlet is always windward and outlet negatively pressurized when open. Such wind-catcher components are commercially available and have been installed in a large number of buildings in the UK. Examples include office buildings (Kirk and Kolokotroni, 2004) and schools (see Figure 4.6) (Parker and Teekaram, 2005).

## Hybrid stack designs

Just as thermal chimneys should be designed without conflict with wind pressure differences, similarly wind chimneys should work with, rather than against, the temperature gradient, where feasible. In general, both are designed to capitalize on any additional boost that wind or stack pressures afford through:

*Source:* Monodraft Ltd

**Figure 4.6** *Abbey and Stanhope School, Deptford, London*

- roof-profiling and terminal/'rain hat' design options for thermal chimneys; and
- low inlet–high outlet configurations for wind chimneys to benefit from the existing buoyancy effects of rising warm air.

A design example where both wind and buoyancy forces are maximized are the stacks on the teaching block of the School of Architecture at Portsmouth University, UK, by Hampshire County Architects (see Figure 4.7) (Bordass et al, 1999). The stacks are partially glazed in order to capitalize on solar gain, while the outlet is located in the lee of a mono-pitch roof profile designed to create a suction pressure for the comparatively steady prevailing wind.

## Ventilation tower design

Computational fluid dynamics modelling was used by Cook and Short (2005) to analyse thermal stack ventilation strategies and wind tunnels (Parker and Teekaram, 2005), and other physical modelling techniques (Cook at al, 2003) have been used to study ventilation towers. However, the more comprehensive fluid flow models are not readily available to, or affordable for, all designers.

Simplified design guidelines and calculations have been produced by the Chartered Institute of Building Services Engineering (CIBSE, 2005) for thermal

*Source:* NatVent (1998)

**Figure 4.7** *Portland Street Building, Portsmouth, UK*

and wind-driven stacks, which depend upon the inlet area, height and temperature differences. A computerized version of these calculations can be found in NatVent (1998) and VentDiscourse (2006).

## Flow enhancement

Simulation work under the NatVent Project (Skåret et al, 1997) on the availability of drivers for wind and stack-driven ventilation (thermal stack of height 10m) found high frequencies of low ventilation drivers (< 15Pa) for Northern European climates. Options for enhancement of stack performance have included cooling of intake air using passive downdraught evaporative cooling (PDEC) systems (Ford, 2002, 2003) in Southern European climates or the use of roof gardens and device shading (Yannas et al, 2005). For low-level supplies, cool air can be supplied during the summer by locating the fresh air inlets on the shaded side of the building. Another option is the heating of extract air: gas-fired thermal chimneys were commonly used to increase ventilation for health in schools, hospitals and other public buildings.

## Flow control and management

In the Queens Building at De Montfort University, UK (DETR, 1999b), flow is controlled by building management system (BMS)-controlled stack dampers. Minimum fresh air supply is ensured by carbon dioxide ($CO_2$) sensors, and cooling ventilation supply is controlled by the temperature differences between the inside and outside. A similar strategy is employed in the main auditorium in the BRE Environmental Building (NatVent, 1998). Downdraughts are avoided through the use of temperature sensors in the stacks, which activate the stack dampers. A night-cooling strategy is employed, controlled by the fabric temperatures.

### Heat recovery of extract air

This area is in its infancy due to the flow resistance of the equipment, with applications restricted to systems with mechanical assistance. A purpose-built heat-recovery system for northern climates was demonstrated under the NatVent Project, while the feasibility of inclusion of low-loss heat-recovery pipes in conventional stacks has been demonstrated through simulation (Sirén et al, 1997) to achieve pressure losses of only a few Pascals, while achieving heat recovery efficiencies of up to 55 per cent.

## Components for ambient air pre-conditioning: Slab cooling

The basic idea behind slab cooling is the exploitation of the thermal inertia of the building mass for the purpose of energy storage; here, air is used as the primary heat transfer medium. This section describes an activated thermal-mass building component that is fully integrated with a ventilation system. In this integrated system, the building envelope, as well as horizontal and vertical partitions of the building, can be constructed from heavyweight materials in the form of slabs. The channels in the slab are used as ducts for the ventilation air. Whenever the temperature of the air passing through the ducts differs from the slab temperature, energy transfer takes place, allowing the building mass to be used as a rechargeable energy store. During the summer, the system can be run at night to store cool energy in the building mass. The technology requires mechanical supply and exhaust ventilation (Klobut and Kosonen, 1995).

An operational and monitored building that uses this system is the Elizabeth Fry Building at the University of West Anglia, UK (DETR, 1998). The system comprises hollow ceiling core slabs (Thermodeck) (see Figure 4.8). The hollow cores in the concrete ceiling slabs comprise the final part of the air supply duct. The ceiling slabs act as a heat sink and provide radiant heating or cooling, and convective heat transfer to the supply air. The overall ventilation system is fresh air supply via four air handling units (AHUs), including heat recovery. The monitored results from the building both in terms of energy consumption and user satisfaction were excellent (Cohen et al, 1996; DETR, 1998).

### Façade-integrated ventilation for cooling

The design of a building façade influences internal thermal and lighting conditions and energy use associated with the provision of these conditions. Key decisions about the building façade are usually taken during the concept design stage of a building, while decisions about the method of providing the environmental conditions are often taken later in the design process. This dilemma is addressed, in many cases, by the development of concept design

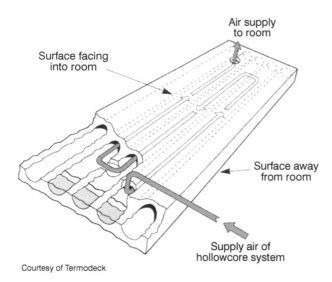

Air supply
to room

Surface facing
into room

Surface away
from room

Supply air of
hollowcore system

Courtesy of Termodeck

*Source:* DETR (1998)

**Figure 4.8** *Illustration of slab cooling component using hollow cores in the concrete ceiling slabs as the final part of the air supply duct*

tools within which ventilation for cooling is one of the main elements since it affects comfort conditions inside the building. Such tools have been developed for residential buildings, such as the RESFEN (University of California, 1999) and OPTI (Gratia and De Herde, 2002) software. More recently, concept tools for the evaluation of façade types for non-domestic buildings have also been published (CIBSE, 2004; Kolokotroni et al, 2004).

Ventilation is a key element of the so-called 'double-skin' façades that are complex systems and are receiving increasing attention. For example, a Pan-European project entitled BEST FAÇADE considers them in detail (BEST FAÇADE, 2005) and the International Energy Agency (IEA) Annex 44 on *Integrating Environmentally Responsive Elements in Buildings* includes studies of double-skin façade facets (IEA, 2006). Some information on double-skin façades is given below based on information from the BEST FAÇADE project.

There are many different principles of how to construct ventilated double-skin façades. These can be classified according to three different criteria, which are independent of one another and are based not only on the geometric characteristics of the façade, but also on its mode of working. The criteria are as follows (BEST FAÇADE, 2005):

1   type of ventilation;
2   partitioning of the façade; and
3   ventilation mode of the cavity.

More information can be found at the website www.bestfacade.com, from where the following is extracted.

## Type of ventilation

The type of ventilation refers to the driving forces at the origin of the cavity's ventilation, located between the two glazed façades. Each ventilated double-skin façade concept is characterized by only a single type of ventilation. One must distinguish between the three following types of ventilation: natural, mechanical or hybrid ventilation (mix between natural and mechanical ventilation).

## Partitioning of the façade

The partitioning of the cavity gives information on how the cavity, situated between the two glazed façades, is physically divided. The partitioning solutions implemented, in practice, can be classified as follows:

- *Ventilated double window.* This is a façade equipped with a ventilated double window and is characterized by a window doubled inside or outside by a single glazing or by a second window. From the partitioning perspective, it is thus a window that functions as a filling element in a wall.
- *Façade partitioned per storey with juxtaposed modules.* This is a ventilated double façade partitioned at each storey, with juxtaposed modules. In this type of façade, the cavity is physically delimited (horizontally and vertically) by the module of the façade, which imposes its dimensions on the cavity. The façade module has a height limited to one storey.
- *Façade partitioned per storey – corridor type.* This is a corridor-type ventilated double-façade partitioned at each storey and is characterized by a large cavity in which it is generally possible to walk. While the cavity is physically partitioned at the level of each storey (the cavities of each storey are independent of one another), it is not limited vertically, and generally extends across several offices or even an entire floor.
- *Shaft-box façade.* The objective of this partitioning concept is to encourage natural ventilation by adapting the partitioning of the façade so as to create an increased stack effect (compared to the naturally ventilated façades, which are partitioned by storey). Thus, it is logical that this type of façade and partitioning is applied only in naturally ventilated double façades. This type of façade is, in fact, composed of an alternation of juxtaposed façade modules partitioned by storey and vertical ventilation ducts set up in the cavity, which extends over several floors. Each façade module is connected to one of these vertical ducts, which encourages the stack effect, thus supplying air via the façade modules. This air is naturally drawn into the ventilation duct and is evacuated via the outlet located several floors above or below.
- *Multi-storey façade.* Multi-storey ventilated double façades are characterized by a cavity that is not partitioned either horizontally or vertically; the space between the two glazed façade layers therefore

forming one large volume. In some cases, the cavity can run all around the building without any partitioning. Generally, façades with this type of partitioning are naturally ventilated; however, there are also examples of façades of this type that are mechanically ventilated. It should be noted that façades of this type generally have excellent acoustical performances with regard to outdoor noise.

- *Multi-storey louvre façade.* The multi-storey louvre naturally ventilated double façade is very similar to a multi-storey ventilated double façade. Its cavity is not partitioned either horizontally or vertically and therefore forms one large volume. Metal floors are installed at the level of each storey in order to allow access to it, mainly for cleaning and maintenance. The difference between this type of façade and the multi-storey façade is that the outdoor façade is composed exclusively of pivoting louvres, rather than a traditional monolithic façade equipped (or not) with openings. This outside façade is not airtight, even when the louvres have all been put in a closed position, which justifies its separate classification. However, the problems encountered with these façades are generally comparable to those encountered in the other ventilated double-skin façades.

## Ventilation mode of the cavity

The ventilation mode refers to the origin and the destination of the air circulating in the ventilated cavity. The ventilation mode is independent of the type of ventilation applied. Not all of the façades are capable of adopting all of the ventilation modes described here. At a given moment, a façade is characterized by only a single ventilation mode. However, a façade can adopt several ventilation modes at different moments, depending upon whether or not certain components integrated within the façade permit it (e.g. operable openings). One must distinguish between the following five main ventilation modes:

1  outdoor air curtain, when the air comes from the outside and is immediately returned to the outside;
2  indoor air curtain, when the air circulates from the inside and returns to the inside;
3  air supply, where fresh air is supplied into the building through the external skin;
4  air exhaust, where the air comes from the inside of the room and is exhausted from the building façade; and
5  buffer zone, where the façade is made airtight; the cavity comprises a buffer zone between the internal and external skin.

Compared to conventional office buildings with large glazed façades, this system provides the follows advantages:

- a thermal buffer zone that reduces heat losses and enables passive thermal gain from solar radiation;

- solar preheating of ventilation air (thus reduced heating demands);
- sound protection (e.g. at locations with heavy traffic) mainly during window ventilation;
- additional shading and protection of shading devices;
- energy savings if the design is well adapted to the climatic conditions;
- natural ventilation – individual window ventilation is almost independent of wind and weather conditions, mainly during sunny winter days and the intermediate season; and
- night cooling of the building by opening the inner windows.

It is very important that the double-skin façade is well designed, with the correct type of ventilation, and is adapted to the climatic region; otherwise, overheating may occur in the occupied spaces. The space within the two skins should be well ventilated either by natural, mechanical or hybrid ventilation.

## Night ventilation

Night ventilation works by using natural or mechanical ventilation to cool the surfaces of the building fabric at night and is more effective where a building includes a reasonably high thermal mass so that heat can be absorbed during the day. Night ventilation can affect internal conditions during the day in four ways:

1 reducing peak air temperatures;
2 reducing air temperatures throughout the day and, in particular, during the morning hours;
3 reducing slab temperatures; and
4 creating a time lag between external and internal temperatures.

A measured typical effect of night ventilation in an office is shown in Figure 4.9 (Kolokotroni, 1998).

A comprehensive review of night ventilation strategies is presented in Santamouris (2004). Night ventilation systems are classified as direct or indirect as a function of the procedure in which heat is transferred between the thermal storage mass and the conditioned space.

In direct systems the cool air is circulated inside the building zones and heat is transferred in the exposed opaque elements of the building. The reduced temperature mass of the building contributes to reduced indoor temperatures for the next day through convective and radiative procedures. In direct systems the mass of the building has to be exposed and the use of coverings or false floors or ceilings has to be avoided.

In indirect systems, the cool air is circulated during the night through a thermal storage medium where heat is stored and recovered during the day period. In general, the storage medium is a slab covered by a false ceiling or floor, while the circulation of the air is always forced. Some components used in these systems were described earlier in the section on 'Building design strategies, systems and components'.

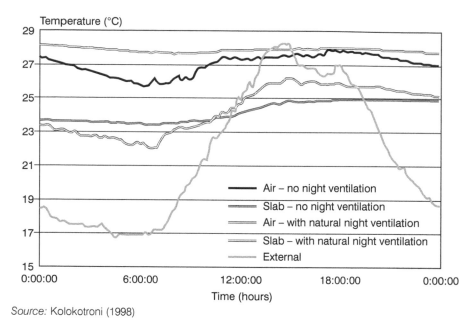

*Source:* Kolokotroni (1998)

**Figure 4.9** *Measured typical effect of night ventilation in an office*

The performance of night cooling systems depends upon three main parameters:

1   the temperature and the flux of the ambient air circulated in the building during the night;
2   the quality of the heat transfer between the circulated air and the thermal mass; and
3   the thermal capacity of the storage medium.

Night ventilation, although a very powerful technique, presents important limitations. Moisture and condensation control is necessary, particularly in humid areas. Pollution, acoustic and fire safety problems, as well as the problem of privacy, are associated with the use of natural ventilation techniques.

The most important limitation of night ventilation techniques is associated with the specific climatic conditions of cities. Increased temperatures due to the heat island effect, as well as the decrease in wind speed in urban canyons, considerably reduces the cooling potential of night cooling techniques.

The effects of the urban environment are discussed in detail in the section on 'Ventilation of urban buildings'. Here, two examples are presented to demonstrate the limitations of night cooling. Geros et al (1999) compared the cooling load of a night-ventilated building when located in an urban canyon or in a non-obstructed site. They found that the relative difference of the cooling loads, between −6 to 89 per cent for the single-sided ventilation, and between

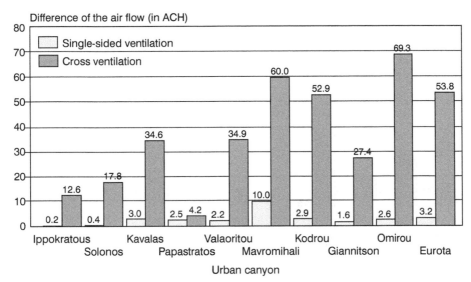

Difference of the air flow (in ACH)

□ Single-sided ventilation
■ Cross ventilation

Ippokratous · Solonos · Kavalas · Papastratos · Valaoritou · Mavromihali · Kodrou · Giannitson · Omirou · Eurota

Urban canyon

*Source:* Group Building Environmental Studies, University of Athens

**Figure 4.10** *Reduction of air change rate (in air changes per hour) for single-sided and cross-ventilated buildings in ten urban canyons*

–18 to 72 per cent for the cross-ventilated building, depends upon the characteristic of the canyon (see Figure 4.10). In parallel, the same comparison was made for the performance of a free-floating night-ventilated building. It has been calculated that the difference between maximum indoor temperatures in the building are between 0°C and 2.6°C for the cross-ventilated buildings and between 0.2°C and 3.5°C for the single-sided ventilation buildings (see Figure 4.11). Kolokotroni et al (2006) carried out a parametric analysis of the effect of external temperature in central London and a reference rural site outside London. The measured ambient temperatures are presented in Figure 4.12 and results of simulations on their effect are presented in Figure 4.13. In general, it was found that during a typical hot week, the rural reference office uses 84 per cent of the energy demand for cooling compared to a similar urban office. An optimized rural office would not need any artificial cooling and would be able to maintain temperatures of below 24°C. An optimized urban office would not be able to achieve this. During an extremely hot week, the rural reference office uses 83 per cent of the energy demand for cooling compared to a similar urban office. An optimized rural office would need only 42 per cent of the cooling demand required for an optimized urban office. Increased night stack ventilation rates would further reduce this amount to 30 per cent.

As a result, optimizing the performance of night ventilation systems requires detailed simulation techniques where all of the energy and environmental parameters are taken into account. Simplified simulation tools designed especially for night ventilation application are also available and can provide very useful information. These are described in the section on 'Models for estimating ventilation cooling potential'.

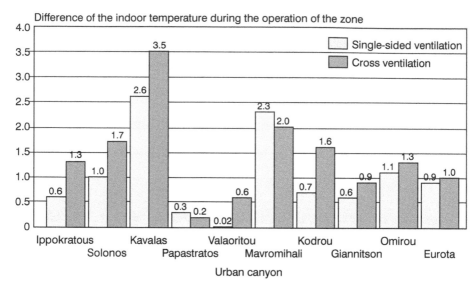

Difference of the indoor temperature during the operation of the zone

*Note:* The analysis refers to ten urban canyons where experiments have been carried out, and results are given for single-sided and cross-ventilated buildings.
*Source:* Geros et al (1999)

**Figure 4.11** *The difference of the maximum indoor air temperature calculated for a night-ventilated building located in a canyon and in a non-obstructed site*

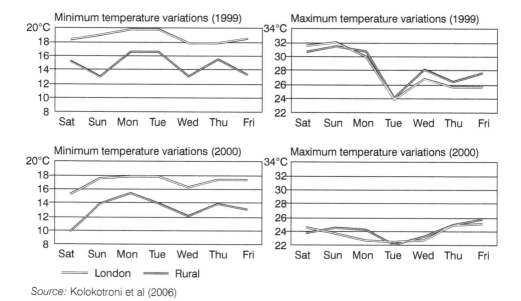

*Source:* Kolokotroni et al (2006)

**Figure 4.12** *Measured ambient temperatures in central London and a rural site just outside London during the night and day for two years; one with an average summer and another with a hot summer*

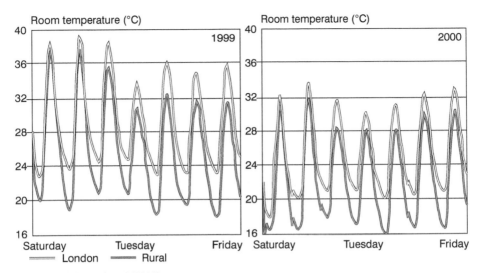

*Source:* Kolokotroni et al (2006)

**Figure 4.13** *Simulation results for internal temperatures in a typical office located in central London and a rural site outside London*

## Documented built examples reported by international projects

This section lists buildings that have been studied by international projects and include measured results of the performance on ventilation, particularly its use for cooling through night operation.

Fifteen buildings incorporating night ventilation strategies with documented monitoring results are described in detail in NatVent (1998) (see Figure 4.14):

1  BRE Environmental Office, Garston, UK – natural ventilation for night cooling through automated openable windows and solar chimneys (which include an extract fan to be used occasionally). Ground cooling through a borehole heat exchanger supplements cooling when required.
2  Canning Crescent Centre, London, UK – natural ventilation for night cooling through dedicated grilles in the façade and ventilation towers (combined solar and wind chimneys).
3  Portland Street Building, Portsmouth, UK: natural ventilation for night cooling through openable windows, ventilation ducts incorporated within the building and connected with staircase towers (a combination of stack chimney with enhanced solar elements).
4  PROBE Office Building, Brussels, Belgium (Figure 4.15): natural ventilation through dedicated ventilation grilles in the façade of the building.
5  Office building along a highway in Aalst, Belgium: natural ventilation for night cooling; mechanical ventilation during the day.

*Source:* NatVent (1998)

**Figure 4.14** *BRG Environmental Office, Garston, UK*

6   Refurbished Office Building, Warengen Belgium: natural (stack) ventilation for night cooling through dedicated ventilation grilles in the façade and chimneys on the roof.
7   Velux Building, Trimbach, Switzerland: natural ventilation for night cooling through BSM-controlled openable windows.
8   EZW Building, Zurich, Switzerland: natural ventilation for night cooling through BSM-controlled openable windows, mechanical ventilation with heat recovery (winter) and refrigerative cooling with ice storage (summer).
9   Basler Versicherung Building, Basel, Switzerland: natural ventilation for night cooling based on user-controlled openable windows.
10  E. Pihl and Son Headquarters, Lynghy, Denmark: natural ventilation through automated multi-positioned ventilation openings and skylights (which include extract fans).
11  BRF-Kredit Headquarters, Lynghy, Demark: natural ventilation through automated multi-positioned ventilation openings and skylights in the atrium.
12  Windowmaster Office Building, Vedbaek, Denmark: natural ventilation through automated openable windows.
13  European Patent Office Building, Rijswijk, The Netherlands: natural ventilation for night cooling through automated openable windows.
14  Zevenhuizen Town Hall, The Netherlands: mechanical (extract) ventilation for night cooling.
15  Enschede Tax Office, The Netherlands: natural ventilation through automated openable windows.

Three buildings incorporating night-time ventilation are described, together with measured data demonstrating the effectiveness of the strategy, in Santamouris (2004). These comprise an office building in Athens, Greece

*Source:* NatVent (1998)

**Figure 4.15** *PROBE office building in Brussels, Belgium*

(Meletitiki Building) (see Figure 4.16); an office building in Kortrijk, Belgium (S.D. Worx), which incorporates an air-to-earth heat exchanger during the day; and a school in Denmark (Sofiendal School) (see Figure 4.17), where a fan is also used to increase flow rates during the night if required.

Three additional built examples with demonstrated good performance are described in Zimmermann and Anderson (1998): one single-family residence, Vila Nova de Gaia, in Porto, Portugal; a design studio in Milton Keynes, UK (both using natural night ventilation); and an office building, IONICA Building, in Cambridge, UK, which used mechanical and natural night ventilation (and also incorporates wind-driven ventilation towers).

Night ventilation is used extensively, together with hybrid ventilation principles, and built examples are reported in Heiselberg (2002):

- B&O Headquarters, Struer, Denmark: night-time ventilation windows used for supply, stairwells used as exhaust stacks and fans located on top of the stack; mechanical displacement ventilation is used during the day.
- Tanga School, Falkenberg, Sweden: stack-assisted natural ventilation with air intake through openings in the façade, an exhaust tower with a solar chimney and manually openable windows.
- Bertolt Brecht Gymnasium, Dresden, Germany: stack-driven ventilation through windows opening in the classrooms and roof openings in the atrium. Mechanical ventilation occurs in winter.
- Palazzina I Guzzini, Recanati, Italy: stack-driven natural night ventilation by windows opening in the façade and atrium. Heating and cooling are by fan coil unit.

*Source:* Santamouris (2004)

**Figure 4.16** *The Meletitiki building, Athens, Greece*

- Fujita Building, Atsugi, Japan: stack-driven natural night ventilation by windows opening in the façade and atrium during unoccupied hours. Air conditioning (variable air volume flow rate, or VAV) occurs during occupied hours.
- Liberty Tower of Meiji University, Tokyo, Japan: stack-driven natural night ventilation by windows opening in the façade and air exhaust through

*Source:* Santamouris (2004)

**Figure 4.17** *Sofiendal School in Denmark*

stairwells and 'wind floors' during unoccupied hours. Air conditioning occurs during occupied hours.

- Tokyo Gas Earth Port, Tokyo, Japan: stack-driven natural night ventilation by windows opening in the façade and atrium during unoccupied hours. Air conditioning (VAV) occurs during occupied hours.

## Simplified models for night-cooling ventilation

Several codes have been developed to calculate the specific performance of night ventilation techniques. These tools are designed to help architects and engineers consider, in a more simplified but accurate way, the sizing of components required for night-cooling techniques. This section provides information on some of the relevant data.

NiteCool (Tindale et al, 1995) was developed under the Energy Related Environmental Issues in Buildings (EnREI) programme of the UK Department of the Environment (DOE) and is designed especially for assessing a range of night-cooling ventilation strategies. The programme is based on single-zone ventilation. This simplified design tool is developed to enable the design team to rapidly explore the effects of a number of key performance parameters and thereby ensure that the basic design concept is workable in terms of the chosen strategy.

A simplified model has been developed by Millet (1997). The model takes into account the thermal inertia of the building and the impact of night ventilation. Attention is paid to the impact of outdoor noise (related to the windows opening at night). This model was validated by comparing its results to a more detailed one (TRNSYS) and was used to produce guidance rules. Used primarily for new buildings, these tools will also be of help in retrofitting.

LESOCOOL is a simple computer tool for evaluating ventilation cooling potential. The small number of input data and the user friendliness of the program help the user to rapidly determine the influence of the main parameters (Roulet et al, 1996). LESOCOOL calculates the cooling potential and the overheating risk in a naturally or mechanically ventilated building, showing the temperature evolution, the airflow rate and the ventilation heat transfer. It can also take into account convective or radiative heat gains.

A detailed methodology to calculate the performance of air-conditioned, as well as free-floating night-ventilated buildings, is presented by Santamouris et al (1996). The method is based on the principle of modified cooling degree days and is extensively evaluated against theoretical and experimental data. The method is integrated within the simulation tool SUMMER (Santamouris et al, 1996) and calculates the variation of the balance-point temperature of a free-floating or air-conditioned night-ventilated building, as well as the overheating hours and the cooling load. In parallel, it performs comparisons with a conventional free-floating or air-conditioned building.

## Example application to low-energy design buildings

In this section, one of the models (NiteCool) is used as an example to assess the suitability of night ventilation cooling for a proposed building. This could be done in the initial concept stage so that particular design options can then be pursued or not. The model has been used to improve the design of a public library in southern England (Rufus and Kolokotroni, 2000).

First, the model was used to predict the measured dry resultant temperatures of another occupied library of similar use in the same topographic area in order to gain some confidence in the model's predictions.

The model was then used to predict internal air temperatures in a proposed design for a new library. It was shown that the original design – which consisted of large areas of single glazing, without a purpose-designed natural ventilation strategy – would suffer from high temperatures even with external maximum temperatures of 25°C. A parametric analysis performed using the model has demonstrated that a substantial reduction in the internal temperature is possible by using day and night natural ventilation. Figure 4.18 presents the effect of day and night ventilation and the exposed thermal mass on the maximum internal temperatures. It can be seen that, for example, a 2.5K reduction could be achieved for the maximum day temperature in a construction with exposed thermal mass to which night ventilation is provided at a rate of 5 air changes per hour (ACH).

It should be noted that calculations for the details of the design cannot be undertaken using the model. Its strength lies in the speed of the predictions so that clients can be shown the results of alternative design strategies very quickly. An experienced engineer or architect will need only a few minutes to access the variations of different strategies for a specific situation. This could be very important in influencing decisions during the design brief stage, which are usually taken during a series of meetings.

Buildings that are cooled using artificial cooling during the day could also benefit from the cooling effect of night ventilation. Benefits could be twofold: a reduction in the energy for cooling required during the day and a reduction in the required size of the air-conditioning plant because of the reduced cooling requirements. An additional benefit is a reduction in peak power demand. Such an application of night ventilation cooling could be described as a hybrid (or mixed-mode) strategy, as outlined in the following section.

A simulation study for an air-conditioned office was carried out for an office located in the hot climate of Kenya (Kolokotroni, 2001). An existing air-conditioned office building was identified in Nairobi and typical data for glazing and opening areas, internal heat gains, operation times and energy consumption figures were obtained. The simulations were carried out for an average month (April: maximum temperature of 26.7°C; minimum temperature of 14.3°C) and the hottest month of the year (February: maximum temperature of 34°C; minimum temperature of 20°C). It was found that by applying night ventilation, 9 per cent less energy would be required during the average month of April to maintain similar internal temperatures as in the

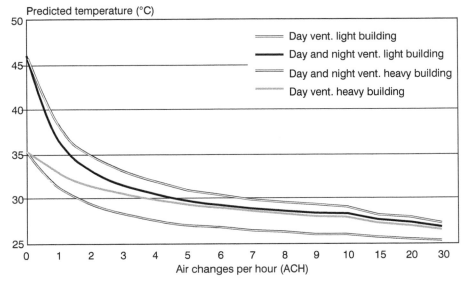

*Source:* Kolokotroni (2001)

**Figure 4.18** *Predicted maximum temperatures for a library building in south-east England as a function of ventilation rate and exposed thermal mass*

reference case. During the hot month of February, a 4 per cent reduction was predicted.

However, if the design of the building is improved according to low-energy good practice principles, further reductions in energy consumption are possible. Figure 4.19 presents the results for daily energy consumption for a typical week during the hottest month of the year (February) in the reference building and in the building optimized in order to benefit most from night-time ventilation. The graph shows energy used in the building in daily values for a typical week during the hottest month. It can be seen that the reference building (i.e. without night ventilation) consumes almost double the electricity to maintain similar internal temperatures as the building optimized for night ventilation.

## Hybrid ventilation

Hybrid ventilation is a term used to describe building-servicing strategies that combine natural ventilation with mechanical ventilation and/or cooling in the most effective manner for a given building application. There are few major sources of information on hybrid ventilation and its applications (Heiselberg, 2002; IEA, 2002; Reshyvent, 2004; Dorer et al, 2005). This section is based on these results, with particular emphasis on the cooling effectiveness of hybrid ventilation.

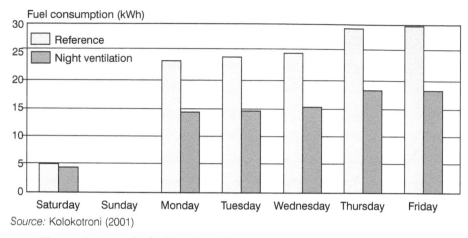

Source: Kolokotroni (2001)

**Figure 4.19** *Daily fuel used in the building during the hottest month in Kenya for the reference office building and a building improved to benefit from night-time ventilation*

## Design principles

The Chartered Institute of Building Services Engineering (CIBSE AM13, 2000) categorizes physical hybrid (mixed-mode) strategies as follows:

- *Contingency designs:* these are usually naturally ventilated buildings that have been carefully planned to permit the selective additions of mechanical ventilation and cooling systems, when this is needed.
- *Complementary designs:* natural and mechanical systems are both present and are designed for integrated operation.
- *Zoned designs:* these allow for differing servicing strategies in different parts of the building.

The International Energy Agency–Energy Conservation in Buildings and Community Systems (IEA-ECSBC) Annex 35 (Heiselberg, 2002) defines the following hybrid ventilation principles:

- *Natural and mechanical ventilation:* this principle is based on two fully autonomous systems where the control strategy either switches between the two systems, or uses one system for some tasks and the other system for other tasks. It covers, for example, systems with natural ventilation in intermediate seasons and mechanical ventilation during midsummer and/or midwinter; or systems with mechanical ventilation during occupied hours and natural ventilation for night cooling.
- *Fan-assisted natural ventilation:* this principle is based on a natural ventilation system combined with an extract or supply fan. It covers natural ventilation systems that, during periods of weak natural driving forces or periods of increased demands, can enhance pressure differences by mechanical (low-pressure) fan assistance.

- *Stack- and wind-assisted mechanical ventilation:* this principle is based on a mechanical ventilation system that makes optimal use of natural driving forces. It covers mechanical ventilation systems with very small pressure losses, where natural driving forces can account for a considerable part of the necessary pressure.

For residential buildings, Dorer et al (2005) identified four systems based on climatic conditions (see Table 4.1).

In the context of Table 4.1, IC-3 and, partly, IC-2 are of interest and are described as follows.

IC-3 offers an improved ventilation system that provides significant good indoor air quality, heating- and cooling-energy savings, and acceptable thermal comfort during the summer by using renewable energy. This concept is a demand-controlled ventilation system based on sensors for presence detection, relative humidity and temperature (see Figure 4.21). Natural driving forces are used whenever possible (thermal stack effect and wind induction), assisted by a photovoltaics-supplied roof cowl fan whose speed is adapted to compensate for natural stack effect deficiency only during the hottest periods. A crucial part of the system is the fan, which has been specially developed to allow natural ventilation when it is switched off. The development of this fan is described in detail in Dorer (2005). The pressure losses due to the fan are lower than 1Pa for airflows of 70 cubic decimetres per second ($dm^3/s$).

IC-2 is a hybrid demand-controlled system with a decentralized supply from the façade and coupled hybrid central mechanical extraction. A characteristic development in this concept is an extremely low-resistance ductwork (< 2Pa at $56dm^3/s$) based on the experiences developed within the EC TIPVENT project (TIPVENT, 2001). A special fan has been developed using 2W at $56dm^3/s$ at 20Pa. Figure 4.20 gives the schematic of the system, which includes $CO_2$ sensors for the living room and bedrooms. A central control unit receives information from these sensors and gives set points to the self-regulating inlets and vents, and, if necessary, starts and controls the fan.

**Table 4.1** *Application fields of the four hybrid ventilation systems developed within the Reshyvent programme*

| | IC-1 | IC-2 | IC-3 | IC-4 |
|---|---|---|---|---|
| Climate | Cold | Moderate | Mild and warm | Severe |
| Building type | Apartments | Dwellings, apartments | Dwellings | Dwellings |
| Renewables | Photovoltaics (PV), wind, heat recovery | PV, wind | PV | Wind, heat recovery |
| Summer comfort | No | Limited | Crucial | No |
| Winter comfort | Important | Important | Important | Crucial |
| Supply | Crucial | Important | Important | Crucial |
| Exhaust | Crucial | Crucial | Crucial | Crucial |

*Source:* Dorer et al (2005)

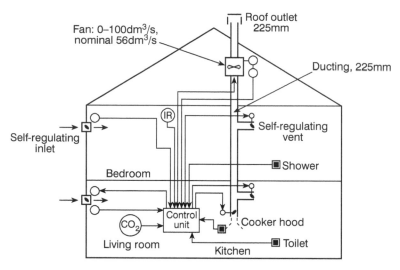

*Source:* Dorer (2005)

**Figure 4.20** *The demand-controlled hybrid ventilation system for moderate climate*

## Control strategies for hybrid ventilation

In all categorizations described above, the importance of appropriately developed and implemented control strategies is highlighted. For this reason, most recent research on hybrid ventilation is complemented by control strategy guidelines. According to Heiselberg (2002), the control strategy for a hybrid ventilation building should at least include a winter control strategy, where IAQ is normally the main parameter of concern, and a summer control strategy, where the maximum room temperature is the main concern. It should also include a spring control strategy to be used in the interval between winter and summer where there might occasionally be a heating demand, as well as excess heat in the building. The control tasks could further be categorized according to the occupancy patterns of the building, as well as integrating controls for solar-shading devices.

To fulfil the control strategy, sensors are needed in the building in order to measure temperature, IAQ and occupancy, as well as actual weather conditions. Box 4.1 provides a list of required sensors.

Further information on sensors for hybrid buildings is available in Hendriksen (2002) and a description of the control strategies in case studies can be found in Aggerholm (2002).

Finally, as in the case of night ventilation, in many instances, detailed simulation techniques are required where all energy and environmental parameters are taken into account. These are described in the section on 'Models for estimating ventilation cooling potential'. Simplified simulation tools designed especially for hybrid ventilation applications are also available

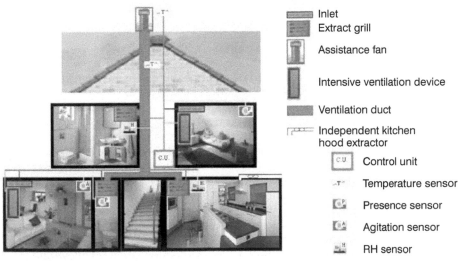

Inlet
Extract grill
Assistance fan
Intensive ventilation device
Ventilation duct
Independent kitchen hood extractor
Control unit
Temperature sensor
Presence sensor
Agitation sensor
RH sensor

*Source:* Dorer (2005)

**Figure 4.21** *Concept of hybrid ventilation systems for warm climate (Belgium/France)*

and can provide very useful information. These are also described in the models section.

## VENTILATION OF URBAN BUILDINGS

### Boundary conditions around urban buildings

The wind patterns and characteristics of dense urban environments differ substantially from the corresponding characteristics in undisturbed conditions. As the air flows from the surroundings to the urban environment, it adjusts to the new boundary conditions defined by the cities. This results in the development of two layers of vertical structures. The so-called 'obstructed sub-layer', or urban canopy sub-layer, extends from the ground surface up to the building's height, while the so-called 'free surface layer', or urban boundary layer, extends above the rooftops. The airflow in the canopy sub-layer is determined by the interaction of the flow field above and the uniqueness of local topography, building geometry and dimensions, streets, traffic and other local features, such as the presence of trees. In a general way, wind speed in the canopy layer is substantially less when compared to the undisturbed wind speed.

Estimations of the wind speed in the urban environment are of great importance for the design of naturally ventilated buildings. Wind speeds measured above the buildings or at airports differ considerably from the speed inside canyons. Since roughness length is greater in an urban area than in the

---

## BOX 4.1 SENSORS IN A BUILDING

Sensors in a building include the following:

- *Temperature:* ordinary room and duct temperature sensors are reliable and are not expensive. Surface temperature sensors exist; but there is not much experience of their use in control systems.
- *Carbon dioxide:* $CO_2$ is an indoor air quality (IAQ) indicator of body odour, but is not harmful to people in the concentrations normally found in buildings. $CO_2$ sensors are quite expensive and need regular calibration.
- *Volatile organic compounds (VOCs):* VOCs are an indicator of IAQ. There is little experience with the use of VOC sensors and it is not clear what they measure and how to calibrate them.
- *Passive infrared (PIR):* infrared presence sensors are reliable and inexpensive. They are easy to test and can also be used for other purposes (e.g. control of artificial light).
- *Air speed:* air speed sensors can be used to measure the airflow rate in ducts. Air speed sensors are quite expensive and require regular cleaning and calibration.

Weather station sensors include the following:

- *External temperature:* external temperature sensors are reliable and inexpensive. The problem is finding a position to install them where the temperature is not influenced by the building or solar radiation.
- *Wind:* traditionally, wind speed is measured with a cup anemometer and wind direction is measured with a wind vane. A new type without moving parts is available where both speed and directions are measured by using the Doppler effect in two directions.
- *Solar radiation:* solar radiation sensors do not need to be very accurate for control purposes. It is preferable to have a sensor on the upper part of each main façade.
- *Precipitation:* precipitation sensors are reliable and inexpensive. They normally only need to produce an on/off signal for overrule purposes.

*Source:* Heiselberg (2002)

---

surrounding countryside, the wind speed at any height is lower in the urban area and is much lower within the obstructed area.

## Experimental work to determine wind speed in urban canyons

Important research work has been carried out to estimate the wind speed in urban areas. This work includes experimental and theoretical studies; as a result, various models that predict the flow in urban canyons have been developed.

The airflow pattern in urban canyons is determined by the characteristics of the canyon, as well as by the characteristics of the wind above the buildings and the thermal conditions along the street. A classification of all flow patterns and characteristics is given in Santamouris (2001).

Most of the work has concentrated in very dense canyons with high height/width (*H/W*) ratios, where *H* and *W* are the height and the width of the canyon, respectively. Important experimental work in various types of urban canyons was carried out during the 1970s and 1980s (see McCormick, 1971; DePaul and Sheih, 1986; Yamartino and Wiegand, 1986; Hoydysh and Dabbert, 1988; Nakamura and Oke, 1989). A full description of the existing experiments is presented in Santamouris (2001).

Since 2000, important experimental work has been carried out in the framework of the European Commission's URBVENT and Reshyvent research programmes. Similar work has also been carried out by the COST programme.

In the framework of the URBVENT programme (Ghiaus et al, 2006), experiments have been carried out in five canyons in Athens, each featuring different urban characteristics. The wind characteristics and the temperature stratification around the building, inside and outside the street canyon, were measured during the summer of 2001. Airflow performance and ventilation rates in each building, as well as indoor air quality, were monitored.

Meteorological data for the building, inside and outside the canyons, were measured. The mobile meteorological station consisted of a vehicle and a telescopic mast, six 3-axis anemometers, five cup anemometers, four thermometers, one infrared thermometer and an infrared camera. The purpose of this equipment was to measure air temperature and wind velocity and direction. Measurements were performed 12 hours per day from morning until night, for three consecutive days, for every canyon. All data were recorded and saved at one-minute intervals (see Figure 4.22).

## First group of measurements

The following measurements were performed:

- *Wind speed and wind direction inside the canyon in the centre of the canyon:* four wind speed and four wind direction anemometers were placed at four different heights (e.g. at 3.5m, 7.5m, 11.5m and 15.5m) in the antenna of the mobile meteorological station.
- *Air temperature inside the canyon in the centre of the canyon:* four thermometers were placed at four different heights (e.g. at 3.5m, 7.5m, 11.5m and 15.5m) in the antenna of the mobile meteorological station.
- *Wind speed inside the canyon near the façades of the canyon:* two 3-axis anemometers were used to measure the three components of the wind speed inside the canyon. Each anemometer was mounted on each exterior façade of the canyon and at distances of 1m to 2m from the wall.
- *Wind direction outside the canyon:* a cup anemometer was also placed on the top of the canyon and at a distance of 6m from its top level to measure the wind speed and direction outside the canyon.

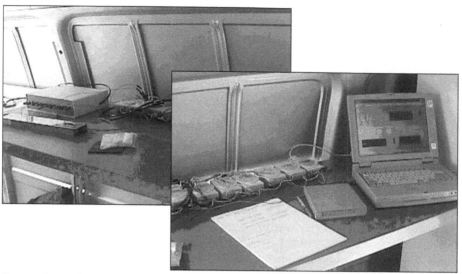

*Source:* Group Building Environmental Studies, University of Athens

**Figure 4.22** *Data acquisition modules and their power supply*

A mobile meteorological station was used for all experiments. The mobile meteorological station consisted of:

- a vehicle; and
- a telescopic mast PT8 combined collar mast assembly, with an extended height of 15.3m, a retracted height of 3.43m and a maximum head load of 15kg. The mast is pneumatically operated for raising and lowering, using the vehicle's power supply (see Figure 4.21).

## Second group of measurements:
## Surface temperature measurements
An infrared thermometer equipped with a laser beam was used. The surface temperatures of the exterior façades of the buildings were measured at different points of the canyon, at different hours during the 12 hours of the experiment. Measurements were taken from the bottom to the top of both canyon façades and along the road between them.

## Third group of measurements:
## Surface temperature measurements using infrared cameras
The distribution of the surface temperatures of all materials inside the canyon was measured using an infrared camera. The surface temperature of the canyon was measured during the 12-hour period.

**Fourth group of measurements: Indoor airflow rate**

Airflow measurements in different buildings were performed in each canyon. Tracer gas techniques were used to estimate the air ventilation rate in the rooms of buildings. The technique is based on the use of a chemically inactive gas: nitrous oxide ($N_2O$) or sulphur-hexafluoride ($SF_6$). These techniques can be used in one-zone or multi-zone spaces inside a building. A detailed description of the measurements performed is given in Santamouris and Georgakis (2003) and Georgakis and Santamouris (2004, 2006). A full analysis of the experiments is given in Georgakis and Santamouris (2005).

A second set of experiments in two urban canyons was performed under the framework of the Reshyvent research project. Similar measurements inside the urban canyons, as previously reported, were performed. In addition, measurements on the efficiency of various hybrid ventilation systems were taken. Analysis of the data is reported in Niachou et al (2007a, b).

The BUBBLE project was a large urban experiment carried out under the auspices of the European COST 715 programme in Basel, Switzerland. Measurements of wind speed and direction in canyons were carried out. An analysis of the data, focusing on the potential of natural ventilation techniques in urban buildings, is given by Germano (2006).

Brown et al (2005) have modelled a street canyon in Oklahoma City, US, with high-density wind sensor instrumentation. The main aim of the experiment was to evaluate the new-generation dispersion models.

## Theoretical methods to calculate wind speed in the urban environment

Two major types of methods exist to calculate the wind speed in urban canyons for natural ventilation purposes:

1  deterministic models based either on simplified calculation assumptions or on the solution of the whole or a reduced set of equations of the flow;
2  data-driven models based on a statistical analysis of existing experiments.

Deterministic techniques are classified in four types of algorithms:

1  simplified models predicting mean speed in a canyon – most of these models are of local use and cannot be used in all cases;
2  analytical models proposing a set of specific equations to calculate wind speed at various locations inside a canyon;
3  network models based on the solution of a reduced set of equations describing flow phenomena;
4  computerized fluid dynamic models based on the solution of the full set of Navier-Stokes and turbulence equations.

Simplified methods to calculate the mean horizontal wind speed in urban canyons have been proposed mainly by Paciuk (1975), Chandra et al (1986) and Nakamura and Oke (1989). Nakamura and Oke (1989) have suggested a simple linear form to calculate the mean horizontal wind speed, $u_h$, inside a canyon:

$$u_h = p \, u_{roof} \tag{1}$$

where $p$ is a decrease factor that depends upon $H/W$ and the measurement level, $u_{roof}$, is the wind speed measured at the roof level, and $H$ and $W$ are the height and width of the canyon, respectively.

A simplified but quite comprehensive method is developed and proposed by Chandra et al (1986). The method proposes a terrain correction factor (TCF) to multiply the design air change rate in order to take into account the effects related to the reduction of wind speed because of the building's location. The same method proposes a reduction coefficient to take into account the effect of neighbouring buildings. The coefficient is based on the wall height of the upwind building, $b$, as well as on the gap between the building and the adjacent upwind building, $g$. The final airflow rate can then be calculated as the product of the design airflow rate multiplied by the neighbourhood correction factor and the terrain correction factor.

Paciuk (1975) has developed a formula predicting relative wind speed based on wind tunnel experiments. The objective of these experiments was to identify the effects of building height and distance between buildings on wind speed in the open spaces between buildings when the buildings are perpendicular to wind direction. Paciuk (1975) proposed the following formula:

$$V_{r(u.h)} = 10 + (66(1 - e - 0.08h))e - 0.18D/W \tag{2}$$

where:

- $V_{r(u.h)}$ is the relative wind speed expressed as the percentage of the wind at the same height well in front of the first line of buildings;
- $D$ is the distance travelled by the wind in metres: $D = n(b+W) - 0.5W$;
- $b$ is the depth of the buildings in metres;
- $n$ is the serial number of the space (downwind);
- $h$ is the height of the buildings in metres;
- and $W$ is the width of the spaces between buildings in metres.

Simplified methods can be used in order to offer a very first assessment of the expected order of magnitude of the wind speed in the urban environment. In most cases, the proposed values are valid inside the limits of the experiments used to develop the method.

Various analytical methods to calculate wind speed in urban canyons have been proposed, mainly by Hotchkiss and Harlow (1973), Nicholson (1975) and Yamartino and Wiegand (1986). The algorithms of these methods are presented in Santamouris (2001).

These methods have been compared against the experimental data collected during the URBVENT experiments. The results of the comparison are given in Bozonnet (2005). The methods may predict wind speed in canyons with acceptable accuracy, especially when the temperature difference-induced flow is not important. The use of zonal models to predict the spatial variation of wind speed in urban canyons has been proposed by Bozonnet et al (2005). The method is fully described in Bozonnet (2005). A comparison of the predicted values of wind speed against data collected during the URBVENT experiments has been carried out, and has demonstrated that the proposed method predicts with sufficient accuracy the wind speed inside canyons, mainly when airflow is due to temperature differences.

Computational fluid dynamic (CFD) models have frequently been applied to predict wind flow in urban canyons (Jicha et al, 2000; Jeong and Andrews, 2002; Assimakopoulos et al, 2003). An extensive comparison between predicted wind speeds using CFD simulations and the experimental data collected during the URBVENT experiment are reported by Assimakopoulos et al (2006). Results from the computations have shown that the wind field in urban areas is quite complex, presenting areas of very low wind speeds and the convergence of vortices. The model underestimated the measured wind speed intensities, which may be partly explained by the uncertainty of specific input parameters, the necessary simplifications for the application of such models, and the geometrical complexity of the area modelled.

A comparison of the measured wind speed during the Reshyvent experiments against another CFD code is reported by Jospisil et al (2005). It has been shown that, under specific boundary conditions, CFD simulations may predict wind speed with sufficient accuracy.

The use of deterministic techniques to calculate wind speed in canyons is accompanied by relatively low accuracy because of the high uncertainty of the input data and the incomplete description of the physical phenomena. The development of very high-speed computers has made data-driven techniques quite popular. Data-driven techniques are based on the analysis of existing experimental data, using advanced mathematical methods, and are used to develop and propose methods of describing the physical behaviour of the system.

Using all data collected during the URBVENT and Reshyvent experiments, a data-driven statistical method to predict the spatial distribution of wind speed in canyons has been proposed (Santamouris et al, 2006). Using fuzzy clustering techniques, clusters of input–output data have been developed using inertia and gravitational forces as criteria. For each cluster of data the more probable wind speed has been calculated. The calculated data together with the corresponding inputs comprise a reduced space input file. This reduced data space has been used to develop three data-driven prediction models. The models are a three-dimensional (3D) graphical interpolation method, a tree-

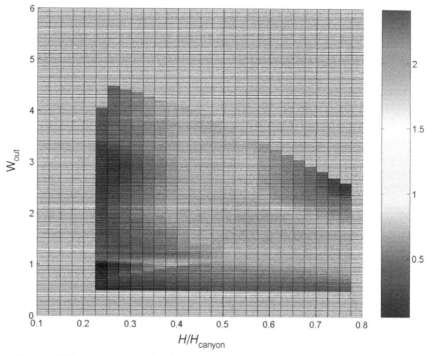

*Source:* Group Building Environment Studies, University of Athens

**Figure 4.23** *Developed graphical data-driven model to predict the more probable wind speed close to the canyon façades when the flow is parallel to the canyon axis*

based model and a fuzzy estimation model. Comparison against the experimental values from the URBVENT and Reshyvent experiments, as well as from other canyon experiments not considered in the development of the method, shows a very good agreement between the theoretical and experimental values (see Figures 4.23 and 4.24).

## The potential of natural ventilation within urban buildings

One method of calculating the suitability of a site and its potential for natural ventilation has been presented by Germano (2006). The method uses multi-criteria analysis, and the criteria for selection are wind-induced and buoyancy-induced pressure, and the levels of noise and atmospheric pollution. A software program has been developed through the URBVENT research programme (see Figure 4.25). The model is validated against existing experimental data.

Using the software to calculate wind speed in canyons that was developed through the URBVENT research project, a method has been created to calculate airflow through openings located in urban canyons. In parallel, a

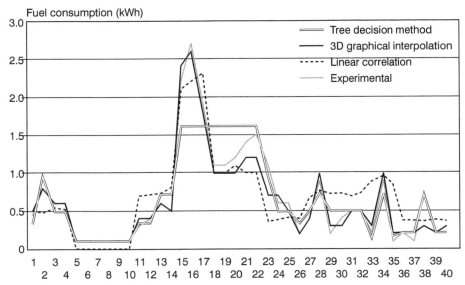

*Source:* Group Building Environment Studies, University of Athens

**Figure 4.24** *Comparison of the predicted against the experimental values of the more probable wind speed inside the canyon at a height* H

method to calculate the necessary openings of an urban building to achieve a specific airflow performance has been developed. The methods are based on the use of a neural network trained with data simulated with the developed urban airflow tool. Two types of configuration are considered:

1  single-side ventilation; and
2  ventilation through chimneys.

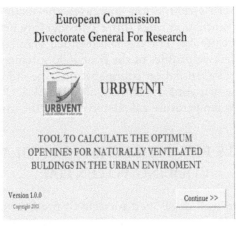

*Source:* Ghiaus and Allard (2005)

**Figure 4.25** *Main page of the URBVENT software*

More information on the tool is given in Ghiaus and Allard (2005). A CD with the tool is also included.

The developed methodology has been validated against experimental data collected through the URBVENT programme (see Figure 4.26 and Table 4.2).

The potential of night ventilation techniques for ten urban canyons, situated in the extended region of Athens, Greece, has been analysed by Geros et al (1999, 2005). Since climatic conditions in the urban environment frequently present important variations between different locations in the same urban area, outdoor air temperature and wind profiles have been measured inside and outside the experimental canyons. The research has studied the impact of the urban environment on night ventilation energy performance by examining a typical room located in the urban domain, under air-conditioned and free-floating operation. During the study, single-sided and cross-ventilation were considered throughout the night period. The influence of the urban microclimate on the efficiency of the technique has been examined by considering the typical zone inside the canyons and under undisturbed conditions.

A comparison of the results evaluated the impact of the urban environment on the effectiveness of night ventilation techniques as also described in the section on 'Night ventilation'. The analysis performed shows that due to the increase in air temperature and the decrease in wind velocity inside the canyons, the efficiency of the techniques studied is significantly reduced when compared with undisturbed conditions, which dominate outside the urban canyons.

It has been found that for single-sided ventilation configurations, the relative decrease of the cooling potential in the urban environment varies from –6 to 89 per cent. For cross-ventilation, this difference varies between –18 and 72 per cent.

Figure 4.26 illustrates the indoor temperature difference during daily operation when the zone is located inside and outside the ten urban canyons, and when the typical zone operates under free-floating conditions. As shown, the application of cross-ventilation increases the examined temperature difference since the ventilation rates differ considerably between the interior and exterior of the canyons. For the single-sided ventilation scenario, the difference between the two profiles of the zone indoor temperature varies from 0°C to 2.6°C. For cross-ventilation configurations, the indoor temperature of the zone inside the canyons is from 0.2°C to 3.5°C higher than the corresponding indoor temperature calculated outside the canyons.

## MODELS FOR ESTIMATING VENTILATION COOLING POTENTIAL

In general, there are well-developed models to predict ventilation rates in buildings, as well as (to some extent) the energy impact of ventilation during the heating season. There is less development within existing models for

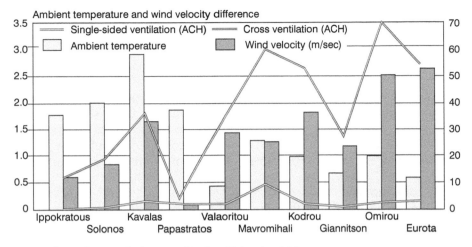

*Source:* Group Building Environment Studies, University of Athens

**Figure 4.26** *The average difference of the ambient temperature, the wind velocity (horizontal component) and the ventilation airflow rate between the two locations of the typical zone, with single-sided and cross-ventilation*

Single-sided ventilation
Experimental ACH (air changes per hour)

| Ermou | Miltiadou | Voukourestiou | Kanniggos | Dervenion |
|---|---|---|---|---|
| 0.2–0.8 | 0.4–1.1 | 0.8–1.2 | 0.2–1.0 | 0.4–1.5 |

Single-sided ventilation
Mean theoretical ACH (air changes per hour)

| Ermou | Miltiadou | Voukourestiou | Kanniggos | Dervenion |
|---|---|---|---|---|
| 0.65 | 1.5 | 1.0 | 1.3 | 1.35 |

**Table 4.2** *Comparison between experimental and theoretical air changes per hour for single ventilation in the five measured canyons*

predicting cooling energy-related issues. This section presents an overview of general ventilation models and a brief reference to their integration with thermal/energy models so that energy demand/consumption calculations can be carried out.

## Types of ventilation models

### Empirical models
Simplified models offer general correlations to calculate airflow rate. These expressions combine airflow with temperature difference, wind velocity and, possibly, a fluctuating term in order to give a bulk evaluation of the airflow rate or the air velocity in a building. These tools are useful because they offer a

rapid initial estimation of the airflow rate, but should always be used within the limits of their applicability.

## Network (zonal) models

Network models use airflow networks to represent all openings in each individual zone. A network is developed for infiltration openings, representing the natural porosity of the building fabric, and for purposely provided openings, such as open windows or air vents in the building envelope, or between individual zones. Each network may consist of many flow paths interconnecting the zones or rooms with differing pressure or temperature.

## Single-zone models

For a single-zone network airflow model, the building considered is treated as a single zone. Airflow paths are represented by infiltration routes and purposely provided openings. This model is an acceptable approximation for open-plan buildings or small family homes. The model assumes that doors in the buildings are left open or are relatively leaky (see Figure 4.27; Liddament, 1996).

## Multi-zone models

A multi-zone network airflow model is more complex than a single-zone model. The building is divided into a number of single zones where internal pressures and temperatures are distinct from one another, separated by internal partitions. This type of model is applicable to commercial and multi-storey buildings where floor space is divided up into separate rooms (see Figure 4.28).

## Computational fluid dynamic (CFD) models

CFD models are based on Navier-Stokes equations, which are solved at all points of a two- or three-dimensional grid that represents the buildings and its

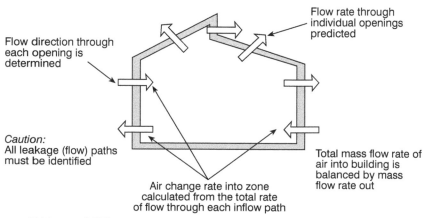

*Source:* Liddament (1996)

**Figure 4.27** *Single-zone airflow paths*

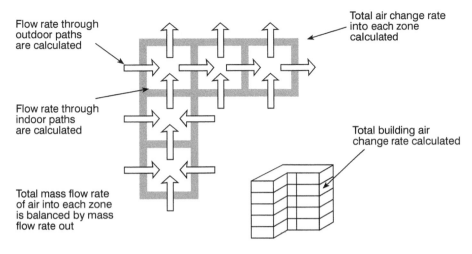

Flow rate through
outdoor paths
are calculated

Total air change rate
into each zone
calculated

Flow rate through
indoor paths
are calculated

Total building air
change rate calculated

Total mass flow rate
of air into each zone
is balanced by mass
flow rate out

*Source:* Liddament (1996)

**Figure 4.28** *Multi-zone airflow paths*

surroundings. The unknown pressure and velocity components are determined for given boundary and initial conditions.

### Effect of mechanical ventilation on modelling

At the most basic level, mechanical ventilation is applied as a fixed-flow rate. One way of accomplishing this is to establish a path in which the flow exponent is set to zero and the flow coefficient is set to the mechanical ventilation rate. This fixed-flow approach is valid provided that the pressure change in the zone can be met by the fan without any change in flow rate. In practical terms, this means that the calculated pressure difference across the fan flow path should be checked with manufacturer's data to ensure that the system can supply the rated airflow at the calculated conditions. A more precise representation of mechanical ventilation is to incorporate the system and fan pressure drop versus airflow rate relationship as if it were any other type of flow path. This requires understanding the flow characteristics of the fan and the impact of associated ductwork.

A comprehensive description of such models giving their theoretical background is included in Allard (1998), and a summary of the available models, including comparisons and availability, is given in Orme (1999).

In many cases, these models have been adapted for specific applications of natural ventilation strategies. For example, the multi-zone model COMIS has been used to develop a simple tool to assess the feasibility of hybrid ventilation systems (Fracastoro et al, 2002).

Recently, new approaches have been developed for predicting air change rates in buildings. A multi-criteria analysis method has been developed especially for applications to buildings located in urban areas, as described in the section on 'Ventilation of urban buildings'. In this method, the driving forces (calculated using the models described above) and the constraints (urban characteristics) are integrated with time, obtaining wind-time, stack-time, pollution-time and noise-time indicators. Then, a multi-criteria analysis is performed by using the Qualifelx method. The result is a classification of the suitability of natural ventilation in the given site for the buildings considered in the evaluation. The method used is described in detail in Germano et al (2005) and Germano and Roulet (2005).

## Energy-related calculations: Integration of ventilation models with thermal models

Thermal models consider the heat transport into and out of buildings, predicting the thermal movements of conduction, convection and radiation, and taking account of heat storage in and transfer through the building fabric, solar gains, and other heat gains and losses. These models, when integrated with an airflow model, can be used for estimating the internal temperature distribution of a building in order to determine the impact of the ventilation, heating and cooling systems on thermal comfort. The energy used by the heating and/or cooling systems for conditioning the air may also be predicted by these models.

There are various methods of integrating the thermal and airflow models. Ideally, the two models would be completely integrated in such a way that the governing equations are solved simultaneously; this method is termed direct coupling (see Figure 4.29). However, a simpler method is to solve the governing equations separately with a feedback link between the two simulations or sequential coupling (see Figure 4.30). In this method, mass flow rates are calculated first, with assumed internal temperatures, and these values are then fed into the thermal model. There is no feedback from the thermal model to the ventilation model. Most commercially available thermal models incorporate ventilation models in this way.

Hensel (1995) describes two additional approaches to coupling thermal and ventilation models: 'ping-pong' and 'onions'. The first involves the airflow rates from the ventilation model being used by the thermal model. The

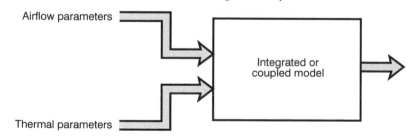

Airflow parameters

Thermal parameters

Integrated or coupled model

*Source:* Liddament (1996)

**Figure 4.29** *Direct coupling simulation technique of thermal and airflow models*

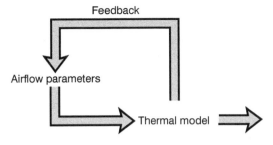

*Source:* Liddament (1996)

**Figure 4.30** *Sequential coupling simulation technique of thermal and airflow models*

calculated air temperatures are then fed back into the ventilation model at the next time step. In the 'onions' approach, airflow rates are passed from the ventilation model to the thermal model, and air temperatures are passed back from the thermal model to the ventilation model. This process is repeated until 'convergence' is reached. Figure 4.31 shows the various types of input data needed to satisfy the airflow models (mass balance) and the thermal models (heat balance), and the feedback links that simulations may use to integrate the two models.

The integration of CFD and thermal models is also possible. A comprehensive review of development in this area is given in Somarathne et al (2005). CFD cannot easily and quickly solve time-dependent thermal interactions across the boundaries of a building model. Hence, time-averaged solutions of boundary conditions (usually in the form of surface temperatures)

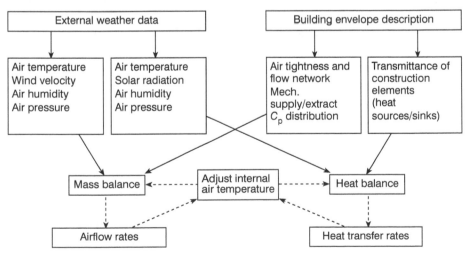

*Source:* Orme (1999)

**Figure 4.31** *Input data and feedback links for coupled simulation of thermal and airflow models*

solved in a dynamic thermal model (DTM) code are manually input into a steady-state CFD code. In such a scenario, if extreme conditions – for example, the hottest day in summer with maximum occupancy rates – are modelled both in CFD and a DTM code, then through the process of combining the solutions of the two tools, representative analysis of extreme thermal conditions is achieved. With current building simulation resources, building systems are commonly designed to function under extreme conditions and, hence, are often over-designed.

In many applications, there is a requirement for a single tool that can provide detailed dynamic thermal simulations with CFD accuracy for ventilation flows. A single tool would enable a building services engineer to closely design systems to the individual traits of a building, taking advantage of the thermal capacity of building fabrics and the interaction of the air that they enclose. To date, the development of such a tool has been severely limited by computational hardware capacity despite extensive research in this field. The fundamental limitation has been the solution process and data storage over vastly different time constants of solids and air. Building fabric responds at a slower rate to thermal effects than air and, hence, requires excessive computational data storage to account for this in CFD.

Work, to date, falls into two main categories that have evolved with the improvement of hardware capacity. The first category was focused upon the development of a more unified building simulation code that could incorporate a variety of computational building services tools. The objective of this main stream of research was to construct a more holistic building simulation package that could account for all aspects of a building environment, such as solar effects and interstitial condensation (Clarke and Irving, 1988).

To build a more holistic building simulation tool, several researchers used a technique where links were created between the various modules of the complete simulation package – for example, Intelligent Integrated System for the Analysis of the Building Thermal Environment, as developed by the Heating, Ventilating and Air-Conditioning (HVAC) Division of Tsinghua University, Beijing. In the case of research conducted by Hong et al (1997), the link between the modules was in the form of a communal database. Multi-iterative processing was used, where results of the separate modules were compared, corrected and updated.

By increasing the overall simulation power of building modelling codes by integrating several different codes, very coarse grids were superimposed over the entire multi-zone geometry of a building. Occasionally, CFD (also an integrated tool in the package) was employed to model likely problematic regions of a zone (i.e. near supply duct exits and extract duct entrances). Some of the earliest publications of the combination of CFD and DTM were by Chen and Jiang (1992) and Holmes et al (1990), where DTM and CFD were used in parallel to swap information with each other.

Maintaining CFD and DTM as separate codes, where information was swapped between them via a third code, appears to be a popular technique of research (Bartak et al, 2002). Similar methods were also employed by Negrao

(1997) and Neilson and TryggVasen (1998), where CFD and DTM packages were run simultaneously and comparisons were made between the solutions generated by them, using an additional process of iteration. The success of their methods relied on the relative lack of detail of each code by using a coarse grid. Larger thermal gradients obtained within a zone using CFD created convergence problems when comparing solutions within DTM.

The second main objective of current work has been to develop a more accurate and detailed tool by focusing on the integration of DTM techniques within CFD. Temperature was a strong link between the two techniques – in particular, the temperatures from the surfaces of the solution domain, which are normally taken from the dynamic thermal model and transferred to a CFD model. In addition, a technique experimented with by Moser et al (1995) and Kato et al (1995) demonstrated that only solutions of the temperature equation were important at the solid–air interface created by the building fabric and the indoor environment.

In a traditional CFD model, accurate representations of the temperatures of the internal and external surfaces of the building fabric over significant lengths of time are essential pieces of information required to produce complete time-dependent simulations. Techniques were developed to do this in Japan by Takeya et al (1998). A 3D transient CFD code was designed to solve coupled heat transfer between airflow and indoor materials (wall envelopes and furniture), including convective heat transfer, radiation exchange among indoor materials, heat conduction through walls, and heat source/sink-like radiation panels.

The research conducted by Takeya et al (1998) was successful in solving transient indoor airflow, but did not take into account the simultaneous external time-varying dynamic conditions, which inevitably occur due to weather effects. Takeya et al (1998) have highlighted that DTM is still required to give initial conditions.

Research work has developed from embedding CFD into DTM, to embedding DTM into CFD; but no independent CFD tool has yet been developed that can provide an accurate dynamic thermal modelling function – research continues in this area.

In the scope of International Energy Agency (IEA) Annex 35 (Heiselberg, 2002), probabilistic methods were applied in thermal building simulation, single- and multi-zones models and CFD (Brohus et al, 2002a). In probabilistic methods, some or all of the input parameters are modelled either as random variables or as stochastic processes, described by statistics (i.e. mean values, standard deviations, auto-correlation functions, etc.), and the rules are the corresponding statistics of the output. A stochastic method is therefore a formulation of a physical problem, where the randomness of the parameters is taken into account. The advantage of a probabilistic method is the possibility of not only designing for peak load and estimating annual energy consumption based on a reference e year, but of examining the range of variation and quantifying the uncertainty. Probabilistic methods can be used as a tool to evaluate the trade-off between economy (cost, energy and environment) and risk (expectation not met, violation of regulations, etc.) on a firm foundation.

## CONCLUSION

This chapter has focused on strategies for providing thermal comfort through ventilation. The functions of ventilation were first described briefly; building design strategies, systems and components that facilitate ventilation for cooling were then explained in more detail. This included ventilation towers, slab cooling, façade ventilation, night ventilation and hybrid ventilation. Urban buildings are a special case and important research results were outlined in the section on 'Ventilation of urban buildings'. Modelling techniques for estimating ventilation cooling potential were described that indicate that there is still a need for development in this area. This is connected to documented performance in operational buildings. Although some information is available, this is an area that requires further work so that realistic parameters can be used for further development of prediction models.

## REFERENCES

Aggerholm, S. (2002) *Characterisation of Hybrid Ventilation Strategies and Control Strategies in the Case Studies*, IEA Annex 35, http://hybvent.civil.auc.dk

Allard, F. (ed) (1998) *Natural Ventilation in Buildings: A Design Handbook*, James and James, London

Assimakopoulos, V. D., ApSimon, H. M. and Moussiopoulos, N. (2003) 'A numerical study of atmospheric pollutant dispersion in different two-dimensional street canyon configurations', *Atmospheric Environment*, vol 37, no 29, pp4037–4049

Assimakopoulos, V., Georgakis, C. and Santamouris, M. (2006) 'Experimental validation of computational fluid dynamic codes to predict the wind speed in street canyons for passive cooling purposes', *Solar Energy*, vol 8, no 4, pp423–434

Bartak, M., Beasoleil-Morrison, I., Clarke, J. A., Denev, J., Drkal, F., Lain, M., Macdonanld, I. A., Melicov, A., Popiolek, Z. and Stankov, P. (2002) 'Integrating CFD and building simulation', *Building and Environment*, vol 37, pp865–871

Bordass, W., Cohen, R., Leaman, A. and Standeven, M. (1999) 'Probe 18: Portland Building', *Building Services Journal*, pp35–40

Bozonnet, E. (2005) 'Impact des microclimats urbains sur la demande energetique', PhD thesis, Université de La Rochelle, France

Bozonnet, E., Belarbi, E. and Allard, R. (2005) ' Modelling solar effects on the heat and mass transfer in a street canyon: A simplified approach', *Solar Energy*, vol 79, no 1, pp10–24

*BRE Digest* (1994) 'Natural ventilation in non-domestic buildings', *BRE Digest*, vol 399, October, p8

Brohus, H., Frier, C. and Heiselberg, P. (2002a) *Qualification of Uncertainty in Thermal Building Simulation by Means of Stochastic Differential Equations*, IEA Annex 35, http://hybvent.civil.auc.dk

Brohus, H., Frier, C. and Heiselberg, P. (2002b) *Stochastic Single and Multi-Zone Models of a Hybrid Ventilated Building: A Monte Carlo Simulation Approach*, IEA Annex 35, http://hybvent.civil.auc.dk

Brown, M. J., Boswell, D., Streit, G., Nelson, M., McPherson, T., Hilton, T., Pardyjak, E. R., Pol, S., Ramamurthy, P., Hansen, B., Kastner-Klein, P., Clark, J., Moore, A., Walker, D., Felton, N., Strickland, D., Brook, D., Princevac, M., Zajic, D., Wayson, R., MacDonald, J., Fleming, G. and Storwold, D. (2005) 'Joint Urban 2003 Street Canyon Experiment', paper for American Meteorological Society, http://ams.confex.com/ams/pdfpapers/74033.pdf

Chandra, S., Fairey, P. W. and Houston, M. M. (1986) *Cooling with Ventilation*, SERI/SP-273-2966, DE86010701, Solar Energy Research Institute, Golden, CO

Chen, Q. and Jiang, Z. (1992) ' Significant questions in predicting room air motion', *ASHRAE Transactions*, vol 98, part 1, pp929–939

CIBSE (Chartered Institute of Building Services Engineering) (2004) *Environmental Performance Toolkit for Glazed Façades*, TM35, CIBSE, London

CIBSE (2005) *Natural Ventilation in Non-domestic Buildings*, Applications Manual 10 (AM10), CIBSE, London

CIBSE AM13 (2000) *Mixed Mode Ventilation*, CIBSE, London

Clarke, J. and Irving, A. D. (1988) 'Building energy simulation: An introduction', *Energy and Buildings*, vol 10, pp157–159

Cohen, R., Leaman, A., Robinson, D. and Standeven, M. (1996) 'Probe 8: Queens Building, Anglia Polytechnic University', *Building Services Journal*, December, pp27–31

Cook, M. J., Ji, Y. and Hunt, G. R. (2003) 'CFD modelling of natural ventilation: Combined wind and buoyancy forces', *International Journal of Ventilation*, vol 1, no 3, p169

Cook, M. and Short, A. (2005) 'Natural ventilation and low energy cooling of large, non-domestic buildings – four case studies', *International Journal of Ventilation*, vol 3, no 4, pp283–294

DePaul, F. T. and Sheih, C. M. (1986) 'Measurement of wind velocities in a street canyon', *Atmospheric Environment*, vol 20, pp455–459

DETR (UK Department of the Environment, Transport and the Regions) (1998) *The Elizabeth Fry Building: Feedback for Designers and Clients*, New Practice Final Report 106, BRE, London

DETR (1999a) *The Ionica Building, Cambridge: Feedback for Designers and Clients*, DETR, New Practice Final Report 115, BRESCU, London

DETR (1999b) *The Queens Building, De Montford University: Feedback for Designers and Clients*, New Practice Final Report 102, BRESCU, London

Dorer, V., Pfeiffer, A. and Weber, A. (2005) *Parameters for the Design of Demand Controlled Hybrid Ventilation Systems for Residential Buildings*, AIVC, Brussels

Ford, B. H. (2002) 'Market assessment of passive downdraught evaporative cooling in non-domestic buildings in Southern Europe', in *Proceedings of the Third European Conference on Energy Performance and Indoor Climate in Buildings (EPIC)*, Lyon, France, October

Ford, B. H. (2003) 'Passive downdraught cooling: Hybrid cooling in the Malta Stock Exchange', in *Proceedings of the PLEA Conference, 2003*, Santiago, Chile, 9–12 November

Fracastoro, G. V., Perino, M. and Mutanin, G. (2002) *A Simple Tool To Assess the Feasibility of Hybrid Ventilation Systems*, IEA Annex 35, http://hybvent.civil.auc.dk

Georgakis, C. and Santamouris, M. (2004) 'On the airflow in urban canyons for ventilation purposes', *International Journal of Ventilation*, vol 3, no 1, June, pp53–66

Georgakis, C. and Santamouris, M. (2005) 'Wind and temperature in the urban environment', in Ghiaus, C. and Allard, F. (eds) *Natural Ventilation in the Urban Environment: Assessment and Design*, James and James, London,

Georgakis, C. and Santamouris, M. (2006) 'Experimental investigation of air flow and temperature distribution in deep urban canyons for natural ventilation purposes', *Journal of Energy and Buildings*, vol 38, pp367–376

Germano, M. (2006) *Qualitative Modelling of the Natural Ventilation Potential in Urban Context*, PhD thesis, EPFL, Lausanne, Switzerland

Germano, M., Ghiaus, C. and Roulet, C. A. (2005) 'Natural ventilation potential', in Ghiaus, C. and Allard, F. (eds) *Natural Ventilation in the Urban Environment: Assessment and Design*, James and James, London

Germano, M. and Roulet, C. A. (2005) 'Multicriteria assessment of natural ventilation potential of a site', in *Proceedings of Passive and Low Energy Cooling for the Built Environment*, PALENC, Santorini Greece, 19–21 May, Heliotopos Conferences, vol II, pp1039–1044

Geros, V., Santamouris, M., Karatasou, S., Tsangrassoulis, A. and Papanikolaou, N. (2005) 'On the cooling potential of night ventilation techniques in the urban environment', *Energy and Buildings*, vol 37, pp243–257

Geros, V., Santamouris, M., Tsangrasoulis, A. and Guarracino, G. (1999) 'Experimental evaluation of night ventilation phenomena', *Energy and Buildings*, vol 29, pp141–154

Ghiaus, C. and Allard, F. (eds) (2005) *Natural Ventilation in the Urban Environment: Assessment and Design*, James and James, London

Ghiaus, C., Allard, F., Santamouris, M., Georgakis, C. and Nicol, F. (2006) 'Urban environment influence on natural ventilation potential', *Buildings and the Environment*, vol 41, pp395–406

Gratia, E. and De Herde, A. (2002) 'A simple design tool for the thermal study of dwellings', *Energy and Buildings*, vol 34, pp411–420

Heiselberg, P. (2002) *Principles of Hybrid Ventilation*, IEA-ECBCS, Denmark

Hendriksen, O. J. (2002) *A Sensor Survey for Hybrid Ventilation Control in Buildings*, IEA Annex 35, http://hybvent.civil.auc.dk

Hensel, J. L. M. (1995) 'Modelling coupled heat and air flow: Ping-pong vs onions', in *Proceedings of the 16th AIVC Conference*, AIVC, Palm Springs, pp253–264

Holmes, M. J., Lan, J. K.-W., Ruddick, K. G. R and Whittle, G. E. (1990) 'Computation of conduction, convection and radiation in the perimeter zone of an office space', in *Proceedings of ROOMVENT 1990*, Oslo, Norway

Hong, T., Zhang, J. and Jiang, Y. (1997) 'IISABRE: An integrated building simulation environment', *Building and Environment*, vol 32, no 3, pp219–224

Hotchkiss, R. S. and Harlow, F. H. (1973) *Air Pollution Transport in Street Canyons*, Report by Los Alamos Scientific Laboratory for US Environmental Protection Agency, EPA-R4-73-029, NTIS PB-233 252, San Francisco, CA

Hoydysh, W. and Dabbert, W. F. (1988) 'Kinematics and dispersion characteristics of flows in asymmetric street canyons', *Atmospheric Environment*, vol 22, no 12, pp2677–2689

IEA (International Energy Agency) (2002) *Annex 35: Hybrid Ventilation in New and Retrofitted Office and Educational Buildings*, IEA–ECBCS, http://hybvent.civil.auc.dk

IEA (2006) *Annex 44: Integrating Environmentally Responsive Elements in Buildings*, IEA, www.annex44.com

Jeong, S. J. and Andrews, M. J. (2002) 'Application of the k–e turbulence model to the high Reynolds number skimming flow field of an urban street canyon', *Atmospheric Environment*, vol 36, no 514, pp1137–1145

Jicha, M., Katolicky, J. and Pospisil, J. (2000) 'Dispersion of pollutants in street canyon under traffic induced flow and turbulence', *Journal for Environmental Monitoring and Assessment*, vol 65, pp343–351

Kato, S., Murakami, S., Shoya, S., Hanyu, F. and Zeng, J. (1995) 'CFD analysis of flow and temperature fields in atrium with ceiling height of 130m', *ASHRAE Transactions*, vol 101, part 2, pp1144–1157

Kirk, S. and Kolokotroni, M. (2004) 'Windcatchers in modern UK buildings: Experimental study', *International Journal of Ventilation*, vol 3, no 1, pp67–78

Klobut, K. and Kosonen, R. (1995) *Slab Cooling with Air in 'Review of Low Energy Technologies'*, IEA Annex 28, IEA, Watford

Kolokotroni, M. (1998) 'Night ventilation for cooling office buildings', Information paper, BRE, Watford

Kolokotroni, M. (2001) 'Night ventilation cooling of office buildings: Parametric analyses of conceptual energy impacts', *ASHRAE Transactions*, vol 107, part 1, pp479–490

Kolokotroni, M., Ge, Y. T. and Katsoulas, D. (2002) 'Monitoring and modelling IAQ and ventilation in classrooms within a purpose designed naturally ventilated school', *Indoor and Built Environment*, vol 11, no 6, pp316–326

Kolokotroni, M., Giannitsaris, I. and Watkins, R. (2006) 'The effect of the London urban heat island on building summer cooling demand and night ventilation strategies', *Solar Energy*, vol 80, no 4, pp383–392

Kolokotroni, M., Robinson-Gayle, S., Tanno, S. and Cripps, A. (2004) 'Environmental impact analysis for typical office façades', *Building Research and Information*, vol 32, no 1, pp2–16

Liddament, M. (1996) *Energy Impact of Ventilation*, AIVC Guide, AIVC, Brussels

McCormick, R. A. (1971) 'Air pollution in the locality of buildings', *Philosophical Transactions of the Royal Society of London A*, vol 269, pp515–526

Millet, J. R. (1997) 'Summer comfort in residential buildings without mechanical cooling', in *Proceedings of the Second International Conference on Buildings and the Environment*, Paris, 9–12 June, vol 1, pp307–315

Moser, A., Schalin, A., Off, F. and Yuan, X. (1995) 'Numerical modelling of heat transfer by radiation and convection in an atrium with thermal inertia', *ASHRAE Transactions*, vol 101, part 2, pp1136–1143

Nakamura, Y. and Oke, T. R. (1989) 'Wind, temperature and stability conditions in an E–W oriented urban canyon', *Atmospheric Environment*, vol 22, pp2691–2700

NatVent (1998) *Natural Ventilation for Offices*, BRE, Watford

Negrao, O. R. (1997) 'Integration of computational fluid dynamics with building thermal and mass flow simulation', *Energy and Buildings*, vol 27, pp155–165

Neilson, P. V. and Tryggvason, T. (1998) *Computational Fluid Dynamics and Building Energy Performances Simulation*, Aalborg University, Aalborg, Denmark. International Conference on Air Distribution in Rooms, Stockholm

Niachou, K., Hassid, S., Santamouris, M. and Livada, I. (2007a) 'Experimental performance investigation of natural, mechanical and hybrid ventilation in the urban environment', *Journal of Buildings and the Environment*, in press

Niachou, K., Livada, I. and Santamouris, M. (2007b) 'Experimental study of temperature and airflow distribution inside an urban street canyon during hot summer weather conditions. Part 2: Air flow conditions', *Journal of Buildings and the Environment*, in press

Nicholson, S. E. (1975) 'A pollution model for street-level air', *Atmospheric Environments*, vol 9, pp19–31

NiRiain, C. and Kolokotroni, M. (2000) 'The effectiveness of ventilation towers in enhancing natural ventilation in non-domestic buildings', in *Proceedings of the PLEA 2000 Conference*, Cambridge, UK, 2–5 July, pp77–82

Orme, M. (1999) *Applicable Models for Air Infiltration and Ventilation Calculations*, Technical Note 51, AIVC, Brussels

Paciuk, M. (1975) *Urban Wind Fields: An Experimental Study on the Effects of High Rise Buildings on Air Flow around them*, MSc thesis, Technion, Haifa, Israel

Palmer, J. (1999) 'Under pressure', *Building Services Journal*, August, pp24–29

Parker, J. and Teekaram, A. (2005) *Wind Driven Natural Ventilation Systems*, BG2/2005, BSRIA, Brussels

Pospisil, J., Jicha, M., Niachou, A. and Santamouris, M. (2005) 'Computational modelling of airflow in urban street canyon and comparison with measurements', *International Journal of Environmental Pollution*, vol 25, pp191–200

Reshyvent (2004) *Demand Controlled Hybrid Ventilation in Residential Buildings with Specific Emphasis on the Integration of Renewables*, http://www.reshyvent.com/index.htm

Roulet, C., van der Maas, A. and Flourentzos, F. (1996) 'A planning tool for passive cooling of buildings', in *Proceedings of the Seventh International Conference on Indoor Air Quality and Climate*, Nagoya, Japan, 21–26 July

Rufus, S. C. and Kolokotroni, M. (2000) 'The role of simplified ventilation modelling for the application of low energy design: A library case study', in *Proceedings of the 19th AIVC Conference*, The Hague, The Netherlands, 26–29 September

Santamouris, M. (2001) *Energy and Climate in the Urban Built Environment*, James and James, London

Santamouris, M. (2004) *Night Ventilation Strategies*, Ventilation information paper no 4, AIVC, Brussels

Santamouris, M. and Asimakoloulos, D. (eds) (1996) *Passive Cooling of Buildings*, James and James, London

Santamouris, M. and Georgakis, C. (2003) 'Energy and indoor climate in urban environments: Recent trends', *Journal of Building Services Engineering Research and Technology*, vol 24, pp69–81

Santamouris, M., Georgakis, C. and Niachou, A. (2006) 'On the use of data driven and fuzzy techniques to calculate the wind speed in urban canyons', in *Proceedings of the 2006 International Symposium on Evolving Fuzzy Systems*, Lancaster, 7–12 September 2006

Santamouris, M., Geros, V., Klitsikas, N. and Argiriou, A. (1996) 'Summer: A computer tool for passive cooling applications', in Santamouris, M. (ed) *Proceedings of the International Symposium: Passive Cooling of Buildings*, Athens, Greece, 19–20 June

Sirén, K., Riffat, S., Alfonso, C., Oliveira, A. and Kofoed, P. (1997) 'Solar-assisted natural ventilation with heat pipe heat recovery', in *Proceedings of the 18th AIVC Conference: Ventilation and Cooling*, Athens, Greece, 23–26 September, pp323–330

Skåret, E., Blom, P. and Brunsell, J. T. (1997) 'Energy recovery possibilities in natural ventilation of office buildings', in *Proceedings of the 18th AIVC Conference: Ventilation and Cooling*, Athens, Greece, 23–26 September, pp312–315

Somarathne, S., Seymour, M. and Kolokotroni, M. (2005) 'Dynamic thermal CFD simulation of a typical office by efficient transient solution methods', *Building and Environment*, vol 40, pp887–896

Takeya, N., Onishi, J., Koga, S. M., Izuno, M. and Kitagawa, K. (1998) *Computer Effort Saving Methods in Unsteady Calculations of Room Airflows and Thermal Environments*, Department of Environmental Engineering, Osaka University, International Conference on Air Distribution in Rooms, Stockholm, Sweden

Tindale, A. W., Irving, S. J., Concannon, P. J. and Kolokotroni, M. (1995) 'Simplified method for night cooling', in *Proceedings of the CIBSE, National Conference 1995*, Eastbourne, 1–3 October, vol 1, pp8–13

TIPVENT (2001) *Towards Improved Performances of Mechanical Ventilation Systems*, EU Programme JOULE IV, Brussels

University of California (1999) *Program Description of RESFEN 3.1*, Windows and Daylighting Group, Lawrence Berkeley National Laboratory, CA

Wind Towers (1996) *The Design of Wind-Driven Naturally Ventilated Buildings: The Calculation Method*, Battle McCarthy, London

Yamartino, R. J. and Wiegand, G. (1986) 'Development and evaluation of simple models for the flow, turbulence and pollution concentration fields within an urban street canyon', *Atmospheric Environment*, vol 20, pp2137–2156

Yannas, S., Erell, E. and Molina, J. L. (2005) *Roof Cooling Techniques: A Design Handbook*, Earthscan, London

Zimmermann, M. and Anderson, J. (1998) *Case Studies of Low Energy Cooling Techniques*, IEA Annex 28, Low Energy Cooling, IEA–ECBCSP, Watford

# WEBSITES

BEST FAÇADE (2005) www.bestfacade.com
BUBBLE www.unibas.ch/geo/mcr/Projects/BUBBLE
VentDiscourse (2006) http://dea.brunel.ac.uk/ventdiscourse

# 5

# Ground Cooling: Recent Progress

*Jens Pfafferott, with Simone Walker-Hertkorn
and Burkhard Sanner*

It has long been known that the ground changes its temperature more slowly than ambient air. The deeper one goes in the ground, the more the ambient temperature is attenuated, and at a certain depth the ground remains at an almost steady temperature level that is slightly higher than the yearly mean ambient air temperature. As a result, the ground can be advantageously used as a heat sink during summer. Its cooling potential can be utilized directly when the building envelope is in contact with the ground (semi-buried buildings in hot summers), through horizontal earth-to-air heat exchangers or water-driven heat exchangers.

## GROUND TEMPERATURE

Due to the high thermal inertia of the soil, temperature fluctuations at the ground surface are attenuated in the ground and the time lag between the surface and the ground temperature increases with depth. Therefore, at a sufficient depth, ground temperature is lower than the outside temperature in summer and is higher in winter.

Since ground temperature depends upon many boundary conditions, its modelling is rather difficult: Mihalakakou (2002) and Mihalakakou et al (1994, 1996a, 1997) use the energy balance equation and a neural network approach to predict ground surface temperatures. Starting from the calculated surface temperature, ground temperature can be estimated.

Short-term and daily temperature fluctuations at the surface even out within the first few centimetres of soil. As a result, the surface temperature is accurately described by a yearly oscillation $\Delta T_{sf}$ around the mean temperature $T_{sf}$, mean. The phase shift $t_{\varphi,sf}$ is caused by the time shift between the solar and the temperature oscillation within the seasonal cycle:

$$T_{sf}(t) = T_{sf,mean} + \Delta T_{sf} \cdot sin\left( 2\pi \cdot (t - t_{\varphi,sf}) / 8760 \right) \tag{1}$$

The mean temperature is a function of the mean ambient air temperature $T_{air,mean}$, the long-wave radiation and the solar albedo, $a_{sf}$ (with the mean solar radiation $I_{g,h,mean}$). The long-wave radiation is calculated with the emissivity of the surface $\varepsilon_{sf}$ and the mean sky radiation, which varies with climate, location and season. The global average for typical Central European climates is 63 watts per square metre (W/m²). The evaporative and convective cooling effect is taken into account by the $\xi$ factors:

$$T_{sf,mean} = \left( \xi' \cdot T_{air.mean} - \varepsilon_{sf} \cdot 63 + (1 - a_{sf}) \cdot I_{g,h,mean} - \xi'' \right) \xi''' \tag{2}$$

The amplitude of the surface temperature is calculated with the yearly temperature oscillation $\Delta T_{air}$ and solar radiation oscillation $\Delta I_{g,h}$, and the conductive heat transfer between the surface and the soil $\lambda/dz$:

$$\Delta T_{sf} = \left( \xi' \cdot \Delta T_{air} + (1 - a_{sf}) \cdot \Delta I_{g,h} - cos(2\pi \cdot t_{\varphi,sf}/8760) \right) (\xi''' + \lambda/dz) \tag{3}$$

The convective and evaporative cooling effects are not independent of each other. These effects are described by so-called $\xi$ factors. $\xi$ corresponds to the convective heat transfer $\alpha_{sf}$ and the mean air humidity $\varphi_{air,mean}$. The dimensionless factor $f_{sf}$ describes the water diffusion – saturated vapour tension with 103 Pascals per Kelvin (Pa/K) and 609Pa, and the parameter 0.0168K/Pa – at the surface:

$$\xi'[W / (m^2K)] = \alpha_{sf} \cdot (1 + 0.0168 f_{sf} \cdot 103 \cdot \varphi_{air,mean}) \tag{4}$$

$$\xi''[W / m^2] = \alpha_{sf} \cdot 0.0168 f_{sf} \cdot 609 \cdot (1 - \varphi_{air,mean}) \tag{5}$$

$$\xi'''[W / (m^2K)] = \alpha_{sf} \cdot (1 + 0.0168 f_{sf} \cdot 103) \tag{6}$$

Only by changing the surface properties can the surface and, hence, the soil temperature be significantly reduced or raised. The solar albedo ranges from 0.2 (white gravel) to 0.1 (shaded by grass) to 0.05 (black gravel), and the water diffusion factor ranges between 0.1 for sealed surfaces, 0.4 to 0.8 for dry soils to moist surfaces covered with vegetation, and 1 for saturated soils.

As a reference value, a grassy, dry surface is usually about 1K warmer than the ambient air (yearly mean temperature) due to the positive radiation balance at the surface, while the monthly maximum surface temperature is similar to the monthly maximum air temperature due to the compensating heat transfer between the soil and the surface.

The ground temperature profile is strongly influenced by the material parameters of the soil. Usually, determination of the heat conductivity, heat capacity and density is difficult. A detailed summary of practical models is given by Dibowski and Rittenhofer (2000). The thermal capacity $\rho c$ ranges between 1400kJ/(m³K) for dry sand, 1800 to 2400kJ/(m³K) for typical soil and 2800kJ/(m³K) for moist clay. The heat conductivity $\lambda$ ranges between 0.7W/(mK) for dry sand, 1.4 to 1.9W/(mK) for typical soil and 2.9W/(mK) for moist clay. According to these properties, the thermal diffusivity ranges from $6 \times 10^7$ to $10 \times 10^7$ square metres per second (m²/s). In addition to the ground cover and the material parameters of soil, the possible thermal influence of a building or groundwater should be taken into account (Pfafferott, 2000).

The undisturbed ground temperature $T_{\mathrm{ground}}(z,t)$ at the depth $z$ is calculated with the surface temperature, the phase shift $t_{\varphi,\mathrm{sf}}$, and the temperature conductivity of the ground as represented by Equation 7:

$$T_{\mathrm{ground}}(z,t) = T_{\mathrm{sf,mean}} + \Delta T_{\mathrm{sf}} \cdot sin\left(2\pi \cdot (t - t_{\varphi,\mathrm{sf}})/8760 - \chi\right) \cdot exp(-\chi)$$

[7]

with $\chi = z \cdot \sqrt{\dfrac{\pi}{a_{\mathrm{ground}} \cdot 8760}}$ and $a_{\mathrm{ground}} = \dfrac{\lambda}{\rho c}$.

## APPLICATIONS

Different ground-cooling technologies are available, and they can be applied to very different building concepts and in various climates. Since ground cooling utilizes environmental energy, an accurate analysis of the microclimatic conditions and the soil around the building is indispensable in order to estimate the ground-cooling potential and to decide which ground-cooling technique should be applied to the building.

### Semi-buried buildings

Semi-buried houses increase their thermal inertia by direct contact with the soil. The outdoor temperature fluctuation is evened out and the house remains cooler than the ambient air during the summer period, provided that the heat gains are small and the ventilation can be enhanced during the night and cool periods. These houses have been built during various historical periods. New projects are located mainly in hot, arid regions. A good introduction to the principles and design of such buildings is given by Argiriou (1996). As the design of semi-buried houses is strongly related to the site, it is difficult to give generally accepted advice. In any case, semi-buried houses need either to be constructed in accordance with local experience, or to be designed with sophisticated design tools and building simulation that take the heat transfer between the soil and the building accurately into account. Semi-buried houses are not discussed in this chapter.

## Horizontal earth-to-water heat exchangers

Horizontal earth-to-water heat exchangers are suitable for heat pump applications in winter. Many of these systems have been installed for residential buildings. However, an earth-to-water heat exchanger close to the soil surface cannot produce a water temperature in summer that could be used for cooling purposes, such as a water-to-air heat exchanger for a ventilation system or thermally activated building components can. Furthermore, these plain absorbers require a large area near the building for heat transfer. Hence, horizontal earth-to-water heat exchangers are not discussed in this chapter.

## Earth-to-air heat exchangers (EAHXs)

When ambient air is directly drawn through buried pipes, the air is cooled in summer and heated in winter. Earth-to-air heat exchangers (EAHXs) are run horizontally at a ground depth of 3m to 5m. Argiriou (1996) describes the underlying thermodynamic models, show some experimental data and provides a calculation method for the design of EAHXs. A cross-section analysis of monitoring data from EAHXs in operation is discussed in the section on 'Earth-to-air heat exchangers'.

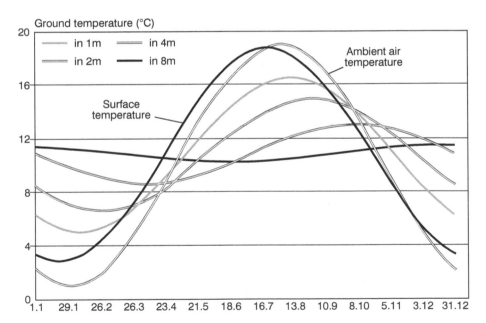

*Note:* For typical mid-European weather with $T_{air,mean} = 10°C$, $\Delta T_{air} = 9K$, $I_{g, h, mean} = 117W/m^2$, $\Delta I_{g, h} = 104W/m^2$ and $\varphi_{air, mean} = 80$ per cent, a clayey soil [$\rho c = 2400kJ/(m^3 K)$ and $\lambda = 2.3W/(m K)$] and a grassy, moist surface [$\varepsilon_{sf} = 90$ per cent, $a_{sf} = 10$ per cent, $\alpha_{sf} = 10W/(m^2 K)$ and $f_{sf} = 0.6$].

**Figure 5.1** *Ambient air, surface and ground temperatures*

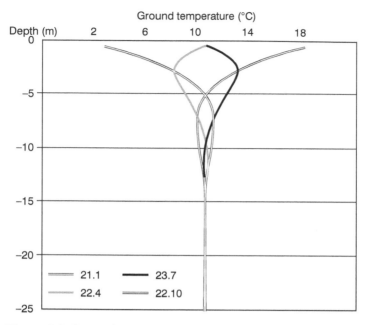

**Figure 5.2** *Ground temperature oscillation in winter (21 January), spring (22 April), summer (23 July) and autumn (22 October) for the boundary conditions given in Figure 5.1*

## Borehole heat exchangers (BHEs)

Borehole heat exchangers (BHEs) are water-driven cooling systems. They are run vertically at depths of up to 100m. The principle of BHEs and some monitoring results are discussed in the section on 'Borehole heat exchangers'.

## Geographic limits of applicability

As a basic principle, there are no geographic limits for the application of ground cooling since the soil works as a heat storage and attenuates the temperature variations of the ambient air temperature. The application of ground cooling, however, is practically restricted to sites where the soil temperature in summer is low and where energy potential can be easily utilized.

An EAHX requires an area around the building that should be unsealed and overgrown. Due to the enhanced heat transport in the soil, groundwater improves the energy transport to and within the EAHX. However, groundwater complicates the construction of EAHXs since water leakage into the EAHX must be avoided.

A BHE requires only a small or, according to its construction, even no area at the building site and functions independently from ground surface properties. Groundwater increases the cooling potential. Ground cooling can also be favourably used in cool, moderate and warm climate zones.

Examples also exist from hot arid climate zones. The ground-cooling potential is limited in cold climates (there is no need for cooling) or tropical and sub-tropical climates (the soil temperature is too high for cooling).

Since ground cooling provides a certain temperature drop, it provides cooling on a certain ambient temperature level. While the temperature drop only depends upon the design of the EAHX or the BHE, the room temperature changes with the ambient temperature. Furthermore, we should consider the different experience and expectations of users concerning indoor environment in different climate zones (see Chapter 1). Consequently, all European climate zones provide a high ground-cooling potential. How, then, can we utilize the ground-cooling potential for cooling purposes?

# EARTH-TO-AIR HEAT EXCHANGERS

In designing an earth-to-air heat exchanger, a decision on design goals has to be made. If the airflow is given by the ventilation system and the construction site is known, the question arises: is it more important to achieve a high specific energy performance based on the surface area of an earth-to-air heat exchanger (EAHX), a high adaptation of air temperature to ground temperature, or a minimal pressure loss?

The following data evaluation deals with the performance of three EAHXs in service. First, the temperature performance is described by plots over time and characteristic inlet–outlet air temperature diagrams. Second, the energy gain is illustrated by monthly graphs as a function of the ambient air temperature. Third, a parametric model is used to provide general efficiency criteria, such as the dynamic temperature behaviour and the energy efficiency.

EAHXs can (pre-)heat the supply air in winter and (pre-)cool it in summer. EAHXs are:

- simple;
- provide a high cooling and pre-heating potential;
- have low operational and maintenance costs; and
- reduce fossil fuel use and related emissions.

Corporations with high investment costs are opposed to these main advantages.

Pre-heated fresh air supports a heat recovery system and reduces the space heating demand in winter. In summer, in combination with a good thermal building design, the EAHX can eliminate the need for active mechanical cooling and air-conditioning units in buildings. EAHXs are therefore a passive cooling option in moderate climates.

The energy performance of EAHXs is described by the interaction of heat conduction in the soil and the heat transmission from the pipe to the air. Figure 5.3 depicts the main boundary conditions for the design and operation of EAHXs.

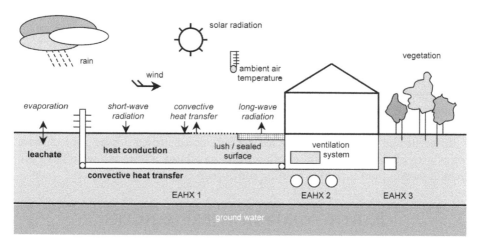

*Note:* The heat transfer between ground and air depends strongly upon the position of the earth-to-air heat exchanger (here: EAHX 1 – 3).

**Figure 5.3** *Boundary conditions for the design and operation of earth-to-air heat exchangers (EAHXs): Climate- and project-specific boundaries (normal), heat flow at the surface (italic) and in the ground (bold)*

## Temperature behaviour

Different parametric and numerical models for EAHXs have recently been published, and some are described below. The simulation models can be classified as models with an analytical or a numerical solution of the ground temperature field, or mixed models.

Albers (1991) developed a parametric analytical model for steady airflow based on a form factor to model the three-dimensional temperature profile. Sedlbauer (1992) published a model based on a heat capacity model in order to take changes in airflow during operation into account. Another approach based on a parametric model using deterministic techniques is given by Mihalakakou (2003).

Numerical simulation models calculate the thermal performance of EAHXs with algorithms describing the coupled and simultaneous heat and moisture transfer in the soil under a three-dimensional temperature gradient (see Pfafferott et al, 1998). This sophisticated model can be applied to any geometry; but its application is time consuming and error prone due to the complex input. Mihalakakou et al (1994) developed a complete numerical model for a single-pipe EAHX. Both models have been validated with long-term measurements and are used to describe the thermal influence of the key variables, such as pipe length, pipe diameter, air velocity and pipe depth (Mihalakakou et al, 1996b). A numerical model for a two-pipe EAHX is described by Bojic (1999). Another numerical model for multiple-pipe EAHXs was validated by Hollmuller and Lachal (2001).

Mixed simulation models are resistance–capacity models based on a numerical solution for the earth temperature near the pipe and an analytical calculation of boundary conditions. Evers and Henne (1999) used such a model to predict the energy performance of EAHXs for different design parameters. In the framework of a European Union (EU) project, a design tool was developed under the guidance of AEE Gleisdorf and Fraunhofer ISE by 15 engineering companies. The simulation model is based on an extended, validated and well-tested resistance–capacity model by Huber (2001).

Some of these calculation models have been implemented in design tools, such as the following:

- WKM, Huber Energietechnik (www.igjzh.com/huber);
- PHLuft, Passivhaus-Institut (www.passiv.de);
- SUMMER, University of Athens (www.cc.uoa.gr/grbes/summer.htm);
- GAEA, Universität Siegen (http://nesa1.uni-siegen.de);
- EWT-Sim, DLR Köln (www.ag-solar.de).

An evaluation of eight simulation models by Tzaferis et al (1992) reached the conclusion that almost all proposed algorithms can predict the outlet air temperature with sufficient accuracy. Furthermore, most commercial building simulation programs have models for EAHXs.

Small EAHXs for residential buildings are sold in complete packages by several suppliers of construction materials and do not need to be planned using simulation tools.

## Design and operation

Various publications deal with the design and operation of EAHXs (see Santamouris and Asimakopoulos, 1996; Zimmermann and Andersson, 1998; Henne, 1999; Pfafferott, 2000; Reise, 2001; and Dibowski, 2004).

EAHXs can be advantageously applied to ventilation systems if cooling is necessary and an area is available for the construction of the buried pipes. Although the EAHX can also be used as a supplementary heat source for pre-heating during the winter, conditions during the summer will determine whether the EAHX can replace investment in active cooling equipment.

Investment costs are mainly determined by the ease of moving earth. In general, EAHXs are cost effective if synergy effects can be realized (e.g. digging the excavation pit, installing a combined air supply system or abandoning air conditioning). Energy-saving costs in summer and winter on their own do not justify the investment. An EAHX reduces the room temperatures in summer and improves the indoor environment in passively cooled buildings without the operation of a chiller. As a result, the investment usually makes sense since the typical total costs (cost of capital, energy and maintenance costs) are around 50 Euros per annum for each worker in office buildings.

In general, EAHXs can be contaminated by fungi and bacteria during summer operation when the warm humid air condenses at the cool pipe surface. Thus, EAHXs must be accurately constructed with a decline of 2° in

**Table 5.1** *Description of evaluated earth-to-air heat exchangers (EAHXs)*

|  | DB Netz AG | Fraunhofer ISE | Lamparter |
|---|---|---|---|
| Number of ducts | 26 | 7 | 2 |
| Length of ducts | 67m–107m | 90m–100m | 90m |
| Diameter | 200mm and 300mm | 250mm | 350mm |
| Depth of ducts | 2m, 3m and 4m, around foundation slab | 4m–5m, partly below foundation slab | 2m–3m, around foundation slab |
| Material | Polyethylene (PE) | PE | PE |
| Mean airflow | 10,300m³/h | 7000m³/h | 1100m³/h |
| Total surface area of ducts | 1650m² | 522m² | 198m² |
| Specific surface area | 0.16m²/(m³/h) | 0.075m²/(m³/h) | 0.18m²/(m³/h) |
| Air speed | 2.2m/s (approx) | 5.6m/s | 1.6m/s |
| Pressure loss at mean airflow | 40Pa (measured) | 166Pa (measured) | 12Pa (calculated) |
| Start of operation | October 1999 | November 2001 | September 2000 |
| Soil type | Dry, rocky | Dry, gravel | Moist, clay |
| Ventilation system | Hybrid ventilation with supply and exhaust air | Hybrid ventilation with supply air | Hybrid ventilation with supply and exhaust air |
| Heat recovery system | Yes: 65% (design) | No | Yes: 80% (measured) |
| EAHX bypass | Yes, but not used | Yes | Yes |
| Control strategy | Time controlled | Temperature controlled (open loop) | Temperature controlled (closed loop) |

*Source:* Fraunhofer ISE, *SolarBau: Monitor*

order to prevent water accumulation in sinks. Filters clean the incoming air continuously in order to prevent fouling. As part of regular maintenance the pipes should be cleaned every two years using a wet cleaning technique. Monitoring results from several projects (Flückinger et al, 1997) have shown that the microbial count in the outlet air is not higher than in the inlet air, even without any cleaning and after several years of operation. Obviously, clean and periodically exchanged filters avoid ongoing fouling during operation.

## Heat and cooling-energy gain

The heat and cooling-energy gain by an EAHX should meet the actual energy demand of the building's ventilation system. If the heating- or cooling-energy gain is lower than the corresponding energy consumption, the supply air has to be heated or cooled. EAHXs in office buildings should meet the complete cooling-energy demand for sufficient thermal comfort in order to substitute a mechanical cooling device.

If the energy gain is higher than the corresponding energy consumption, the supply air has to be cooled (in winter) or heated (in summer). This unwanted energy dissipation could be avoided by an optimized control strategy.

If a heat recovery system with high efficiency is operated, only one third of the heat energy gains can be used to reduce the heating energy demand for ventilation in typical low-energy buildings. But in heat recovery systems with a high efficiency, there is the danger of freezing because the humid exhaust air is cooled below the freezing point at very low inlet-air temperatures. If an EAHX is operated in series with a high-efficiency heat recovery system, it should be large enough to prevent freezing at low ambient-air temperatures.

## Energy efficiency

Besides the heat gain, electricity demand should also be taken into account. De Paepe and Janssens (2002) developed a method to optimize the energy efficiency of an EAHX by reducing the pressure drop for a given thermal efficiency. According to this method, the dissipation energy caused by the pressure loss – and not the electricity demand – is used to calculate the coefficient of performance (COP): the ratio of thermal energy supplied to mechanical dissipation energy.

## Comparative analysis

In spite of many publications and available design tools, there is a lack of comparative analysis. In the following sub-sections, three EAHXs are evaluated according to their energy gain, temperature behaviour and energy efficiency using standardized performance criteria.

The thermodynamic model used for this evaluation reduces the parameters to several characteristic values in order to clearly show the relationship between operation-, climate- and project-specific boundaries.

The model picks up the main features from existing calculation tools and is based on an analytical solution of the undisturbed three-dimensional temperature field using a form factor. This steady-state model is combined with a heat-capacity model, which takes the operation time into account. This transient model is described by Pfafferott (2000).

### Description of evaluated earth-to-air heat exchangers

The EAHX at Fraunhofer ISE in Freiburg, Germany, is located to the left and right of the foundation slab of the building (see Figure 5.4a) and was designed for an airflow of 9000 cubic metres per hour (m³/h). The design idea was to achieve a high-energy performance to assist the ventilation system with cooled or pre-heated air. The ventilation system for the office building of DB Netz AG

*Note:* While the design criteria are different, the demand on operating time is similar in each building. The characteristic attribute is the surface area per airflow, which varies from 0.075 to 0.18m²/(m³/h). The key information is given in Table 5.1.
*Source:* (a) Fraunhofer ISE (b) University of Karlsruhe, (c) Archtektengemeinschaft Nürtingen

**Figure 5.4** *EAHXs under construction: (a) Fraunhofer ISE (Freiburg, Germany); (b) DB Netz AG (Hamm, Germany); and (c) Lamparter (Weilheim, Germany)*

is designed for an airflow of 12,000m³/h. For cooling in summer, a maximum outlet temperature of 19°C is required at an ambient air temperature of 36°C. The total surface area was designed to be quite large in order to cool the air almost to the undisturbed earth temperature, which reaches approximately 18°C in late summer. The large register had to be positioned in the small excavation at the construction site in Hamm, Germany (see Figure 5.4b). The EAHX for the Lamparter office in Weilheim, Germany, was designed for an airflow of 1200m³/h and for small pressure losses. It is installed around the office building (see Figure 5.4c). These projects are presented in detail in *SolarBau: Monitor* (2006).

**Analysis of temperature behaviour**
Temperature behaviour is described separately for every project according to plots over time and characteristic curves.

*Ground temperature* In the design process, the undisturbed ground temperature is a main input parameter. However, undisturbed ground temperature is more of a hypothetical value. It is influenced by the building, but not by the EAHX, and is defined for mean soil properties and, hence, cannot be measured directly during operation of an EAHX. Therefore, undisturbed ground temperature has to be derived from measurements through regression analysis. Figure 5.5 indicates that measured and calculated 'undisturbed' ground temperature differ strongly.

*Air temperatures: Time variation curves and characteristic lines* Figure 5.6 depicts the operation of an EAHX for two days. The temperature difference is –14K in summer and +12K in winter at an air volume flow of 10,000m$^3$/h to 11,000m$^3$/h. The outlet temperature is nearly constant during daily operation. The thermal fatigue is 1K per day in summer and winter.

Due to the buffered oscillation of the ground temperature, the outlet air temperature is higher than the inlet air temperature in winter and is lower in summer. Figure 5.7 (left) shows hourly mean air temperatures during operation for a whole year. These plots illustrate the working principle of EAHXs.

The thermal behaviour of an EAHX can be taken from its temperature characteristic. In Figure 5.7 (right), vertical lines divide the temperature field into three virtual zones for comparison: the heating period for low-energy office buildings with inlet temperatures below 12°C, the cooling period with inlet temperatures above 22°C and the passive period in between, without heating and cooling.

## Operation of earth-to-air heat exchangers
According to the building energy demand for heating or cooling, at a specific ambient temperature the inlet air should be either heated or cooled. However, occasionally there is an unwanted temperature decrease in winter or an increase in summer due to non-ideal control (see the monitoring data in Figure 5.7). In addition to the design and operation, the control strategy has a strong impact on the temperature performance of an EAHX.

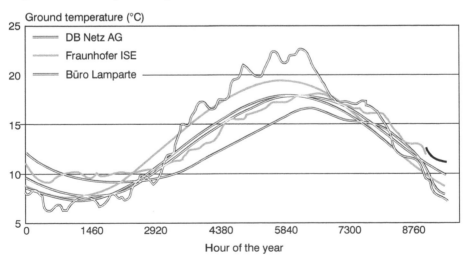

**Figure 5.5** *Ground temperature profile at the (average) depth of each EAHX (hourly data); the monitored data (dashed lines) do not correspond to the calculated undisturbed ground temperature (drawn lines)*

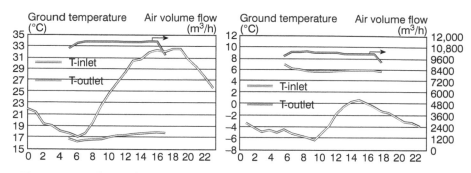

**Figure 5.6** *Inlet and outlet air temperature at Fraunhofer ISE for a summer (27 July 2001) and a winter day (17 January 2001)*

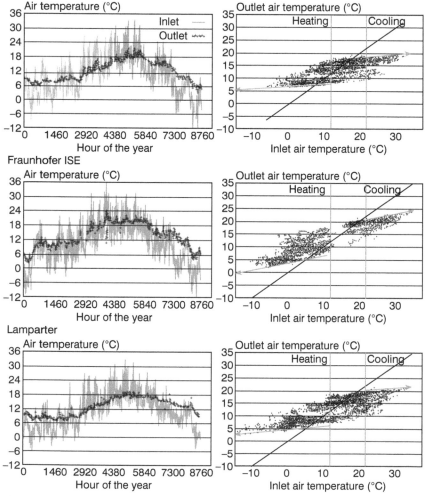

**Figure 5.7** *Ambient and outlet air temperatures at DB Netz AG, Fraunhofer ISE and Lamparter; the right-hand graphs also show the regression lines (grey arrows) for high and low ambient air temperatures*

The EAHX at DB Netz AG is operated during working days between 8 am and 5 pm. Since the EAHX operates without controls, there is a wide temperature range with unwanted (below 12°C and above 22°C) or unusable (between these temperature limits) heating and cooling.

Since the airflow at Fraunhofer ISE is controlled by the inlet air temperature, there is no operation between 12°C and 16°C during working hours. Due to the open-loop control, there is only a very small temperature overlap with unwanted heating or cooling.

The offices at Lamparter are ventilated with a constant air temperature supply of 22°C during working hours. Accordingly, the EAHX warms or cools the incoming air to 22°C. Since the closed-loop control is realized without a temperature hysteresis, the EAHX is operated at (almost) every ambient temperature.

Figure 5.8 illustrates two special aspects. The graph for Fraunhofer ISE shows the variation of outlet temperature with inlet temperature. Due to the high thermal inertia, the ground temperature lags behind the ambient air temperature. As a result, the ground is cool at the beginning of summer with the first warm days. But by the end of summer, the ground has warmed up and

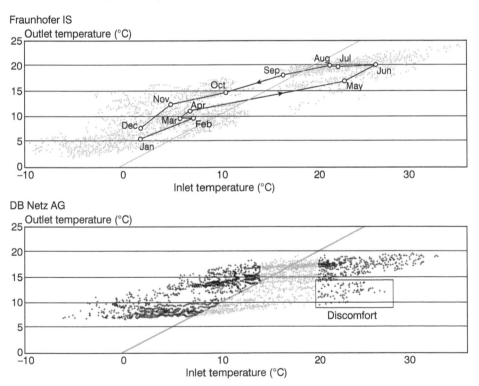

**Figure 5.8** *Inlet and outlet air temperatures at Fraunhofer ISE (with monthly mean temperature and time lag between ground and ambient air temperature) and at DB Netz AG (with optimized control; suboptimal operation in the grey field)*

the inlet air during the last summer days (e.g. with temperatures of 20°C) cannot be cooled as much as during the first summer days (e.g. with temperatures of 20°C). This is valid analogously for the winter. The graph for DB Netz AG indicates how an optimized control can reduce the operation time to a minimum. The limits for the inlet temperature are 12°C and 20°C, and the EAHX is not operated when the air is heated during the summer or cooled during the winter. Besides the reduced energy consumption for the fan, an advantage of closed-loop control over operation that is only time controlled is that the ground can regenerate thermally when the EAHX is not in use.

Taking the energy saving for the fan operation and the transient thermal behaviour into account, the mean COP can be enhanced by optimizing the control strategy without any additional investment costs. If the EAHX is operated only at ambient air temperatures below 15°C (heating) and above 18°C (cooling) and for temperature differences of at least 1K, the COP is improved by:

- 50 per cent at DB Netz AG (only time controlled → ideal control);
- 25 per cent at Fraunhofer ISE (open-loop control → ideal control); and
- 40 per cent at Lamparter (closed-loop control with small hysteresis → ideal control).

Ambient air drawn through buried pipes in summer is typically cooled during the daytime, but may be heated at night when ambient air cools down substantially. Since this may happens at the end of summer when the soil has been heated and the days are becoming cooler, a control algorithm should be applied to the earth-to-air heat exchanger, which, at least, avoids these suboptimal operating points.

## Energy gain and supply

The energy gain is illustrated by time variation curves and is sorted by the ambient temperature. However, the energy gain from an EAHX is not identical to its utilized energy gain. The interaction between the EAHX and the ventilation concept can be evaluated only by a coupled building simulation.

*Yearly energy gain* The energy performance $\dot{Q}_{air}$ of an EAHX is calculated from the airflow $\dot{V}_{air}$, the volumetric heat capacity $\rho c_{air}$ and the air temperature difference between the inlet and outlet:

$$\dot{Q}_{air} = \dot{V}_{air} \cdot \rho c_{air} \cdot (T_{out} - T_{in}) \qquad [8]$$

Taking the operation time into account, the yearly heat and cooling-energy gain can be calculated from Equation 8. The specific energy gain (in relation to surface area from Table 5.1) is compared for different EAHXs in Table 5.2. As expected, the specific energy gain increases with operation time and the available temperature difference between the earth and air, but with decreasing

**Table 5.2** *Heating- and cooling-energy gain*

|  | DB Netz AG | Fraunhofer ISE | Lamparter |
|---|---|---|---|
| Period of measurement | 1 January 2001 to 31 December 2001 | 7 November 2001 to 6 November 2002 | 1 January 2001 to 31 December 2001 |
| Hours of operation | 3701 hours | 4096 hours | 3578 hours |
| Heating-energy gain | 27,700kWh/yr | 26,800kWh/yr | 3,200kWh/yr |
| Specific (pipe surface) | $16.8kWh/(m^2_{EAHX}yr)$ | $51.3kWh/(m^2_{EAHX}yr)$ | $16.2kWh/(m^2_{EAHX}yr)$ |
| Cooling-energy gain | 22,300kWh/yr | 12,400kWh/a | 2,400kWh/yr |
| Specific (pipe surface) | $13.5kWh/(m^2_{EAHX}yr)$ | $23.8kWh/(m^2_{EAHX}yr)$ | $12.1kWh/(m^2_{EAHX}yr)$ |

specific surface area. Thus, the EAHX at Fraunhofer ISE has the highest specific energy gain. Since the energy gain is dependent upon climate and the EAHX's operation time, the specific energy gain is suitable to evaluate the energy-saving potential of an EAHX, but is not suitable as a general efficiency criterion.

*Monthly energy gain* Depending upon the operation time, ground temperature and inlet-air temperature, there is a heating-energy gain in winter and a cooling-energy gain in summer, which is illustrated for Fraunhofer ISE in Figure 5.9. Compared to typical weather conditions, June 2001 was warm and November 2001 was cool. Thus, there are comparatively high energy gains

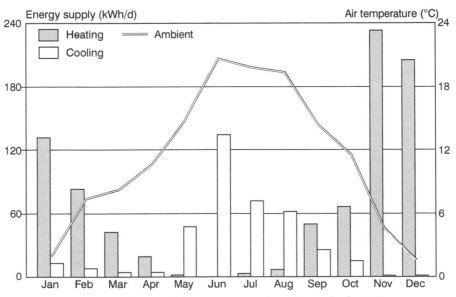

**Figure 5.9** *Monthly energy supply at Fraunhofer ISE (November 2001–October 2002)*

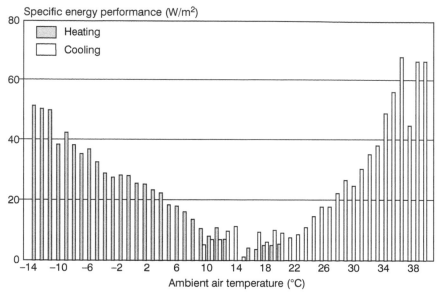

**Figure 5.10** *Heating- and cooling-energy performance at Fraunhofer ISE (November 2001–October 2002)*

during these months because of the high temperature difference between the inlet air and ground, as the ground is – contrary to the air – cool in June and warm in November.

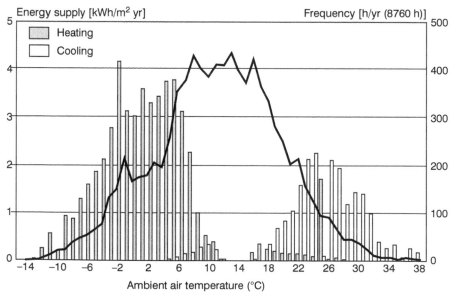

**Figure 5.11** *Specific energy supply sorted by ambient air temperature for Fraunhofer ISE (November 2001–October 2002)*

*Energy supply sorted by ambient air temperature* Each EAHX is integrated within a ventilation system. By approximation, there is either a heating- or a cooling-energy demand at a given ambient air temperature. Depending upon the ambient temperature, the supply air should be either heated or cooled. If the energy performance of an EAHX is sorted by the ambient temperature, the thermal performance at a specific temperature can be calculated (see Figure 5.10).

The EAHX at Fraunhofer ISE is time and temperature controlled (open-loop control). A total of 96 per cent of heating energy is supplied below 12°C and 92 per cent of cooling energy above 18°C (see Figure 5.11).

## Comparison of efficiency

A parametric model is used to provide general efficiency criteria. Time variation curves for both inlet and outlet air temperatures and ground temperature, $T(t)$, can be analysed with a regression function using a mean temperature, $T_{mean}$, a temperature amplitude, $\Delta T$, and a phase shift, $t_\varphi$:

$$T(t) = T_{mean} - \Delta T \cdot sin(2\pi \cdot (t + t_\varphi) / 8760) \qquad [9]$$

$T$ can be replaced by the inlet air, $T_{in}$, the outlet air, $T_{out}$, or the undisturbed ground temperature, $T_{ground}$.

The outlet air temperature, $T_{out}$, can be calculated from the ground temperature, $T_{ground}$, the inlet air temperature, $T_{in}$, and the dimensionless $NTU$ (number of transfer units). The temperatures are time dependent and $NTU$ is a function of the operation time $t_{op}$:

$$T_{out}(t) = T_{ground}(t) + \left[ T_{in}(t) - T_{ground}(t) \right] \cdot exp \left[ - NTU(t_{op}) \right] \qquad [10]$$

$NTU$ is the quotient of the convective heat transfer at the surface $A_{EAHX}$ (with the pipe diameter $d_{EAHX}$ and the length $l_{EAHX}$) and the heat transport.

In this model, $NTU$ takes into account both the thermal capacity of the ground and the heat transfer from the ground to the air. During operation, the ground near the pipe is discharged thermally. Starting from the undisturbed ground temperature, the effective temperature difference between the earth and the air decreases with operation time. Thus, the heat transfer coefficient $h_{EAHX}$ and, consequently, $NTU$, decreases with operation time. However, a constant mean $NTU$ is used for data evaluation:

$$NTU(t_{op}) = \left[ h_{EAHX}(t_{op}) \cdot A_{EAHX} \right] / (\dot{V}_{air} \cdot cp_{air}) = const \qquad [11]$$

with $A_{EAHX} = \pi d_{EAHX} \cdot l_{EAHX}$

These equations summarize the complex thermal interrelations from Figure 5.3. Using the least-squares method, these parameters can be calculated using a non-linear regression analysis. The results are summarized in Table 5.3.

**Table 5.3** *Regression analysis for ground, inlet and outlet air temperatures and number of transfer unit (NTU) sets from parameter identification for one year*

|  | DB Netz AG | Fraunhofer ISE | Lamparter |
|---|---|---|---|
| **Ground** |  |  |  |
| Ground temperature ($T_{ground}$) | 12.8°C | 13.8°C | 11.9°C |
| Amplitude ($\Delta T_{ground}$) | 5.1K | 5.9K | 5.3K |
| Phase shift ($t_{\varphi, ground}$) | 55 days | 21 days | 50 days |
| **Inlet** |  |  |  |
| Inlet air temperature ($T_{in}$) | 11.78°C | 12.94°C | 11.35°C |
| Amplitude ($\Delta T_{in}$) | 8K | 11.21K | 6.05K |
| Phase shift ($t_{\varphi, in}$) | 26 days | 2 days | 10 days |
| **Outlet** |  |  |  |
| Outlet air temperature ($T_{out}$) | 12.80°C | 13.81°C | 11.90°C |
| Amplitude ($\Delta T_{out}$) | 5.27K | 6.77K | 5.30K |
| Phase shift ($t_{\varphi, out}$) | 52 days | 21 days | 38 days |

The profiles of undisturbed ground temperatures are similar. However, due to the construction under the foundation slab, the undisturbed ground temperature at Fraunhofer ISE is higher and shows a smaller phase shift. Since the ground is slightly warmer than the ambient air, the mean outlet temperature is also higher than the mean inlet temperature.

Due to the thermal inertia of the soil, the temperature fluctuations are strongly buffered. While the yearly inlet air temperature oscillates between 6.1K and 11.2K, the yearly outlet temperature oscillates only between 5.3K and 6.8K.

The temperature ratio $\Theta$ is a design criterion (see Mihalakakou et al, 1995) and can be calculated using the results from the regression analysis:

$$\Theta = \frac{T_{in} - T_{out}}{T_{in} - T_{ground}} \qquad [12]$$

The energy gain of an EAHX is associated with an energy demand for the fan. The energy efficiency of an EAHX is the ratio between its energy gain and the electricity demand for the fan to draw the air through the EAHX. Since the mechanical dissipation energy – and not the fan – is the characteristic parameter of an EAHX, the COP is calculated with the overall energy gain (kilowatt hours, thermal energy; $kWh_{th}$) supplied by the EAHX and the mechanical dissipation energy (kilowatt hours, mechanical energy; $kWh_{mech}$) during operation time:

$$COP = \frac{\sum\limits_{t_{operation}} (Q_{heat} + Q_{cool})}{\sum\limits_{t_{operation}} (\Delta p \cdot V)} \qquad [13]$$

The pressure difference in an EAHX is measured without filters, since filters are already used in the ventilation system. The pressure loss is only 12Pa at Lamparter due to the large pipe diameter and low air speed. Due to the slightly higher air speed and more blends, the pressure loss at DB Netz AG is 40Pa. In contrast, the pressure loss of 166Pa at Fraunhofer ISE is much higher since the EAHX had to be integrated within the air inlet, resulting in more flow resistances. The pressure loss results in a mechanical dissipation power at mean airflow rate of 114W at DB Netz AG, 322W at Fraunhofer ISE and only 20W at Lamparter.

Of course, the usable heat and cooling-energy supply is the most important result from the data. However, the energy supply depends upon both the hours of operation and the climate. In order to characterize the thermal efficiency of an EAHX, its main characteristics should be independent of these changeable parameters. In addition to the energy gain in Table 5.2, Table 5.4 summarizes the four main characteristics of EAHXs, which have been calculated from a regression analysis starting with the results from Table 5.3.

Since the specific surface area at DB Netz AG is higher (small mass flow rate), its *NTU* is higher than at Fraunhofer ISE. The *NTU* at Lamparter is smaller than that at DB Netz AG because of its comparatively large pipe diameter (small convective heat transfer coefficient).

The overall heat transfer coefficient, *h*, at Lamparter is smaller than at DB Netz AG and Fraunhofer ISE, although a better heat transfer was expected at Lamparter than at DB Netz AG or Fraunhofer ISE due to the higher heat conductivity of the moist earth (see Table 5.1). Obviously, the convective heat transfer between the piping and the air due to the higher air velocities at

Table 5.4 *Main characteristics for earth-to-air heat exchangers*

|  | DB Netz AG | Fraunhofer ISE | Lamparter |
|---|---|---|---|
| Mean number of transfer units (*NTU*) | 2.59 | 1.57 | 1.74 |
| Heat transfer ($h_{mean}$) | 5.5W/(m²K) | 5.0W/(m²K) | 3.2W/(m²K) |
| Efficiency (Θ) | 0.944$K_{real}$/$K_{ideal}$ | 0.766$K_{real}$/$K_{ideal}$ | 0.804$K_{real}$/$K_{ideal}$ |
| Coefficient of performance (COP) | 88kWh$_{th}$/kWh$_{mech}$ | 29kWh$_{th}$/kWh$_{mech}$ | 380kWh$_{th}$/kWh$_{mech}$ |
| COP (primary energy) | 20.5kWh$_{th}$/kWh$_{prim}$ | 6.7kWh$_{th}$/kWh$_{prim}$ | 89kWh$_{th}$/kWh$_{prim}$ |

Fraunhofer ISE and DB Netz AG dominates the overall heat transfer from the ground to the air.

The temperature ratio $\Theta$ at Fraunhofer ISE is smaller than at Lamparter and DB Netz AG because the specific surface area is smaller; as a result, the total heat transfer is lower. $\Theta$ at Lamparter is smaller than at DB Netz AG, though the specific surface area is similar: since the heat transfer is worse (small $h_{EAHX}$), the temperature ratio is smaller in spite of the similar specific surface area.

Although there are large differences in COP, it should be mentioned that each COP is high enough to save both end and primary energy. The energy saving can be calculated from the mechanical dissipation power and the thermal energy gains. Taking a typical fan efficiency of around 70 per cent into account, the end energy saving is 70 per cent lower than the COP. Starting from this end energy saving and taking a primary energy conversion factor for electric energy into account (i.e. $3\mathrm{kWh}_{prim.energy}/\mathrm{kWh}_{end\ energy}$), the primary energy saving is one third that of the end energy saving.

## Long-term monitoring and short-term measurements

In each EAHX, the air temperatures were measured during several weeks at distances of 10m (DB Netz AG and Fraunhofer ISE) and 9m (Lamparter). These short-term measurements can be used to validate the results from the parameter model. Starting from Equations 10 and 11, the local air temperature at $x_{EAHX}$ can be calculated:

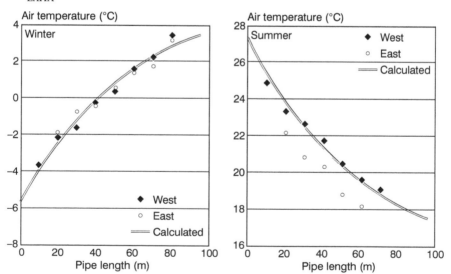

**Figure 5.12** *Local temperature variation at Fraunhofer ISE from short-term measurements in the west and east segments during operation hours on 4 January and 16 May 2002; the calculated temperature variation is valid for the mean value from the two segments*

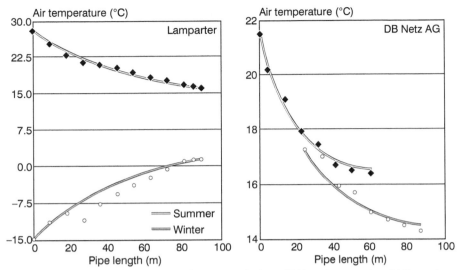

*Note:* The Lamparter EAHX is analysed for data on 27 June 2001, at 1 pm, and on 24 December 2001, at 8 am. The two segments of DB Netz AG EAHX are analysed for data on 23 July 2001, at 2 pm. Since the airflow is higher through the short south segment (smaller NTU), the local temperature gradient is smaller than in the long north segment.

**Figure 5.13** *Local temperature variation at Lamparter and DB Netz AG from short-term measurements*

$$T_{air}(x_{EAHX}) = T_{ground} + (T_{in} - T_{ground}) \cdot exp\left[ - NTU \cdot \frac{x_{EAHX}}{l_{EAHX}} \right] \qquad [14]$$

Figure 5.12 shows the results from the EAHX at the Fraunhofer ISE building for a mean winter and summer day. Figure 5.13 shows results from a comparison at a particular time for the EAHX at the DB Netz AG and the Lamparter buildings.

The measured temperature variation (short-term measurement) is predicted accurately by the calculation model (long-term monitoring: *NTU* from a yearly data analysis) under different conditions (summer/winter, geometric data, airflow or mean/current temperature).

## Summary

The thermal performance (temperature behaviour and energy efficiency) of earth-to-air heat exchangers has been calculated using four different approaches:

1  specific energy supply: kWh/(m²yr) and W/m²;
2  heat transfer: *NTU* and $h_{mean}$;
3  temperature behaviour: $\Theta$; and
4  energy efficiency: COP.

But which EAHX is the best concerning thermal performance? Due to different design criteria and certain demands on the ventilation systems, an EAHX should be evaluated for each project according to project-specific criteria. Each of the evaluated EAHXs at DB Netz AG, Fraunhofer ISE and Lamparter is the best from a specific point of view:

- The outlet air temperature of the EAHX at DB Netz AG is close to the undisturbed Earth temperature.
- The EAHX at Fraunhofer ISE supplies the highest specific energy gain based on the total surface area.
- Taking the lower heat transfer into account, the EAHX at Lamparter has the highest COP.

The method for evaluating the data derived from EAHXs in operation provides noteworthy considerations for the design and operation of EAHXs:

- The influence of earth parameters (soil and surface) and of the building on the earth temperature is as important as the pipe diameter on the thermal efficiency.
- In spite of a high heat flow density, the influence of the density of a pipe register on the energy gain is small. Obviously, thermal regeneration is sufficient to prevent a reduced energy performance.
- Pipe lengths of up to 100m and pipe diameters of around 250mm are profitable. If the EAHX aims at achieving a high specific energy performance, a small specific surface area should be provided using fewer pipes. If the EAHX aims at achieving a high temperature ratio, a high specific surface area should be produced using more pipes. However, the construction site itself often determines the dimensions of an EAHX.
- In operation, the control strategy plays a decisive role in the usable energy supply provided by the EAHX. Temperature control is important in order to prevent unwanted heating in summer and cooling in winter. Of course, more efficient utilization of energy supply is achieved by a closed-loop control; but its programming is difficult because of long dead times in EAHXs. An open-loop control runs robustly; but usually its programming should be adjusted after the first year of operation, when the temperature behaviour is known. Although the operation of each evaluated EAHX could be improved, each EAHX supplies much more heating and cooling energy than the primary energy that it uses for fans.
- In their role of assisting other passive cooling techniques, EAHXs replace active mechanical cooling systems during the summer.

## BOREHOLE HEAT EXCHANGERS

The nearly constant temperature of deep ground (depths of up to 100m) can be favourably utilized by borehole heat exchangers (BHEs). During the

*Source:* (a) VIKA Ingenieure, (b) Druckerei Engelhardt & Bauer, (c) solares bauen GmbH

**Figure 5.14** *On the construction site: (a) drilling machine, (b) borehole heat exchangers and (c) thermally activated ceiling*

summer, cool water can be used directly as a heat sink for cooling with thermally activated building systems (TABS). During winter, warm water can be used as a heat source for heat pumps.

The knowledge of underground thermal properties is crucial in order to design BHEs. In small plants (residential houses), these parameters are usually estimated. However, for larger plants, thermal conductivity should be measured. A useful tool for this is a thermal response test (TRT), carried out on a BHE in a pilot borehole (later to be part of the borehole field). The thermodynamic modelling of BHEs is explained by means of the TRT method in the following section.

## Basic principles and thermal response test

In order to conduct a thermal response test, a defined heat load is put into the borehole and the resulting temperature changes of the circulating fluid are measured. Since the late 1990s, this technology has become more and more popular, and today is used routinely in many countries for the design of larger plants with BHEs, allowing sizing of the boreholes based upon reliable underground data.

The TRT is a suitable method to determine effective underground thermal conductivity, as well as the borehole thermal resistance or the thermal conductivity of the borehole filling. A temperature curve is obtained, which can be evaluated with different methods. The resulting thermal conductivity represents a value for the total heat transport underground, noted as thermal conductivity (see Gehlin, 1998, 2002; Gehlin and Hellström, 2000). Other effects, such as convective heat transport (e.g. in permeable layers with groundwater; see Hellström, 1994) or further disturbances, are automatically included; therefore, it may be more correct to speak of an 'effective' thermal conductivity, $\lambda_{eff}$. The test equipment can easily be transported to the site.

## The thermodynamic model and application of tests

The theoretical basis for the TRT has been determined over several decades (see Choudary, 1976; Mogensen, 1983; Claesson et al, 1985; Claesson and Eskilson, 1988; Hellström, 1991; van Gelder et al, 1999; Cruickshanks et al, 2000; Sanner et al, 2000; Spitler et al, 2000; Mands and Sanner, 2001a, 2001b; Eugster et al, 2002). During the 1990s, the first practical applications were made (e.g. for the investigation of borehole heat storage in Linköping, Sweden; see Hellström, 1997). In 1995, mobile test equipment were developed at Luleå Technical University, Sweden, in order to measure the ground thermal properties for BHEs between depths of 10m to over 100m (see Eklöf and Gehlin, 1996; and Gehlin and Nordell, 1997). A similar development has been occurring independently since 1996 at Oklahoma State University, US (Austin, 1998). The first thermal response tests in Germany were performed in the summer of 1999 (Sanner et al, 1999).

**Figure 5.15** *The mobile TRT-test rig of UbeG GbR on site in Langen, Germany*

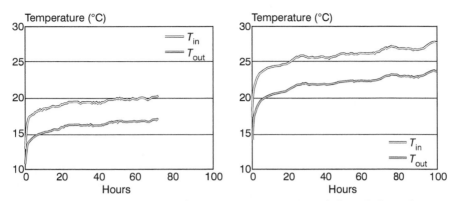

**Figure 5.16** *Measured temperature curves with low (left) and strong (right) climatic influence*

In order to achieve good results, it is crucial to set up the system correctly and to minimize external influences (see Figure 5.15). This is achieved more easily by heating the ground (with electric resistance heaters) than by cooling (with heat pumps). However, even with resistance heating, the fluctuations of voltage in the grid may result in fluctuations of the thermal power injected into the ground. Climatic influences are another source of deviation, primarily affecting the connecting pipes between the test rig and the BHE, the interior temperatures of the test rig, and occasionally the upper part of the BHE in the ground. Heavy insulation is required to protect the connecting pipes; sometimes even air conditioning for the test rig is required, as was the case at Oklahoma State University. With open or poorly grouted BHEs, rainwater intrusion may also cause temperature changes. A longer test duration allows for statistical correction of power fluctuations and climatic influence, and results in a more trustworthy evaluation. Typical test curves with strong and low climatic influence are shown in Figure 5.16.

### TRT duration and its evaluation with simplified models

With the increasing commercial use of TRTs, the desire for a shorter test duration has become apparent, particularly in the US. A recommendation for a minimum of 50 hours was given by Skouby (1998) and Spitler et al (1999a), which is compatible with the International Energy Agency (IEA) recommendations (IEA, 2006); but there is also scepticism (Smith, 1999). A test time of approximately 12 hours is desired, as this would prevent having the test rig on site overnight. In general, there are physical limits for shortening the measuring period because a somewhat stable heat flow has to be achieved in the ground. During the first few hours, temperature development is primarily controlled by the borehole filling and not by the surrounding soil or rock. A time of 48 hours is considered the minimum test period. In the evaluations made of German tests, the minimum duration criterion, $t_b$, as noted by Eklöf and Gehlin (1996), proved helpful:

$$t_b = \frac{5R^2}{\alpha} \qquad\qquad [15]$$

with:

- $R$ = borehole radius (m);
- $\alpha$ = thermal diffusivity ($\alpha = \lambda/(\rho c)$) with estimated values (m²/s).

However, visual cross-checking is recommended because the measured data may deviate from the theoretical assumptions. It is also worthwhile recalculating the minimum duration criterion with the thermal conductivity resulting from the first evaluation.

The easiest way to evaluate TRT data is to make use of line source theory. This theory was used as early as the 1940s to calculate temperature development in the ground over time for ground-source heat-pump plants (Ingersoll and Plass, 1948). An approximation is possible with the following formula, given in Eklöf and Gehlin (1996):

$$k = \frac{Q}{4\pi H\lambda_{eff}} \qquad\qquad [16]$$

with:

- $k$ = inclination of the curve of temperature versus logarithmic time (Ks);
- $Q$ = heat injection/extraction (J);
- $H$ = length of borehole heat exchanger (m);
- $\lambda_{eff}$ = effective thermal conductivity (W/(mK)); $\lambda_{eff}$ can be directly calculated from Equation 16.

The first test in Germany was performed for a large office building in Langen during the summer of 1999 (Seidinger et al, 2000). This example explains the above mentioned procedure.

Figure 5.17 depicts the regression curve of the mean fluid temperature from 6.9 to 50 hours, on a logarithmic timescale. The slope of the curve after 7 hours is 1.41, and using Equation 10 and the values given in Table 5.5, the thermal conductivity can be calculated thus:

Table 5.5 *Parameters of the first thermal response test in Langen, Germany (1999)*

| | |
|---|---|
| Test duration | 50.2 hours |
| Ground temperature | 12.2°C |
| Injected heat | 4.90kW |
| Depth of borehole heat exchanger (BHE) | 99m |
| Borehole diameter | 150mm |

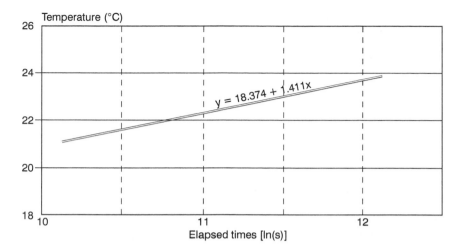

**Figure 5.17** *Regression curve of mean fluid temperature in the thermal response test in Langen, Germany (1999)*

$$\lambda_{eff} = \frac{4900}{4\pi 99 \cdot 1.411} = 2.79 \qquad [17]$$

A second value that can be determined by a response test is the borehole thermal resistance. For the office building in Langen, it was calculated as $r_b = 0.11 K/(W/m)$. This value gives the temperature drop between the natural ground and the fluid in the pipes. It is also possible to calculate $r_b$ from the dimensions and materials used (e.g. with the program Earth Energy Designer (EED); see www.buildingphysics.com). The result in the Langen case is $r_b = 0.115 K/(W/m)$ and matches the measured value well.

## More sophisticated models for TRT data evaluation

A more complicated method to evaluate a thermal response test is parameter estimation using numerical modelling, as was conducted, for instance, at a duct store in Linköping, Sweden (Hellström, 1997). Further work on parameter estimation was done at Oklahoma State University (Bose et al, 2002) and Oak Ridge National Laboratory (Shonder and Beck, 1999), among other locations.

Spitler et al (1999b) found a deviation of ±5 per cent in thermal conductivity between different methods of evaluation if data over 50 hours were used, but ±15 per cent when using only the first 20 hours of data. A comparison of four different evaluation methods is reported in Gehlin and Hellström (2003), and, in summary, the inclusion of data with less than 30 hours of continuous evaluation is not recommended. Busso et al (2003) compared three evaluation methods with data from a nine-day test in Chile and conclude that:

*Application of the classical slope determination and/or two-variable parameter fitting can be used as a fast and reliable tool for data evaluation. Accuracy of the evaluation depends on the care taken when performing the test.*

As a result, more advanced evaluation methods (parameter estimations through numerical simulation) can enhance accuracy and give additional information, but can reduce test time only slightly.

The TRT, meanwhile, is used routinely for the commercial design of BHE systems. The exact knowledge of ground thermal properties enables necessary safety margins to be reduced when estimating parameters; thus, the TRT is economical for systems comprising approximately ten BHEs or more.

## TRT and groundwater flow

A limitation to TRT is the amount of groundwater flow. Because the thermal conductivity obtained includes convection effects, with high groundwater flow the directly measured thermal conductivity becomes masked, and the values cannot be used for the design of BHE plants. The groundwater flow is not characterized by the velocity (in metres per second), which describes the time during which a water particle travels from one point to another. We consider the Darcy velocity ($(m^3/s)/m^2$), which describes the amount of water flowing through a given cross-section in a certain time. The Darcy velocity thus depends upon the porosity and the velocity of groundwater flow.

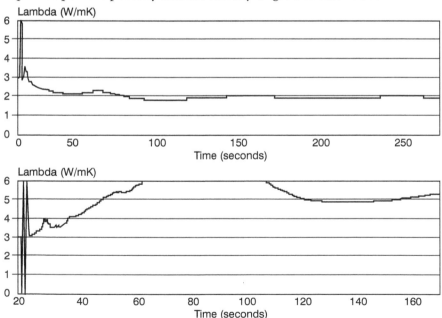

**Figure 5.18** *Step-wise evaluation showing perfect convergence (above) and test with high groundwater flow (below) and unreasonably high thermal conductivity value*

A useful method for checking excessive groundwater flow in the standard line-source evaluation is step-wise evaluation, with a common starting point and increasing length of data series. The resulting thermal conductivity for each time span can be calculated and plotted over time. Usually, in the first part of such a curve, the thermal conductivity swings up and down, converging to a steady value and a horizontal line in the case of a perfect test. If this curve continues to rise, a high groundwater flow exists and the test results may be useless (see Figure 5.18). This method also shows whether other external factors (e.g. weather and unstable power for heating) are disturbing the measurement.

## Repeatability of TRT

Results from the TRT can be reproduced, and different rigs on the same site yielded similar results. On a site in Mainz, Germany, two tests were made in the same underground conditions. Table 5.6 shows a very close match between ground thermal conductivity.

**Table 5.6** *Results of two tests on the same site in summer 2003*

|         | *Thermal conductivity* | *Borehole thermal resistance* |
|---------|------------------------|-------------------------------|
| Mainz 1 | 1.43W/(mK)             | 0.16K/(W/m)                   |
| Mainz 2 | 1.41W/(mK)             | 0.20K/(W/m)                   |

## The impact of grout on borehole thermal resistance

A comparison of three different TRT rigs was conducted in October 2000 at the site for a new borehole storage system in Mol, Belgium. Three BHEs with different grouting were available for the test:

1  single-U, grouted with sand produced while drilling;
2  single-U, grouted with specially graded sand; and
3  single-U, standard bentonite/cement grout.

**Table 5.7** *Results of the thermal response test comparison in Mol, Belgium (October 2000)*

|                | *Rig 1*            | *Rig 2*            | *Rig 3*            |
|----------------|--------------------|--------------------|--------------------|
| Produced sand  | $\lambda = 2.47$   | –                  | $\lambda = 2.47$   |
|                | $r_b = 0.06$       | –                  | $r_b = 0.05$       |
| Graded sand    | $\lambda = 2.40$   | –                  | $\lambda = 2.51$   |
|                | $r_b = 0.1$        | –                  | $r_b = $ n/a       |
| Bentonite      | –                  | $\lambda = 2.49$   | –                  |
|                | –                  | $r_b = 0.13$       | –                  |

*Notes:* Values for $\lambda$ are in W/(mK) and for $r_b$ are in K/(W/m); n/a = not available.

**Table 5.8** *Influence of thermal conductivity on borehole thermal resistance, calculated with Earth Energy Designer*

| Type of borehole heat exchanger (BHE) | λ grout | $r_b$ |
|---|---|---|
| Single-U, polyethylene (PE) | 0.8W/(mK) | 0.196K/(W/m) |
| | 1.6W/(mK) | 0.112K/(W/m) |
| Double-U, PE | 0.8W/(mK) | 0.134K/(W/m) |
| | 1.6W/(mK) | 0.075K/(W/m) |

*Note:* borehole diameter: 150mm; pipe size: 32mm; shank spacing: 70mm.

All tests resulted in a thermal conductivity of the ground of between 2.40W/(mK) and 2.51W/(mK), while the borehole thermal resistance differed between 0.05K/(W/m) and 0.13K/(W/m), according to the various backfill materials (see Table 5.7). In the saturated underground site in Mol, simple sand had the lowest thermal resistance, while the standard bentonite grout did not perform well.

A parameter where engineering can help to increase the efficiency of a BHE is borehole thermal resistance. By increasing the thermal conductivity of the borehole filling (grout), borehole thermal resistance is decreased. Table 5.8 shows the theoretical improvement for some examples.

The TRT enables theoretical assumptions to be checked in practice. In Figure 5.19, the borehole thermal resistance obtained in a number of TRTs performed by just one company in Germany is plotted against borehole diameter. As expected, borehole thermal resistance increases with increasing borehole diameter. However, two fields of data can be identified: for standard and for thermally enhanced grout. The TRT, hence, verifies the positive effect of decreased borehole thermal resistance.

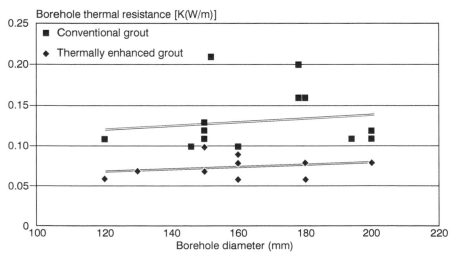

**Figure 5.19** *Borehole thermal resistance versus borehole diameter for a number of thermal response test (TRT) results*

## Conclusions and further development

The TRT has developed into a routine tool for investigating ground thermal parameters when designing BHE plants. The concept has proven reliable. A prerequisite, therefore, is high accuracy in temperature sensing, a diligent test set-up and operation, and a sufficiently long test time. The standard line source-based evaluation method is, in most cases, sufficient and can be enhanced by step-wise evaluation. Parameter estimation with numerical modelling can yield additional accuracy and information.

Further development of the TRT points in two directions:

1  'quick and dirty' tests with reduced accuracy for routine checking in quality control during the construction of BHE fields, or for the design of small systems in residential houses; and
2  more sophisticated tests with additional information (e.g. vertical thermal conductivity distribution along the BHE; see Heidinger et al, 2004; and Rohner et al, 2004).

Guidelines for TRTs are required to prevent inadequate testing and to ensure the necessary accuracy for a given task. A draft guideline was published as an appendix to Eugster and Laloui (2001).

# Design of borehole heat exchangers

There are three main parameters governing the thermal performance of a BHE:

1  thermal conductivity of the geological formation;
2  groundwater flow; and
3  the material and type of the BHE and the filling of the annulus (borehole thermal resistance).

To a lesser extent, and mainly for systems storing heat and/or cold underground, the specific heat (per volume) also has an influence. The influence of groundwater flow is usually overestimated; only in aquifers with high porosity and a substantial groundwater volume in motion (Darcy velocity) can a significant impact be found.

The thermal parameters of the geological formation, as well as the groundwater regime, cannot be influenced artificially. The only parameter that engineering can improve is the borehole thermal resistance. Methods include spacers, thermally enhanced grouting material and optimized geometry.

The German guideline *Thermische Nutzung des Untergrunds* (VDI 4640, 2006) and *IEA Heat Pump Annex 8: Advanced In-Ground Heat Exchange Technologies* (IEA, 2006) summarize design guidelines for BHEs. There are several software tools for the design of BHEs, such as EED (see www.buildingphysics.com).

A typical borehole heat exchanger (depth 100m, location mid-Europe) supplies a specific thermal power of up to 50W/m and a specific yearly energy

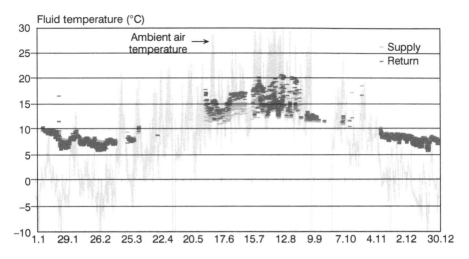

**Figure 5.20** *Return and supply temperature in 2004 (the ambient air temperature is shown for comparison, indicating cooling and heating load)*

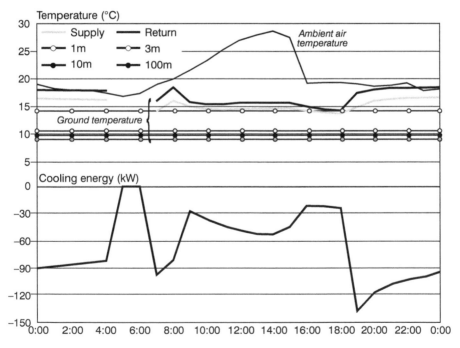

*Note:* Supply temperature between 14°C and 17°C; stable ground temperatures.

**Figure 5.21** *Operation on Friday, 23 July 2004, during the whole day (with the exception of two hours in the morning), with a higher fluid volume flow during the night than during the day*

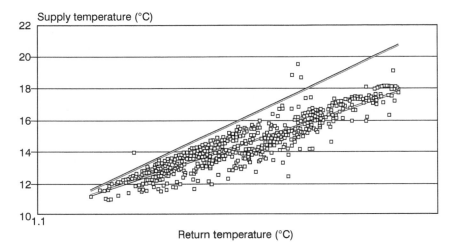

*Note:* At high return temperatures, the supply temperature is 3K below the return temperature. At low return temperatures, the temperature difference is minimal since the ground temperature is similar.

**Figure 5.22** *Supply and return temperature for the summer period of June to August 2004*

gain of approximately 30kWh/(m yr) for both heating and cooling.

## Monitoring results

Unfortunately, there is only limited high-quality monitoring data available from BHEs in operation. As a result, we cannot draw detailed conclusions from a cross-section analysis, as for the EAHXs. Nevertheless, some monitoring data from a German project demonstrate the potential of BHEs.

The ENERGON Building in Ulm, Germany, has been designed as a passive house with a net ground area of 6900 square metres, of which 5000 square metres are heated and cooled by thermally activated ceilings (see *SolarBau: Monitor*, 2006). Forty BHEs with a length of 100m each provide the heat sink for passive cooling (and a heat source in winter for pre-heating of supply air). The heat sink (primary cycle) is separated from the concrete slab cooling (secondary cycle) by a heat exchanger. Figure 5.20 shows the fluid temperatures in 2004 during regular operation of the primary cycle. The supply temperature ranges between 7°C and 18°C.

Figure 5.21 focuses on a typical working day during summer. The concrete slab cooling works during the whole day (with the exception of two hours in the morning), with a higher fluid volume flow in the secondary cycle during the night than during the day. The supply temperature is a function of volume flow, return temperature and ground temperature, and ranges between 14°C and 17°C. The ground temperature is stable during the day and changes with depth from 9°C to 14°C.

During summer, the BHE should not only provide a heat sink, but also a sufficiently low temperature level for cooling since no additional system may lower the fluid temperature. Figure 5.22 shows how the BHE utilizes this cooling potential. As the temperature difference between return and supply temperature (here: closed loop) decreases at low fluid temperatures in the (uncontrolled) primary cycle, the volume flow should be controlled by the temperature in order to avoid a reduced heat transfer at small temperature differences, and to reduce the electricity consumption for the pumps.

The BHE at ENERGON supplied 46,700kWh/yr heating and 68,300kWh/yr cooling energy in 2004, corresponding to 12kWh/(m yr) and 17kWh/(m yr).

## Conclusion

1   In recent years, BHE has become a main cooling source for low-energy cooling concepts. A third of new office buildings in Germany use groundwater or BHE for cooling today.
2   Thermal response tests are an important step during the design phase since these tests provide the main design parameters which can be used by thermal simulation programs.
3   Ground cooling with BHE is very energy efficient, provided that the hydraulic system is well designed and has a low pressure drop.

## REFERENCES

Albers, J. (1991) *Untersuchungen zur Auslegung von Erdwärmeübertragern für die Konditionierung der Zuluft für Wohngebäude*, Universität Dortmund, Germany
Argiriou, A. (1996) 'Ground cooling', in Santamouris, M. and Asimakopoulos, D. (eds) *Passive Cooling of Buildings*, James and James, London
Austin, W. (1998) *Development of an In-Situ System for Measuring Ground Thermal Properties*, MSc thesis, Oklahoma State University, Oklahoma City, OK
Bojic, M., Papadakis, G. and Kyritsis, S. (1999) 'Energy from a two-pipe earth-to-air heat exchanger', *Energy*, vol 24, pp519–523
Bose, J. E., Smith, M. D. and Spitler, J. D. (2002) 'Advances in ground source heat pump systems – an international overview', in *Proceedings of the Seventh IEA Heat Pump Conference*, IEA Heat Pump Programme, Beijing, China, pp313–324
Busso, A., Georgiev, A. and Roth, P. (2003) 'Underground thermal energy storage – first thermal response test in South America', in *International Congress RIO 3: World Climate and Energy Event*, Rio de Janeiro, Brazil, pp189–196
Choudary, A. (1976) *An Approach to Determine the Thermal Conductivity and Diffusivity of a Rock In Situ*, PhD thesis, Oklahoma State University, Oklahoma City, OK
Claesson, J., Eftring, B., Eskilson, P. and Hellström, G. (1985) *Markvärme, en Handbok om Termiska Analyser*, vol 3, SCBR T16-18, Stockholm
Claesson, J. and Eskilson, P. (1988) 'Conductive heat extraction to a deep borehole: Thermal analysis and dimensioning rules', *Energy*, vol 13, no 6, pp509–527

Cruickshanks, F., Bardsley, J. and Williams, H. (2000) 'In-situ measurement of thermal properties of cunard formation in a borehole, Halifax, Nova Scotia', in *Proceedings of Terrastock 2000*, Universität Stuttgart, Stuttgart, pp171–175

De Paepe, M. and Janssens, A. (2002) 'Thermo-hydraulic design of earth-air heat exchangers', *Energy and Buildings*, vol 35, pp389–397

Dibowski, G. (2004) *Luft-Erdwärmetauscher, AG Solar Nordrhein-Westfalen*, www.ag-solar.de/projekte/berichte/LEWT_PLF1_Kurzfassung.pdf, 3 April

Dibowski, G. and Rittenhofer, K. (2000) *Über die Problematik der Bestimmung thermischer Erdreichparameter*, HLH 51, VDI-TGA, Springer Verlag, Berlin, pp32–41

Eklöf, C. and Gehlin, S. (1996) *TED: A Mobile Equipment for Thermal Response Test*, MSc thesis, University of Technology, Luleå

Eugster, W. J. and Laloui, L. (eds) (2001) *Proceedings of the Workshop Geothermische Response Tests Lausanne*, GtV, Ecole Polytechnique Fédérale de Lausanne, Lausanne

Eugster, W. J., Sanner, B. and Mands, E. (2002) 'Stand der Entwicklung und Anwendung des Thermal-Response-Test', in *Proceedings 7: Geothermische Fachtagung*, Geothermische Vereinigung e.V., Wareu, pp304–314

Evers, M. and Henne, A. (1999) *Simulation und Optimierung von Luftleitungs-Erdwärmeüber-tragern*, TAB 12/99, Technik am Bau, Berlin

Flückinger, B., Wanner, H. and Lüthy, P. (1997) *Mikrobielle Untersuchungen von Luftansaug-Erdregistern*, Institut für Hygiene und Arbeitsphysiologie, ETH Zürich

Gehlin, S. (1998) *Thermal Response Test: In-Situ Measurements of Thermal Properties in Hard Rock*, University of Technology, Luleå, Sweden

Gehlin, S. (2002) *Thermal Response Test: Method Development and Evaluation*, PhD thesis, University of Technology, Luleå, Sweden

Gehlin, S. and Hellström, G. (2000) 'Recent status of in-situ thermal response tests for BTES applications in Sweden', in *Proceedings of Terrastock 2000*, Universität Stuttgart, Stuttgart, pp159–164

Gehlin, S. and Hellström, G. (2003) 'Comparison of four models for thermal response test evaluation', *ASHRAE Transactions*, vol 109, pp1–12

Gehlin, S. and Nordell, B. (1997) 'Thermal response test – a mobile equipment for determining thermal resistance of borehole', in *Proceedings of Megastock 1997*, Sapporo, Japan, pp103–108

Heidinger, G., Dornstädter, J., Fabritius, A., Welter, M., Wahl, G. and Zurek, M. (2004) 'EGRT – enhanced geothermal response test', in *Proceedings 8: Geothermische Fachtagung*, Geothermische Vereinigung e.V., London/Pfalz, pp316–323

Hellström, G. (1991) *Ground Heat Storage: Thermal Analysis of Duct Storage Systems, I. Theory*, Luuds Tekniska Högskola, Luud

Hellström, G. (1994) 'Fluid-to-ground thermal resistance in duct ground heat storage', in *Proceedings of Calorstock 1994*, Espoo, Finland, pp373–380

Hellström, G. (1997) 'Thermal response test of a heat store in clay at Linköping, Sweden', in *Proceedings of Megastock 1997*, Sapporo, Japan, pp115–120

Henne, A. (1999) *Luftleitungs-Erdwärmeübertrager (Grundlegendes zum Betrieb)*, TAB 10/99, Technik am Bau, Berlin

Hollmuller, P. and Lachal, B. (2001) 'Cooling and preheating with buried pipe systems: Monitoring, simulation and economic aspects', *Energy and Buildings*, vol 33, pp509–518

Huber, A. (2001) *WKM Version 2.0 – PC-Rechenprogramm für Luft-Erdregister*, Huber Energietechnik, Zürich

IEA (International Energy Agency) (2006) *IEA Heat Pump Annex 8: Advanced In-Ground Heat Exchange Technologies*, www.heatpumpcentre.org/, 15 May 2006

Ingersoll, L. R. and Plass, H. J. (1948) 'Theory of the ground pipe heat source for the heat pump', *Heating, Piping and Air Conditioning*, vol 20, no 7, pp119–122

Mands, E. and Sanner, B. (2001a) 'In-situ-determination of underground thermal parameters', in *Proceedings of IGD Germany 2001 Bad Urach*, Supplement, IEA Heat Pump Center, Bad Urach, pp45–54

Mands, E. and Sanner, B. (2001b) 'Erfahrungen mit kommerziell durchgeführten Thermal Response Tests in Deutschland', *Geothermische Energie*, vol 32, no 33/01, pp15–18

Mihalakakou, G. (2002) 'On estimating soil surface temperature profiles', *Energy and Buildings*, vol 34, pp251–259

Mihalakakou, G. (2003) 'On the heating potential of a single buried pipe using deterministic and intelligent techniques', *Renewable Energy*, vol 28, pp917–927

Mihalakakou, G., Lewis, J. O. and Santamouris, M. (1996a) 'The influence of different ground covers on the heating potential of earth-to-air heat exchangers', *Renewable Energy*, vol 7, pp33–43

Mihalakakou, G., Lewis, J. O. and Santamouris, M. (1996b) 'On the heating potential of buried pipes techniques – application in Ireland', *Energy and Buildings*, vol 24, pp19–25

Mihalakakou, G., Santamouris, M. and Asimakopoulos, D. N. (1994) 'Modelling the thermal performance of earth-to-air heat exchangers', *Solar Energy*, vol 53, pp301–305

Mihalakakou, G., Santamouris, M., Asimakopoulos, D. N. and Tselepidaki, I. (1995) 'Parametric prediction of the buried pipes cooling potential for passive cooling applications', *Solar Energy*, vol 55, pp163–173

Mihalakakou, G., Santamouris, M., Lewis, J. O. and Asimakopoulos, D. N. (1997) 'On the application of the energy balance equation to predict ground temperature profiles', *Solar Energy*, vol 60, pp181–190

Mogensen, P. (1983) 'Fluid to duct wall heat transfer in duct system heat storages', *Building Research*, vol 16, pp652–657

Pfafferott, J. (2000) *Auslegung und Betrieb von Erdwärmetauschern*, HLH Heizung – Lüftung/Klima – Hautechnik 51, Springer VDI Verlag, Berlin, pp46–52

Pfafferott, J., Gerber, A. and Herkel, S. (1998) *Erdwärmetauscher zur Luftkonditionierung (Anwendungsgebiete, Simulation und Auslegung)*, gi Haustechnik – Bauphysik – Umwelttechnik 119 Heft 4, Eigenverlag, pp201–213

Reise, C. (2001) *Planning Tool for Earth-to-Air Heat Exchangers*, EU-Contract JOR3-CT98-7041, Fraunhofer ISE, Freiburg, Germany

Rohner, E., Rybach, L. and Schärli, U. (2004) 'Neue Methode zur in-situ-Bestimmung der Wärmeleitfähigkeit für die Dimensionierung von Erdwärmesonden-Feldern', in *Proceedings 8: Geothermische Fachtagung*, Geothermische Vereinigung e.V., London/Pfalz, pp324–328

Sanner, B., Reuss, M. and Mands, E. (1999) 'Thermal Response Test – eine Methode zur In-Situ-Bestimmung wichtiger thermischer Eigenschaften bei Erdwärmesonden', in *Geothermische Energie*, vol 24–25, no 99, pp29–33

Sanner, B., Reuss, M., Mands, E. and Müller, J. (2000) 'Thermal Response Test – experiences in Germany', in *Proceedings of Terrastock 2000*, Universität Stuttgart, Stuttgart, pp177–182

Santamouris, M. and Asimakopoulos, D. (eds) (1996) *Passive Cooling of Buildings*, James and James, London

Sedlbauer, K. (1992) *Erdreich/Luft-Wärmetauscher zur Wohnungslüftung*, Fraunhofer Institut für Bauphysik, Stuttgart, Germany

Seidinger, W., Mornhinweg, H., Mands, E. and Sanner, B. (2000) 'Deutsche Flugsicherung (DFS) baut Low Energy Office mit größter Erdwärmesondenanlage Deutschlands', *Geothermische Energie*, vol 28–29, no 00, pp23–27

Shonder, J. A. and Beck, J. V. (1999) 'Determining effective soil formation: Thermal properties from field data using a parameter estimation technique', *ASHRAE Transactions*, vol 105, no 1, pp458–466

Skouby, A. (1998) 'Thermal conductivity testing', *The Source*, 11 December 1998

Smith, M. (1999) 'Comments on in-situ borehole thermal conductivity testing', *The Source*, 1 February 1999

*SolarBau: Monitor* (2006) 'Internet platform', www.solarbau.de/english_version/ 2006-15 May

Spitler, J. D., Rees, S. and Yavuzturk, C. (1999a) 'More comments on in-situ borehole thermal conductivity testing', *The Source*, 3 April

Spitler, J. D., Yavuzturk, C. and Jain, N. (1999b) *Refinement and Validation of In-Situ Parameter Estimation Models*, Report, OSU, www.mae.okstate.edu/Faculty/spitler/pdfs/insitu.pdf

Spitler, J. D., Yavuzturk, C. and Rees, S. J. (2000) 'In situ measurement of ground thermal properties', in *Proceedings of Terrastock 2000*, Universität Stuttgart, Stuttgart, pp165–170

Tzaferis, A., Liparakis, D., Santamouris, M. and Argiriou, A. (1992) 'Analysis of the accuracy and sensitivity of eight models to predict the performance of earth-to-air heat exchangers', *Energy and Buildings*, vol 18, pp35–43

van Gelder, G., Witte, H. J. L., Kalma, S., Snijders, A. and Wennekes, R. G. A (1999) 'In-situ-Messung der thermischen Eigenschaften des Untergrunds durch Wärmeentzug', in *Proceedings of OPEC Seminar Erdgekoppelte Wärmepumpen*, pp56–58

VDI 4640 (2006) 'Thermische Nutzung des Untergrunds', *Blatt*, vol 1–4, Beuth Verlag, Berlin

Wärmeentzug', in *Proceedings of OPET-Seminar Erdgekoppelte Wärmepumpen*, Geothermische Vereinigung e. V., Geeste, pp56–58

Witte, H. J. L., van Gelder, G. and Spitler, J. D. (2002) 'In-situ thermal conductivity testing: A Dutch perspective', *ASHRAE Transactions*, vol 108, no 1, pp263–272

Zimmermann, M. and Andersson, J. (1998) *Case Studies of Low Energy Cooling Technologies*, IEA Energy Conservation in Buildings and Community Systems, Annex 28: Low Energy Cooling, EMPA ZEN, Dübendorf

# 6

# Evaporative Cooling

*Evyatar Erell*

## INTRODUCTION

Evaporative cooling is employed all over the world to improve thermal comfort in buildings. The most common application of evaporative cooling is found in desert coolers (sometimes referred to as swamp coolers), in chillers installed as part of medium- or large-scale cooling plants for offices and warehouses, and in cooling towers used in industrial facilities.

So-called direct evaporative cooling works by introducing into the conditioned space air that has been exposed to liquid water. The process of evaporation results in the conversion of sensible heat to latent heat at a constant wet bulb temperature; as a result, the air supplied is not only cooler, but is also more humid. An example of a direct evaporative cooler is the desert cooler, which draws fresh outdoor air through wet pads and supplies the cooled moist air directly to the building interior. Its design has changed very little since it was commercialized during the 1920s, and it will not be discussed in this chapter.

Indirect evaporative cooling reduces the dry bulb temperature of the air without increasing its moisture content: The air supplied to the conditioned space is fed through a heat exchanger that contains air or water that has been cooled separately by a direct evaporative cooler. Many commercial buildings employ large-scale indirect evaporative chillers, which supply a heat-exchange fluid for space cooling by means of fan-coil units installed in the building interior. The design of such systems is beyond the scope of this book, which focuses on passive or hybrid cooling and does not deal with mechanical engineering. Cooling towers designed to remove excess heat from a coolant used in industrial plant will also not be discussed for the same reason.

This chapter provides an overview of progress in research on two types of applications that feature evaporative cooling and that are not yet applied extensively: roof ponds and downdraught evaporative cool towers (DECTs). Both techniques have attracted considerable research in recent years. However, neither method is applied, in practice, in conventional buildings, partly because there has been relatively little published information available to engineers and architects on issues of practical application.

This chapter will attempt to address this shortcoming. It begins with a brief overview of the principles governing evaporation of water as a foundation for following sections that deal with recent research on the application of evaporative cooling to buildings. Readers interested in a more comprehensive discussion of the underlying mechanisms may refer to previous publications on this subject, such as Yellot (1989) or Argiriou (1995); to general texts on heat transfer, such as Incropera and De Witt (1990); or to the chapter on psychrometrics in the *ASHRAE Handbook: Fundamentals* (ASHRAE, 2005). The second part of the chapter deals with ways of assessing the degree to which climatic conditions affect the applicability of evaporative cooling in a specific location. The third and fourth sections, which constitute the main part of this chapter, include a detailed discussion of specific issues relating to the construction and operation of downdraught cool towers and of roof ponds, respectively. The chapter concludes with a brief discussion of thermal comfort and health issues that have a particular bearing on evaporative cooling systems.

# BASIC PRINCIPLES

Evaporation is the phase change of water from liquid to gas (vapour). It is accompanied by the absorption of sensible heat from the air that comes in contact with the water being evaporated or from a wet solid surface where evaporation takes place. This results in the cooling of the surface and of the surrounding air, and an increase in the moisture content of the air. The energy, present in the form of latent heat, may be released again if the water vapour condenses.

Evaporation takes place when the vapour pressure at the surface of the water is greater than that of the adjacent atmosphere. The energy required for the phase change of water from liquid to gas is the latent heat of vaporization. When the change in air temperature occurs without the addition or extraction of energy from the system, the process is adiabatic. Evaporation of water may be a very potent cooling mechanism because the amount of energy required to evaporate water is very large – about 2.44MJ/kg (at a temperature of 25°C and an atmospheric pressure of 100 kilopascals (kPa). For example, evaporating just 1 litre of water would cool about 200 cubic metres of air – comparable to the volume of a modest apartment – by 10°C.

Dry air consists primarily of nitrogen (78.1 per cent) and oxygen (20.9 per cent), in addition to small proportions of other substances, notably carbon dioxide. However, atmospheric air also contains some water vapour in an amount that varies from zero to a maximum that depends upon temperature and pressure. When this maximum occurs, the air is said to be saturated. The amount of water vapour in air, even at saturation, is minute in terms of weight per unit of dry air, but is of vital importance in atmospheric processes.

There are several methods of expressing atmospheric humidity. Among the most common are relative humidity, which is the ratio (percentage) of water vapour present in a volume of air to the amount of water vapour that would result in saturation (at the same temperature and pressure); the dew point, which is the temperature to which a given volume of moist air would have to be cooled to reach saturation (at a constant pressure); and wet bulb temperature, which is the temperature at which a given volume of air may reach saturation purely through a process of evaporation of water (i.e. adiabatically, without adding or removing heat).

The relationship between the temperature of air and its moisture content is presented graphically in the psychrometric chart, first introduced by Carrier in 1911 (see Figure 6.1). The Carrier chart, shown here in a simplified format, plots the humidity ratio W as the ordinate versus dry bulb temperature as the abscissa, both on linear scales. The humidity ratio, also referred to as the moisture content, is expressed in International System of Units (SI) as grams of water per kilogram of dry air, and dry bulb temperature in degrees Centigrade. The curved lines on the chart are lines of equal relative humidity, and the diagonal ones are lines of constant wet bulb temperature.

Air cannot be cooled by evaporation to a temperature that is lower than its wet bulb temperature. In practice, even this theoretical limit is rarely attained, and the output of most evaporative coolers is at least 2°C warmer than the ambient wet bulb. This implies that even in very dry conditions, where evaporation can cool the air substantially, extremely high ambient

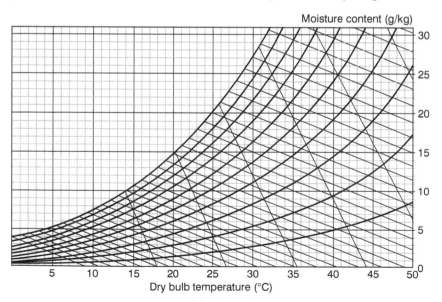

**Figure 6.1** *Simplified psychrometric chart*

temperatures may result in wet bulb temperatures that are too high for human thermal comfort.

Evaporative cooling may be used to cool the air supplied to a building, or it may be used to cool the building structure, improving the thermal comfort of building occupants indirectly. The application of evaporative cooling on rooftop installations has historically concentrated mostly on structural cooling; evaporative cooling of supply air is generally carried out by so-called desert coolers, which may be installed on the roof or on any other exterior building surface with equal success. The exceptions to this rule are passive downdraught evaporative cooling (PDEC) towers, which rely on thermal buoyancy and wind forces to supply cool air without mechanical intake fans (Alvarez et al, 1991; Pearlmutter et al, 1996).

## APPLICABILITY

For a given volume of air, the potential for evaporative cooling is the difference between its moisture content when brought to saturation adiabatically, $\omega_s$, and its actual moisture content, $\omega_a$. The potential for cooling varies among different regions and over time at a specific location. These variations must be assessed in order to decide upon the utility of installing evaporative cooling in a proposed building project.

*Source:* Yannas et al (2006)

**Figure 6.2** *Map of Europe showing the total wet bulb temperature depression during the summer in degree hours*

## Spatial variations

The benefits of evaporative cooling are very limited in warm humid conditions, where the potential for evaporation is low. A common measure of the potential for evaporative cooling is the wet bulb temperature depression, which is the difference between the ambient dry bulb temperature and the wet bulb temperature. Figure 6.2 illustrates this parameter for typical daytime conditions during summer around the Mediterranean.

## Temporal variations

The moisture content of air at a given location is not fixed, but changes over the course of time. It is affected by advection, reflecting the moisture availability of the surface over which incoming air has travelled and, thus, the complex action of a multitude of factors that control wind direction. However, it also displays distinct diurnal and seasonal patterns.

The diurnal pattern of the wet bulb temperature is controlled by the diurnal variations of net radiation, which, in turn, drives the typical pattern of variations in the other components of the surface energy balance. In hyper-arid areas where there is little vegetation or soil moisture, net radiant exchange is transformed into large sensible heat fluxes, leading to rapid heating during the daytime and rapid cooling at night. In humid locations, part of the net radiant flux is channelled into latent heat, thus reducing the diurnal temperature range.

The variation of specific humidity depends upon the diurnal course of evapotranspiration and condensation, surface temperatures, turbulence and the thickness of the boundary layer (Arya, 1988). Large diurnal changes in surface temperature lead to large variations of specific humidity because the saturation vapour pressure and air temperature are linked. The daily minimum and maximum values of moisture content typically coincide with the corresponding extremes of air temperature. Givoni (1994) suggested that the diurnal range of the wet bulb temperature is typically about one third of the amplitude of the dry bulb temperature of the air at any given location. However, this ratio should be viewed only as a fairly coarse approximation, and the diurnal pattern of the wet bulb temperature depression may vary substantially independently of the changes in the dry bulb temperature of air. According to Costelloe and Finn (2003), long-term climatic averages show that the diurnal variation in the average summer ambient adiabatic saturation temperature (AST) in Milan is 3K to 4K, compared to 2K to 3K in Dublin, which is both cooler and more humid (AST is, in practice, almost identical to the wet bulb temperature). During experiments on a DECT at Sde Boqer in Israel (Erell et al, 2005b), diurnal variations in the wet bulb temperature ranged from 2K to as much as 7K. In all cases, the lowest wet bulb temperatures were recorded just before sunrise, while daily maxima were typically measured in the afternoon.

Although the wet bulb temperature depression is a convenient means of comparing the potential for evaporative cooling in two locations with the same dry bulb temperature, it is the absolute value of the wet bulb temperature that determines whether atmospheric conditions support useful evaporative cooling. This is because the lower limit for evaporative cooling is the wet bulb temperature, or, in practice, about 2K to 3K above this temperature. In order to absorb heat from the space being cooled, the wet bulb temperature must be substantially lower. Therefore, simply noting that there is a large wet bulb temperature depression is not in itself an indication that conditions at a given place and time favour evaporative cooling: the wet bulb temperature must be sufficiently low.

Analysis of conditions at a given location must be based on long-term climatic averages, such as the Typical Meteorological Year (TMY). Once the required output temperature of the cooling system is specified (based on desired comfort levels and the characteristics of the system), the evaporative cooling potential can then be established by calculating one or more of the following indicators (Costelloe and Finn, 2003):

- total annual availability;
- average monthly cooling temperatures; and
- analysis of unavailability, particularly during warm but humid periods.

If one assumes that the output of an evaporative cooling system (chilled air or water) is 3K above the ambient wet bulb temperature, a chart can be drawn

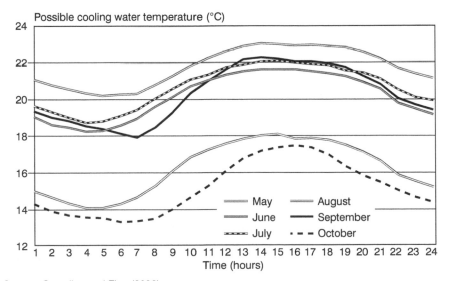

*Source:* Costelloe and Finn (2003)

**Figure 6.3** *Cooling water temperature obtained from an evaporative cooling system, assuming water is cooled to 3K above ambient adiabatic saturation temperature (AST); data for Milan, May–October*

showing the hourly variation in this output at a given location, based on TRY data for monthly periods. Figure 6.3 illustrates such a calculation for the city of Milan, Italy (Costelloe and Finn, 2003).

Note that the total number of degree hours of wet-bulb temperature depression during the cooling season, although a simpler datum, does not in itself provide an indication of the applicability of evaporative cooling at a given location because it is not correlated with any specific output temperature produced by the cooling system.

# DOWNDRAUGHT EVAPORATIVE COOL TOWERS (DECTs)

## Background

Wind towers or wind catchers have been used in traditional architecture throughout the Middle East and Central Asia for centuries to promote natural ventilation, sometimes incorporating evaporative cooling. They are known variously as the *malqaf* in Egypt or as the *badgir* in Iraq and Iran, and are found as far east as Pakistan and Afghanistan (Al-Megren, 1987). In a traditional wind tower, air entering through the windward opening at the top with positive wind pressure leaves the tower through any of a number of openings that have a pressure coefficient that is lower (Bahadori, 1985). In traditional *badgirs*, the air may be cooled if it flows over moist surfaces or a shallow pond of water at the base of the tower. However, such wind catchers have several significant drawbacks:

- They are not effective where wind speed is low.
- The cooling output of the towers is low because contact between the air and moist surfaces is insufficient.
- The aerodynamic design of the towers does not maximize airflow to the building.

Bahadori (1985) proposed an innovative design that incorporated several novel modifications to the traditional design (see Figure 6.4). First, the inlets to the tower were equipped with gravity-shut dampers to reduce losses from the leeward opening at the top of the tower. Second, Bahadori proposed that water should be sprayed at the top of the tower, rather than near the bottom. Finally, he suggested that vertical clay conduits be installed beneath the sprayers to increase the area of moist surfaces and, thus, improve cooling.

Bahadori (1985) assumed constant density for air flowing through the tower, and that the driving potential creating air movement is the pressure difference between the inlet section of the tower and the outlet (door or window) through which air leaves the building that is being cooled. Wind tunnel experiments on a reduced-scale model provided pressure coefficients for a variety of openings in an isolated tower, a tower joined to a building,

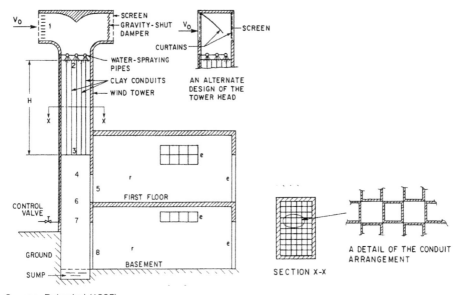

*Source:* Bahadori (1985)

**Figure 6.4** *Cross-section of wind tower proposed by Bahadori*

and a tower and building surrounded by a courtyard (Karakatsanis et al, 1986).

Cunningham and Thompson (1986), who built and monitored an experimental downdraught cool tower attached to a test building in Tucson, Arizona, in the US, realized that while wind might assist in maintaining airflow through the tower, air movement would be generated by differences in air density between warm dry air and air cooled by evaporation of water. In their experimental design (see Figure 6.5), air was drawn into the top of the tower through wet porous mats similar to those used in desert coolers attached across vertical openings on all sides. The velocity of air in the tower varied between 0.25 to 0.7 metres per second (m/s), depending upon the magnitude of the cooling effect. Givoni (1993), using the same data, later found that if the pads were kept moist, the temperature reduction equalled approximately 87 per cent of the wet bulb temperature depression in all conditions. A temperature reduction equalling between 85 and 90 per cent of the wet bulb temperature depression was also found in other types of downdraught cool towers (Pearlmutter et al, 1996; Erell et al, 2005b) and may be assumed to be a property of all such systems, assuming that sufficient moisture is available and that air and water are in good contact.

A downdraught cool tower with a different design was installed in the Blaustein International Centre for Desert Studies building at Sde Boqer, Israel, constructed during 1989 to 1991 (see Figure 6.6). In this tower, water was introduced into the air stream by means of sprayers, instead of being drawn in through moist pads. Air movement was generated by the density difference

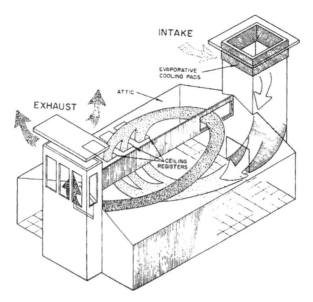

Source: Cunningham and Thompson (1986)

**Figure 6.5** *Airflow path through a test house with a wind tower and solar chimney*

Source: W. Motzafi-Haller (2006)

**Figure 6.6** *View of downdraught cool tower in the atrium of the Blaustein International Centre for Desert Studies building, Sde Boqer, Israel*

*Source:* E. Erell (2001)

**Figure 6.7** *View of downdraught cool towers at the Avenue of Europe, Expo 1992, Seville, Spain*

between inlet and outlet, and was supported by an electric fan. The sprayers produced relatively coarse drops of water, akin to raindrops, and excess water not evaporated was collected in a shallow pool of water that was constructed for this purpose. The advantage of this configuration over a design that employs wetted pads is that aerodynamic resistance is lower, and airflow through the tower was expected to be greater. The performance of this tower was monitored in a variety of configurations, and the best performance was achieved with a combination of sprayers producing both coarse drops of water and a fine mist (Pearlmutter et al, 1994; Pearlmutter et al, 1996).

Evaporative cooling was employed extensively to promote thermal comfort of outdoor spaces at the World Exposition held at Seville, Spain, in 1992. The application of PDEC was demonstrated in the Avenue of Europe, where two parallel rows of six 30m high towers were constructed (Alvarez et al, 1991). These towers (see Figure 6.7) employed micronizers to create a very fine mist as a means of optimizing contact between air and water, while minimizing loss of air pressure. The operation of the water-spraying mechanisms was designed to ensure full evaporation of water, thus allowing pedestrian traffic beneath the towers, while maximizing cooling output. The primary driving force generating air movement through these towers was the density difference between warm ambient air and cooler moist air in the tower. However, the tapered, slightly conical towers also had a diagonal opening at the top acting as a directional inlet, favouring flow from the direction of the prevailing winds.

## Principle of operation

The main forces that move the air through a downdraught evaporative cool tower are:

- the wind pressure at its top inlet;
- the increase in the specific weight of the cooled air in the upper section of the tower, which is slightly heavier than its surroundings and, thus, descends, generating air movement through the full height of the tower (Rodriguez et al, 1991a); and
- momentum transfer from the water to the air if large drops of water are sprayed that are not fully evaporated in the tower.

The first of these forces, investigated by Bahadori (1985), may act even in the absence of evaporative cooling. The following generic expression was proposed to estimate the magnitude of the driving force:

$$\Delta p_a = (C_{pi} - C_{pe})\frac{1}{2}\rho V_\infty^2 \qquad [1]$$

where $\Delta p_a$ is the pressure difference between the inlet of the tower and the outlet from the space being cooled, $C_{pi}$ and $C_{pe}$ are the wind pressure coefficients at the inlet and outlet, $\rho$ is air density, and $V_\infty$ is ambient wind speed. The wind pressure coefficients depend upon the geometry of the tower inlet and the building outlet, upon wind direction relative to these openings, upon the aerodynamic effect of adjacent structures, etc., and must be obtained empirically. A series of wind tunnel tests were conducted to obtain indicative values for typical configurations (Karakatsanis et al, 1986).

Evaporative cooling may produce air movement in a downdraught tower even in the absence of wind, due purely to the effect of negative buoyancy. In the free atmosphere, and neglecting friction generated by the motion of air itself, the velocity of air resulting from buoyancy is given by the following expression:

$$w = \left(2gz\frac{T_e - T_p}{T_e}\right)^{1/2} \qquad [2]$$

where $w$ is the vertical velocity (m/s), $g$ is the acceleration due to gravity (m/s²), $z$ is the difference in height (m), $T_e$ is the temperature of the environment (K) and $T_p$ the temperature of the parcel of air cooled by evaporation (K).

Equations 1 and 2 are sufficient to describe the forces creating air movement in a tower in which water evaporates within a negligible height – either because it is present in the liquid form only on the surface of wet pads, or because the radius of drops sprayed is so small evaporation is practically instantaneous. If water is introduced in the form of relatively coarse drops, they will not be fully evaporated near the inlet and will descend through the tower in liquid form. Since the air near the inlet becomes almost saturated within a very short distance (Pearlmutter et al, 1996), very little evaporation will occur further down the tower and these drops will therefore remain approximately constant in size. As the water drops descend down the tower, aerodynamic drag will result in a transfer of momentum to the air, increasing its downward velocity. The transfer of momentum is governed by the principle of conservation of momentum, which for a characteristic drop may be written as follows (Guetta, 1993):

$$m_d \frac{dv}{dt} = C_D \rho_a (U - v) \left| U - v \right| \frac{\pi d_d^2}{8} + m_d g \qquad [3]$$

where $m_d$ is the drop's mass, $d_d$ is its diameter, $\rho_a$ is the density of air, $U$ is the velocity of air, $v$ is the velocity of the drop, $g$ is the acceleration due to gravity and $C_D$ is the coefficient of drag between the drop and the surrounding air. The value of $C_D$ has been shown empirically to be approximately $C_D = 24/Re$ for low values of the Reynolds number (Re) (Re < 0.5).

Air velocity at the outlet of a particular downdraught cool tower depends upon the relative importance of each of the three mechanisms described above and, in addition, upon the pressure loss resulting from the aerodynamic design of the tower itself. This includes pressure losses at the inlet of the tower, at its outlet and the effects of friction with its interior surfaces. The sum of the pressure loss coefficients for the small tower at the Environmental Research Laboratory in Tucson, Arizona, was estimated to equal about 6 (Cunningham and Thompson, 1986), effectively reducing air velocity by about 60 per cent.

## Design factors

### Aerodynamic design
Most traditional wind catchers have a regular cross-section that does not vary with height. Al-Megren (1987) classified wind catchers according to the following characteristics:

- *Flow concept:* a wind catcher may be either uni-directional, if wind enters through the top and air is allowed to escape into the house at ground level, or it may be bi-directional, if the tower is divided along its vertical axis so that air enters through an opening on the windward side, travels down to ground level in one passage and then up again through a second passage, to be exhausted through an opening on the lee side of the tower.
- *Cross-sectional shape:* traditional wind towers are constructed of brick and are generally either square or rectangular. Modern towers are typically circular.
- *Orientation with respect to wind:* rectangular wind catchers generally have their long axis normal to the prevailing winds. Bi-directional towers are

usually oriented diagonally with respect to the prevailing wind and have a square section.

- *Shape of the top end of the catcher:* uni-directional catchers have an inclined roof with a slope of 30° to 45°. Bi-directional towers may have either flat tops or inclined ones (a 'butterfly' configuration).
- *Height:* traditional wind towers range in height from about 5m to 15m, and generally project at least one storey above roof height.

## Height of tower

As Equation 2 shows, the potential for pure thermodynamic convection in a tower is related to its height. However, the height of a tower built for space cooling may be limited by considerations of structural strength, expense or architectural form. In a building-integrated tower, reducing tower height also causes an increase of outlet areas (to the building mainly on the top floor), required to keep constant mass fluxes; a less uniform temperature and velocity distribution, leading to uneven performance for different building floors; and a risk that some of the water drops may not be fully evaporated (Alvarez and Sanchez, 2000). Furthermore, if the wind catcher is not sufficiently removed from the roof of the building, turbulence caused by the roof may reduce airflow into the tower. Alvarez and Sanchez (2000) suggested that a height of 6m above the roof was a reasonable compromise.

## Water-spraying system

There are several methods of introducing moisture to the air entering a building through a wind catcher tower:

- *Wetted pads:* air is drawn through cellulose pads that are kept moist by trickling water over their entire surface. The tower at the Environmental Research Laboratory in Tucson, Arizona, is an example of this type of tower (Cunningham and Thompson, 1986).
- *Atomizer nozzles:* a very fine mist of water is generated at the top of the tower, which evaporates rapidly and cools the air around it. In the 'mist DECT', all water drops are evaporated before reaching ground level, so the space directly below it is suitable for pedestrian activity. The towers in the Avenue of Europe at Expo 1992 in Seville, Spain, are an example of this type of tower (Alvarez et al, 1991).
- *Coarse sprayers:* relatively large drops of water are sprayed at the top of the tower and are not fully evaporated by the time they reach ground level. The cool tower at Sde Boqer, Israel, is an example of such a tower (Pearlmutter et al, 1996). Several small experimental towers of this type have also been monitored by Givoni in different climates (Givoni, 1997; Yajima and Givoni, 1997). Because the volume of water introduced is larger than the potential for evaporation, a shower DECT produces chilled water in addition to cooling the air in the tower (Rodriguez et al, 1991b). The excess water is collected into an operational reservoir beneath the

tower and may be circulated through heat exchangers to cool non-adjacent spaces.

*   *Wet internal surfaces:* the internal surfaces of the tower are kept moist by a system of drip irrigation. In the simplest configuration, this involves keeping the perimeter of the tower wet. However, the total surface area exposed to the air may be increased by the addition of porous clay surfaces near the centre of the cross-section, as suggested by Bahadori (1985). The drawback inherent to this approach is that friction with the additional solid surfaces reduces the total airflow rate through the tower. More recently, experiments with ceramic bricks have been carried out in Greece (Papagiannopoulos and Ford, 2003) and the UK (Ibrahim et al, 2003), and a building system based on such bricks, EVAPCOOL, has been developed (Ford and Schiano-Phan, 2003). The porous ceramic bricks are intended to produce more even wetting of the surface than clay tiles and to avoid the need for collecting excess water that may not have evaporated.

In 'mist DECTs', total evaporation can occur only if water supply to the spraying system is adjusted periodically in response to changing environmental conditions (Alvarez et al, 1991). Alternatively, a conservative approach may be adopted where water supply is restricted to a rate that ensures full evaporation at all times. This strategy results in suboptimal performance in hot dry conditions, where the potential for evaporation in the tower exceeds the water supply to the sprayers. In 'shower DECTs', where total evaporation of water spray is not required, spraying excess water effectively ensures that the rate of evaporation will always be the maximum possible in the given environmental conditions.

The water-spraying system in shower DECTs is usually simpler and more reliable than that in mist DECTs: the spray heads do not require a pressurized water supply, are less susceptible to clogging than the micronizers incorporated in mist DECTs, and are cheaper. On the other hand, the operation and maintenance of the operational reservoir incorporated in shower DECTs requires care: recycled water must be filtered to protect the pumping system and the reservoir itself must be cleaned periodically to remove dust particles washed out of the air by the water drops (Etzion et al, 1997). Shower DECTs are also more exposed to the risk of legionnaire's disease.

To conserve pumping energy, the rate at which water is sprayed into the incoming air should be exactly equal to the amount required to bring its entire volume to saturation. This water flow rate can be determined from the moisture deficit of the ambient air $(g/m^3)$ if the volumetric flow rate of air is known.

In practice, neither the moisture content of the ambient air nor the volumetric flow rate is constant with respect to time. This means that the rate at which water is sprayed into the airflow must be regulated to compensate for these variations. Insufficient water provision results in the loss of cooling potential, whereas excess water leads to a waste of pumping energy.

Regulation of the water-spraying rate may be achieved by varying the pressure in the supply system or by changing the number of active sprayers. However, changes in the water pressure supplied to the sprayers lead to changes in the characteristics of the drops produced, both with respect to the initial trajectory of the drops and to the drop size distribution. Incremental changes to the number of active sprayers can provide a consistent drop size. However, simply adding sprayers to those already active at any given time may result in a non-uniform distribution of the water droplets throughout the cross-section of the tower, at least some of the time. This problem may be overcome by the installation of separate water supply circuits for each step change in the flow rate, each of which provides a uniform water distribution pattern. This adds to the cost and complexity of the system.

A more practical approach may be to spray an excess amount of water, thus guaranteeing that there is sufficient water to bring the air to saturation in any environmental conditions likely to occur at the given location for the expected volumetric airflow. This approach requires the installation of a water collection system to recycle the excess water. It should also be noted that increasing the water mass ratio requires more pumping power; but it has diminishing returns with respect to the cooling output obtained (Guetta, 1993).

## Drop diameter

Drop size has an important effect on the evaporation process: for a given volume of water, smaller drops result in a larger total surface area, leading to better contact with the air and, thus, to faster evaporation. However, the following discussion will demonstrate that fine spray is not necessarily required for efficient operation of a cool tower.

Drop size is a by-product of atomization – the process of generating drops. The process of atomization begins by forcing liquid through a nozzle. The potential energy of the liquid (measured as liquid pressure for hydraulic nozzles), along with the geometry of the nozzles, causes the liquid to emerge as small ligaments. These ligaments then break up further into very small 'pieces', which are usually called drops, droplets or liquid particles. Drop size refers to the size of the individual spray drops that comprise a nozzle's spray pattern.

Water drops within a given spray vary in size, and may be described by a mathematical drop size distribution from which a collection of characteristic or mean diameters can be extracted. These diameters are single values that express the various mean sizes in the spray.

The volume mean diameter, $D_{30}$, which is the drop diameter below or above which lies 50 per cent of the volume of the drops, is defined for a discrete distribution as:

$$D_{30} = \left( \frac{\sum N_i \delta_i^{\,3}}{\sum N_i} \right)^{1/3} \qquad [4]$$

Presuming we have a set of experimental data with drop diameters, we determine an arbitrary number of bins, $n$, with a certain size range. $N_i$ is thus the number of droplets in bin $i$ and $\delta_i$ is the middle diameter of its size range.

Analogous to $D_{30}$ (see Equation 4), we can define a surface mean diameter $D_{20}$ as the drop diameter of a distribution with surface area equal to the mean surface area of all the droplets in the spray:

$$D_{20} = \left( \frac{\sum N_i \delta_i^{\,2}}{\sum N_i} \right)^{1/2} \qquad\qquad [5]$$

By combining volume mean diameter $(D_{30})$ and surface mean diameter $(D_{20})$, the Sauter mean diameter, or volume–surface mean diameter $(D_{32})$, can be obtained:

$$D_{32} = \frac{D_{30}^{\,3}}{D_{20}^{\,2}} = \frac{\sum N_i \delta_i^{\,3}}{\sum N_i \delta_i^{\,2}} \qquad\qquad [6]$$

It is defined as the diameter of a drop having the same volume-to-surface ratio as the entire spray.

Other than the effects of the specific liquid being sprayed, the three major factors affecting drop size are nozzle type, capacity and spraying pressure. Lower spraying pressures provide larger drop sizes, while higher spraying pressures yield smaller drop sizes. The smallest drop sizes are achieved by air atomizing nozzles, which may require pressure as high as 25 bar. Within each type of spray pattern the smallest capacities produce the smallest spray drops, and the largest capacities produce the largest spray drops.

Sprayers may be supplied with water from the mains. In this case, however, excess water cannot be recycled directly into the tower system, and the system requires a means of regulating water pressure. An alternative solution is to feed the sprayers from a reservoir acting as a buffer, collecting excess water. In such a system, water is supplied to the sprayers by an electric pump. The type of sprayers installed affects the operating parameters of this pump and requires careful consideration.

Sprayers should be selected with the following considerations in mind:

- Is full evaporation of the water drops required? In this case, the drop diameter must be small enough to ensure that the time required to fully evaporate the largest drops produced by the sprayer is shorter than the residence time of the drops as they fall down the tower.
- If full evaporation is not required, is it nonetheless desirable to minimize drift of airborne drops beyond the perimeter of the tower? Preventing drift allows full recovery of excess water and avoids the wetting of the floor

**Table 6.1** *Properties of several commercial sprayers considered for the experimental cool tower at Sde Boqer, Israel*

| Type | p (bar) | $D_{32}$ (μm) | A ($m^2/s$) | $Q_w$ (l/min) | n | P (W) |
|------|---------|-----------|---------|---------|-----|-------|
| PJ32   | 4 | 106 | 0.775 | 0.82 | 24.4 | 180.5 |
| LN20   | 4 | 123 | 1.234 | 1.52 | 13.2 | 180.5 |
| HHSJ07 | 3 | 199 | 2.774 | 5.51 | 3.6  | 141.3 |
| TF6    | 3 | 156 | 3.536 | 5.53 | 3.6  | 141.3 |

*Notes:* $p$ = pressure at inlet of sprayer; $D_{32}$ = Sauter diameter; $A$ = total surface area of drops produced per second; $Q_w$ = water flow rate; $n$ = number of sprayers required to provide a total water flow rate of 20l/min; $P$ = required pump power per sprayer.
*Source:* Tambour and Guetta (2003)

surface adjacent to the tower. It also reduces the risk of legionnaire's disease.

* Minimizing the power required to supply the spraying system with the required pressure not only reduces operational costs, but can also result in lower equipment costs.
* Reduce maintenance requirements, especially with respect to treatment of the water in order to prevent clogging of the spray nozzles by particulate matter or by scaling. Nozzles with very fine apertures are more likely to be clogged than nozzles with coarse ones.

The rate at which air in the tower is cooled by evaporation depends upon the total surface area of the water drops. A specified surface area may be achieved by supplying a given volume of very fine droplets, or by a larger volume of coarser drops. Because the creation of very fine drops requires water to be forced through a narrow aperture at high pressure, there is a trade-off between drop size and volume.

Table 6.1 shows the results of a study carried out to select sprayers for a test installation at Sde Boqer, Israel (Tambour and Guetta, 2003), where the required water flow rate was 20 litres per minute (l/min). The table demonstrates that for a given pumping power, the largest total surface area is not necessarily provided by nozzles that produce the finest drops. The TF6 nozzles produced the largest total surface area per nozzle, yet required less power than the PJ32 nozzles, which produced the smallest drops.

## Implementation of DECT in buildings

The implementation of downdraught evaporative cool towers in buildings requires careful analysis of the ventilation and cooling requirements of the specific building, and appropriate planning of the airflow trajectories. The experience gained in a handful of public buildings where DECTs have been installed is illuminating and illustrates the complexity of the issues.

A detailed simulation of a hypothetical office building cooled by DECTs was carried out for a site in Seville, Spain, as part of a European Union (EU) Joule III programme (Cook et al, 2000). In addition to a theoretical framework for modelling flow in the towers, the study highlighted the difficulties posed by the desire to provide effective cooling for each of several occupied floors in a multi-storey building. Since the height of the column of evaporatively cooled air was smaller in the upper stories of the building than in the lower floors, the airflow generated in the upper levels of the building according to the model was proportionately smaller. A similar problem was observed at the Torrent Research Centre building in Ahmedabad, India, a multi-storey building equipped with PDEC towers (Ford et al, 1998): Air change rates of about 9 air changes per hour were measured at ground level, 6 air changes per hour at the first floor and much lower rates at the top floor. Further computational fluid dynamic (CFD) simulations of the hypothetical Seville building showed that the problem could be reduced by separating the tower from the adjacent spaces requiring cooling, allowing airflow only through small openings near the floor at each level. The size of these openings was varied in (inverse) proportion to the effective height of the tower above the floor in question, so that lower-level floors were served by smaller openings than upper-level floors, where the driving pressure of the cool air column was smaller.

The possibility of providing PDEC to a building by means of a large central atrium was explored in another hypothetical study supported by the EU (Francis, 2000). Atria are often integrated in deep-plan office buildings to provide additional daylighting and ventilation. The objective of this study was to evaluate whether the atrium could also be used to distribute air cooled by evaporation to adjacent office spaces. An advantage of this approach is that relatively high mass flow rates may be achieved by passive means alone because although the air velocity is low – typically less than 0.5m/s – the entire cross-section of the atrium may be utilized (Ford and Diaz, 2003). However, this approach also has a number of drawbacks, because the entire volume of air must be cooled even at times when the cooling demands of the building are low and because it is difficult to control the movement of air into office spaces at different levels. A configuration with a number of separate towers may be preferable (Francis, 2000):

- Individual PDEC towers can be designed and programmed to cool different areas of the building at different times, according to needs.
- A more accurate building management strategy can be applied, thanks to increased sensitivity of the system.
- Individual towers provide better adaptation to irregular building typologies.
- Individual towers are more conducive to the detailed design of openings at different levels of the building in order to ensure uniform air velocities and specific flow rates.
- Individual towers allow better integration with wind catchers.

*Source:* Webster-Mannison (2003)

**Figure 6.8** *Interactive Learning Centre of the Charles Sturt University campus at Dubbo, New South Wales, Australia, showing four cool towers*

An additional complication is the necessity to allow adequate openings for the release of indoor air to environment. Each of the spaces to be cooled by air supplied from the cool tower must be equipped with a release vent. If this vent opens directly to the outdoors, wind pressure on the external surface may cause ingress of outside air: increased air pressure in the room may therefore prevent movement of cool air from the tower. This problem was reported in the Interactive Learning Centre of the Dubbo Campus of Charles Sturt University, New South Wales, Australia (see Figure 6.8), which is cooled by four shower towers (Webster-Mannison, 2003). Wind deflectors and baffles were subsequently fitted to exterior windows, and tests showed a decrease in airflow from the outside, allowing cooled air supplied by the towers to flow from the central space to perimeter offices. The building was also equipped with electric fans to support airflow between the central space and these offices, allowing either fully passive or fan-supported operation of the cooling system.

Although the Interactive Learning Centre has not been analysed comprehensively, the thermal performance of the central space, which is cooled by the PDEC towers, appears to be the equivalent, if not better than, conventional evaporative cooling systems (Webster-Mannison, 2003). The design of the building is innovative not only through its use of PDEC, but because it adopts an integrated approach to all aspects of its interaction with the environment, including rainwater harvesting and indoor air quality. Its iconic design demonstrates that although incorporating evaporative

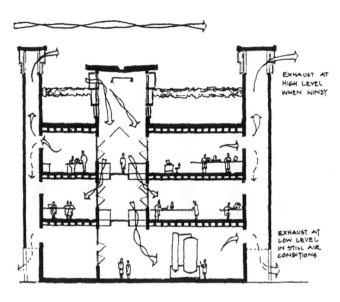

Source: Ford et al (1998)

**Figure 6.9** *Cross-section of the Torrent Research Centre building, Ahmedabad, India, showing downdraught cool towers and exhaust vents*

downdraught cool towers requires an unconventional building design and might generate some unforeseen engineering and building management problems, it also creates opportunities for exciting architecture.

An alternative solution, adopted at the Torrent Research Centre in India, is to construct exhaust chimneys that protrude above roof level, into which vents from individual rooms are connected (see Figure 6.9). A similar configuration was modelled analytically and using a perspex scale model for different heights of the inlet and outflow stacks, confirming that the generic solution is a viable one (Woods et al, 2003).

The design of the air inlet at the top of the tower is sensitive to the roof design of the specific building and to the prevailing wind conditions. In the absence of an appropriate wind catcher to deflect wind-generated airflow downwards, a vertical tower open at the top may act as a chimney, with air being sucked upwards by the low pressure zone created by the wind. This counteracts the thermal downdraught created by the evaporation of water and reduces the supply of cooled air available from the tower (Etzion et al, 1997). Experiments on a reduced-scale tower with a variety of wind catchers suggest that a fixed, curved deflector oriented perpendicularly to the direction of the wind is the most efficient design (Pearlmutter et al, 1996), evaluated on the basis of providing the maximum airflow through the tower for a given wind speed. Similar conclusions were obtained from a wind tunnel study of a hypothetical design for an office building in Catania, Italy, carried out as part of an EU-supported research project on PDEC (Francis, 2000). All designs

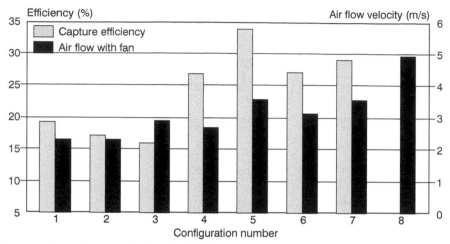

*Source:* adapted from Pearlmutter et al (1996)

**Figure 6.10** *Comparison of the wind capture efficiency of several generic configurations of wind catchers and airflow velocity obtained in the same test facilities with forced draught*

incorporating louvres are less efficient, particularly at low wind speeds, because some of the wind energy is absorbed to open the louvres against the forces of gravity and hinge friction. Fixed designs with other geometries were also shown to be marginally less efficient (see Figure 6.10).

Airflow in downdraught cool towers equipped with wind catchers may be characterized by an asymmetrical airflow pattern. Such asymmetry was, in fact, predicted in the CFD study of the hypothetical Seville office building (Cook et al, 2000), and was expected to result in uneven flow rates into the interior spaces being cooled. A further problem created by varying and uneven airflow in the tower is that the interaction with water sprayed near the inlet may be less than optimal, resulting in imperfect mixing of the water–air mixture, reduced cooling, and drift of water drops not evaporated into the spaces being cooled. The only practical means of overcoming this problem may be to add a baffle at the top of the tower, just below the inlet, to straighten flow irrespective of external wind.

A baffle may also be required to create uniform flow in the vicinity of the sprayers (Francis, 2000), especially in cool towers that combine wind-driven airflow with fan-assisted operation. The patterns of airflow in an experimental tower at Sde Boqer, Israel, monitored in both modes of operation, were, in fact, quite different (Erell et al, 2005a). An axial fan installed in the tower produced a strong centrifugal effect, resulting in a greater downward vector near the perimeter of the tower than in the middle, in addition to the tangential vector. This was in contrast to the flow pattern observed in wind-driven mode, which was characterized by a stronger vertical vector at the middle of the tower than near its perimeter. The implication of these findings, which are hardly surprising, is that unless flow in the tower can be homogenized by means of a baffle, sprayer locations and operational characteristics need to be modified when the fan comes into operation to create greater airflow than in purely passive operation.

In addition to the inlet of a downdraught tower, careful attention may have to be given to the aerodynamic design of its outlet, unless its cross-section area is much greater than that of the tower itself and little turbulence is created. This is because turbulence at the outlet can reduce flow through the tower substantially, especially if air speed is high. The velocity reported in several purely passive downdraught towers, such as the small tower described by Cunningham and Thompson (1986), is generally too low to require a sophisticated outlet design: air speed in this tower was less than 1m/s. However, in much taller towers, or if flow is assisted by an electric fan, air speed in the tower may be more substantial, and an aerodynamic deflector may reduce turbulence at the outlet and therefore increase airflow. A curved conical deflector was tested at the reduced-scale experimental tower at Sde Boqer, and increased flow through the tower by about 15 to 20 per cent in fan-assisted operation when the downward vector at the outlet was greater than 1.5m/s (Erell et al, 2005a).

# ROOF POND SYSTEMS

## Background

Cooling of building roofs by evaporation may be classified into two generic systems: roof ponds, in which a substantial mass of water cooled by evaporation is supported on an essentially flat roof; and spray or trickle systems, in which water is supplied as needed to cool exterior building surfaces exposed to a hot dry environment. The first category may include fixed or moveable insulation floating on the water or suspended above it.

The term 'roof pond' is commonly used to denote a system incorporating a pool of water that acts as a means of heat storage for a building and as a heat exchanger. In the cooling mode the roof pond acts as interim heat sink with heat from occupied spaces rising naturally and transferred to the water through the structural ceiling. The large volumetric heat capacity of water, about 4.18 kJ/m³/K, and the fact that it is cheap, widely available and non-toxic, make it an ideal material for storing heat. The pond cools at night by long-wave radiation to the sky or by evaporation if the water is not enclosed in watertight bags. Evaporation may be enhanced by spraying water using sprinklers or sprayers. During daytime, it is essential to protect the water from solar radiation and from high outdoor temperatures by means of a fixed or moveable protective cover, or by thermal insulation. In locations with fairly mild winter conditions, roof ponds may also be employed to provide some heating. On sunny days, exposure of the pond to solar radiation will warm up the water and can contribute heat to the spaces below. At night, and at times of low solar radiation or low heat demand, use of a cover can insulate the pond from the environment, thus reducing heat loss from occupied spaces.

Historically, the wetting of flat roofs by spraying or flooding has been used regularly in hot dry regions where direct evaporative cooling can provide relief on hot summer days. More recent scientific investigation of roof ponds began with the experiments and applications directed by Harold Hay since the 1950s in India, and during the 1960s and 1970s in the US. The concept patented by Harold Hay as 'Skytherm' (Hay and Yellot, 1969) combined the traditional functions of a roof with an effective natural heating and cooling system. The Skytherm system was applied successfully in several buildings in different parts of the US. In a typical application, a corrugated metal ceiling deck was used for structural support for the pond. The entire ceiling area of the building was treated with roof ponds. The water was contained in plastic bags with typical water depths of 100mm to 300mm. A watertight liner or coating was placed over the metal deck as additional protection from water leakage. The moveable insulation panels were commonly of 50mm to 75mm rigid foam and covered with a protective skin. The panels were operated manually or by automatic control, and a mechanical guidance and drive system physically relocated the panels in either position. In the cooling mode, the panels remain closed during daytime to protect the roof and building interior from solar radiation and from transmission heat gains. The panels are retracted at night to allow the pond to cool by thermal radiation to the sky and by convection to the outdoor air.

The following discussion focuses only on roof ponds cooled by evaporation. Cooling by long-wave radiation is the subject of Chapter 7.

## Roof pond components

The main components of a roof pond system are described as follows.

### Pond support

The building elements in contact with the roof pond must have a high thermal conductivity to provide close thermal coupling between the water and the occupied spaces. This can be provided by a metallic ceiling or a reinforced concrete slab. Metals have higher thermal conductivity and can be used in a very thin layer, thus ensuring very good thermal coupling without the additional thermal inertia provided by a thicker element, such as a concrete slab.

### Water container

The water may be contained in plastic bags, in which case it will not be exposed to evaporative fluxes and will not require replenishment; alternatively, water may be contained within the roof parapet over a watertight lining and its surface exposed to ambient air. In this case, water will be lost by evaporation, and it will be necessary to replenish it periodically to maintain the pond's level.

## Protective cover

In the cooling mode, use of a cover over a roof pond protects the water from unwanted solar gains. The cover may be fixed or moveable, made of cloth or plastic. An alternative to a fixed cover is the use of floating insulation (commonly polystyrene panels).

## Spraying and circulation

Spraying nozzles may be used to circulate water from the roof pond, injecting droplets to the air above the pond, cooling the water by evaporation.

# Design guidelines

The following design guidelines have been drawn up on the basis of parametric studies carried out using the ROOFSOL and PDEC Tool (RSPT) software (Salmeron et al, 2003; Yannas et al, 2006). They are backed up by experimental results obtained during the Roof Solutions for Passive Cooling (ROOFSOL) project.

A detailed mathematical model was developed for the calculation of energy balances and water temperatures in a pond (Rodriguez et al, 1998). The model assumes that water temperature is uniform throughout the pond – a condition that usually applies, in practice, because of natural convection. This was applied to the studies summarized below. The special case of roof ponds where water temperature is not uniform and temperature stratification is preserved will be discussed later.

Roof ponds cooled by evaporation may have insulating covers to reduce daytime solar gains, and water sprays to enhance the cooling rate. The design parameters considered, therefore, comprised the following: depth of water and solar absorptivity of the floor of the roof pond; thermal and solar-optical properties of the pond cover; and characteristics of the spraying system.

*Pond depth* determines the volume of water that it can contain per unit roof area – and, thus, the heat storage capacity and thermal inertia of the system. This is also a measure of the structural loads imposed by the pond. Recommended values for pond depth for the purpose of cooling are 25cm to 50cm, with the upper limit more appropriate for uncovered ponds and the lower for protected ponds. Sensitivity studies showed that for ponds that are shaded with a cover, and/or those cooled by sprays, pond depth has a smaller effect on cooling performance. In such cases, depth may be determined by water availability for circulation or from structural loading considerations. Ponds without a cover are exposed to higher heat gains and, thus, require additional water to prevent overheating.

The *colour of the pond floor* affects the solar absorptivity of this surface. Clearly, light colours should be preferred to reflect radiation. However, sensitivity analysis showed that most of the incoming radiation is absorbed by the water mass before it reaches the bottom of the pond. Thus, this parameter has a negligible effect – especially in the case of shaded ponds.

The *pond support structure* should provide good thermal contact between the roof pond and the building interior. It should therefore be constructed of a highly conductive material, such as a metal sheet, and be as thin as possible.

A roof pond may be *covered* to provide solar protection, reducing unwanted heat gains during the daytime. An opaque cover is preferred; but where total opacity is not possible, a material with solar transmissivity of no higher than 0.2 should be selected. It is preferable that the external surface of the cover is of a light colour in order to reflect solar radiation; this is of lesser importance where the cover material is a good insulator.

The main considerations relating to the operation of the cover are as follows:

- *Moveable cover:* opening the cover at night and closing it during daytime may allow a pond to achieve a good performance without the need for a spraying system.
- *Fixed cover:* with a cover permanently in place and separated from the water level by a ventilated air layer, the nocturnal radiant cooling potential is lost, but the convective and evaporative effects apply; in some locations, these could be sufficient to remove all thermal loads without need of a spraying system.
- *Fixed floating insulation:* this is cheap and easy to apply, and the thickness of the insulation can be specified as needed in order to minimize the heating effect of solar radiation; the insulation does, however, inhibit heat dissipation at night, and a spraying system becomes essential.

A roof pond may incorporate a spraying system to increase evaporative cooling. The design and operational characteristics of the spraying system have a strong influence on the performance of the system, and they are summarized as follows:

- *Droplet radius:* very small droplets may evaporate entirely, leading to excessive water consumption, while larger droplets may not produce a sufficient cooling effect. The optimum diameter is about 1mm, assuming a spray height of about 1m.
- *The height of spray:* typical spray heights in the range of 0.5m to 1m are sufficient to guarantee enough time for the droplets to cool.
- *Horizontal distribution of nozzles:* nozzles should be distributed uniformly around the pond surface; the arrangement must avoid overlap and waste of water.
- *Pond water circulation:* for a pond coupled to a single space, the entire volume of water in the roof pond should be circulated once every hour. Where an imposed thermal load is acting on the pond (e.g. if water from the pond is circulated through a heat exchanger, cooling panel, etc.), the spray flow rate should be increased in proportion to the thermal load.
- *Operational strategy:* the spray system should be turned off if the ambient wet bulb temperature rises above that of the pond temperature since in such cases spraying would lead to a warming-up of the pond.

Recommendations for each of the possible combinations of these features are given below (Yannas et al, 2006).

### Uncovered pond, no sprays

The water pond is permanently exposed to ambient air without a cover. There is no spraying system. This is the simplest roof pond configuration.

Water depth should be at least 30cm to reduce temperature fluctuations resulting from daytime solar gains. Water temperature will increase due to solar gains until compensated by spontaneous evaporation; typical water temperature fluctuation is around 10K.

The colour of the pond floor should be light in order to reduce solar absorptance. However, the effect of this factor is relatively small since even in a pond that is only 30cm deep, the water absorbs most of the incoming radiation.

### Uncovered pond, with spray

The spraying system operates continuously day and night over an uncovered pond to provide a cooling effect. The design of the pond is similar to that with no sprayers. In addition, the design and operation of the spraying system should consider the following factors:

- The water-spraying system should circulate the equivalent of about 1.0 to 1.5 volumes of the roof pond per hour.
- The minimum height of the spray should be 0.5m in order to create a sufficient time of fall for the drops to cool to the ambient wet bulb temperature.
- Average droplet radius should be 0.5mm to 1mm. Generating finer drops requires higher pressure at the sprayers and thus requires more pumping energy. Fine drops are also more likely to drift beyond the perimeter of the roof pond.

Limiting spray operation to night-time can conserve water. However, daytime operation may be required to maintain a stable water temperature in shallow ponds (< 30cm depth). For deeper ponds, the increase in water temperature during the daytime may be less than 7K to 8K, even in warm and sunny conditions.

### Fixed cover, no spray

The pond is shaded at all times. There is no spraying. In this configuration, the pond cover prevents overheating of water, while spontaneous evaporation lowers water temperature below the average ambient air temperature. The design and installation of the cover should be as follows:

- The cover should be installed at a sufficient height to allow unrestricted airflow above the surface of the water, typically about 30cm.

- If airflow is unrestricted, the main function of the cover is to shade the surface of the pond. It should therefore be opaque to solar radiation. Its solar absorptance, thermal conductance and the emissivity of its lower surface have a negligible effect on the temperature of the roof pond.

### Operable cover, with spray

The pond is covered during the daytime only, while the spraying system operates at night. In this configuration, the pond cover reduces fluctuation in pond temperature due to solar absorption during daytime, while spraying lowers pond temperature at night. The cooling output of such a pond is higher compared to an uncovered pond with sprays.

The thermal properties of an operable cover should conform to the recommendations for a fixed one. However, removal of the cover at night increases the structural complexity and requires a mechanical system and control mechanism.

The design of the spraying system is the same as that in an uncovered pond. The spraying system, furthermore, should be operated only at night in order to conserve water. However, changing environmental conditions may create a situation in which ambient wet bulb temperature increases above the temperature of the water in the pond. In this case, water spraying would lead to an increase in water temperature and should be stopped.

## Thermally stratified roof ponds

The analysis in the previous section was based on the assumption that water in the roof pond is well mixed and has negligible thermal stratification. While this is generally the case, a roof pond with substantial thermal stratification may have an advantage over a well-mixed one: water in contact with the structural support is colder, so heat loss from the building interior is greater. At the same time, warmer surface temperature increases the rate at which heat is emitted to the environment.

Natural convection between the surface layer and the rest of the pond can be inhibited, to a certain extent, by the insertion of a layer of fabric just below the surface of the water or floating on it. The fabric absorbs much of the solar radiation during the daytime. This, in addition to heat absorbed from the adjacent air by convection, creates a very thin layer of water at the surface that may be warmer by several degrees than the rest of the pond. Thermal equilibrium is maintained by heat loss through evaporation.

Comparative experiments on small (117cm × 117cm), shallow (17.6cm) ponds showed that the addition of fabric not only improved the performance of an exposed pond, but resulted in overall performance similar to that of a pond that was equipped with an insulating cover that was kept in place during the daytime and removed at night (Tang et al, 2003). Further improvement in performance was obtained by shading the pond with a fixed cover about 50cm above the surface. The main advantage of this technique lies in its simplicity:

no sprayers are required, and no mechanism for moving the insulation panels. However, it should be noted that this type of roof pond, like most other variants, is not appropriate where winter conditions require thermal insulation in cold weather.

## MULTIPLE-COMPONENT SYSTEMS

In many locations, evaporative cooling, whether in the form of PDEC or roof ponds, may be insufficient to provide acceptable levels of thermal comfort. A detailed study of thermal conditions in a hypothetical office building in Seville concluded that 'it is highly unlikely that PDEC will be viable without some form of mechanical support in Seville, irrespective of control efficiency' (Robinson et al, 2004). Nevertheless, the same study suggested that energy savings of up to 83 per cent were possible, depending upon occupancy and set point. However, the cost of installing a full-scale cooling system in order to guarantee thermal comfort in extreme conditions, combined with the complex control mechanisms required for operating PDEC effectively, requires careful examination of the economics of such a building. In order to be accepted more widely, evaporative cooling systems – like most other passive cooling techniques – should allow modular operation of components in order to deal with successively higher loads. Ideally, all of the components should be integrated in such a way that building operation progresses seamlessly from one mode to the next.

The Palenque Entertainment Centre, constructed in Seville for Expo 1992, provides an example of this approach (Velazquez et al, 1991). The centre includes a stage and seating for 1500 people near the centre of the 8000 square metre complex, and additional public spaces surrounding it. Because analysis showed that evaporative cooling of air supplied to the public areas would be insufficient in extreme weather conditions, a multi-tiered approach was adopted. First, undesirable heat gains from the environment were reduced by several methods: by spraying water on the exterior surface of the white polyvinyl chloride (PVC) tent-like structure serving as the roof, using evaporative cooling to lower surface temperature; by restricting the flow of hot dry ambient air through the covered area by means of a dense wall of vegetation; and by cooling this air somewhat using micronizers arranged around the perimeter of the building. Chilled air supplied to the public spaces was cooled by a three-stage system incorporating both evaporation and conventional compression cooling. Supply air was first pre-cooled by means of a coil circulating water from open pools integrated within the complex (which were themselves cooled by evaporation using sprayers). It was then cooled further by evaporation in a second section. If the temperature of the supply air could not be maintained at 19°C or less, due to the combination of internal loads and environmental conditions, this evaporative cooler was replaced by a conventional auxiliary compression chiller. Estimates made during the design of this plant indicate that the auxiliary chiller would supply only 25 per cent of the cooling requirements of Palenque, and its installed capacity is only 43 per cent of the capacity required for a standard conventional cooling system (Velazquez et al, 1991).

# HEALTH AND THERMAL COMFORT

## Thermal comfort

Evaporative cooling differs from other modes of cooling in that heat is not necessarily emitted to an environmental sink. Instead, as has been noted above, sensible heat is converted to latent heat. This has important implications with respect to system design and to human thermal comfort.

Conventional heating, ventilating and air-conditioning (HVAC) design standards require that air temperature and moisture levels in the building remain within fairly narrow limits. To reduce the energy consumption required to maintain the specified conditions, building envelopes are designed to be as airtight as possible, and the minimum air change rate is determined from health considerations. Direct evaporative cooling systems, such as PDEC, cannot usually cool the supply air to the same extent that compression chillers do. To compensate for this, the volume of air supplied is large, and air change rates are typically an order of magnitude greater than in buildings with conventional HVAC. The combination of environmental parameters – air temperature, moisture content and air speed – is therefore likely to be somewhat different in a building cooled by direct evaporation, compared to one with standard HVAC. Furthermore, these conditions are likely to vary over time in response to changing ambient conditions and to internal loads. It is therefore of interest to ascertain that perceptions of thermal comfort in PDEC buildings are not inferior to similar buildings cooled by conventional means. Although the issue of thermal comfort is the subject of a separate chapter, it may be useful to review several studies devoted specifically to comfort in such buildings.

A detailed model of human physiology was used to study dynamic thermal sensation in conditions likely to be encountered in a building cooled by PDEC (Fiala et al, 1999). The study concluded that elevated relative humidity of up to 80 per cent had little effect on comfort perception and thermal acceptability for dry bulb temperatures in the range of 24°C to 27°C. At higher air temperature, increasing the humidity level led to a substantial increase in the percentage of people likely to be dissatisfied. As expected, increasing air speed from 0.3m/s to 0.8m/s improved perceptions of comfort at high temperature and humidity. However, the study also found that rapid changes in air speed were found to have a detrimental effect on the overall comfort perception, and should therefore be avoided. A subsequent study by the same authors (Martinez et al, 2000) found that adaptive behaviour – in particular, modifying clothing worn – could extend the comfort envelope to as much as 28.8°C at low levels of humidity (17 per cent). Increasing air speed could extend this envelope further; but the additional benefits of inducing air speeds greater than 0.9m/s were marginal.

Givoni (1992) proposed that analysis of local climatic conditions plotted on the psychrometric chart could provide an indication of the suitability of specific passive heating and cooling strategies with respect to provision of thermal comfort in the building interior. The suitability of his guidelines to PDEC buildings was assessed by conducting a detailed dynamic thermal simulation of a hypothetical office building in Seville, Spain, cooled by a combination of night ventilation and PDEC (Lomas et al, 2004). The study proposed modifications to the original building bioclimatic charts (BBCCs) that retain the benefits of the original scheme, but introduced additional parameters: the climatic limit of thermal comfort (CLTC), which defines the ambient conditions beyond which indoor thermal comfort is virtually never achieved, and the lower limit of thermal discomfort (LLTD), which defines ambient conditions below which there is a very low risk of summertime thermal discomfort. The study further proposed that modelling the functionality and control strategies of a PDEC system should be improved to allow the boundaries fixing the CLTC and LLTD to be established more accurately.

## Health

Evaporative cooling towers may provide an ideal environment for the proliferation of the bacteria responsible for legionnaire's disease. The disease, first identified in 1976, is a rare but life-threatening form of pneumonia caused by inhalation of fine water droplets contaminated with legionella bacteria. These bacteria are common and may be found in natural water sources, such as lakes and rivers, and in water services within buildings, particularly services that are lukewarm, corroded or have organic debris.

Brundrett (2002) recommends four methods of minimizing the risk of infection:

1  Prevent multiplication of the bacteria in the water. This may be done by temperature control – the bacteria die at water temperature in excess of 60°C and remain inactive at temperatures below 20°C. The storage time of bacteria should be kept to no more than one day. Nutrient supply in the water should be minimized. Biocides may be used, although the latter method is not always effective.
2  Prevent aerosols of water escaping the plant because the bacteria cannot survive outside of water. Cooling towers must be fitted with drift eliminators at the outlet if full evaporation of water cannot be guaranteed.
3  Prevent the aerosol from reaching people.
4  Protect susceptible people, especially smokers, the elderly and those with an existing underlying illness.

Although the characteristics of PDEC systems make them particularly susceptible to legionnaire's disease, careful engineering and meticulous maintenance can meet all health standards, as demonstrated by the Malta Stock Exchange building (Ford and Diaz, 2003).

## CONCLUSION

The application of evaporative cooling to buildings requires a thorough understanding of local climatic conditions. However, energy savings from incorporating evaporative cooling may be substantial in most climates. Indirect evaporative cooling systems are already available that can provide the same level of comfort and user control as conventional compression cooling in some buildings. The challenge is to attain similar levels of engineering and confidence in the performance of PDEC and roof ponds. As the examples discussed in this chapter show, application of evaporative cooling may pose unexpected problems, especially with respect to control of temperature and airflow. However, the experience gained through installation and operation of PDEC in buildings such as the Interactive Learning Centre at Charles Sturt University, Dubbo, in New South Wales, Australia (Webster-Mannison, 2003), shows that innovative design can produce not only exciting architecture but also effective, low-cost and energy-efficient cooling. Since the underlying physical process of evaporation is well understood, research efforts should focus on overcoming problems of implementation, such as those highlighted in some of the case studies described in this chapter. As the use of computer simulation of HVAC systems in buildings becomes more widespread and confidence in its results grows, it may be possible to benefit from evaporative cooling to a greater extent than is possible today.

## ACKNOWLEDGEMENTS

The experimental work that underpins parts of this chapter was carried out at Sde Boqer, Israel, over many years. Professor Yair Etzion played a leading role in all of this research, which would not have been possible without his contribution.

## REFERENCES

Al-Megren, K. (1987) *Wind Towers for Passive Ventilated Cooling in Hot-Arid Regions*, PhD thesis, Department of Architecture, University of Michigan, MI
Alvarez, S., Rodriguez, E. and Molina, J. L. (1991) 'The Avenue of Europe at Expo '92: Application of cool towers', in *Architecture and Urban Space: 9th PLEA International Conference*, Seville, Spain, 24–27 September
Alvarez, S. and Sanchez, F. (2000) *Thermally Driven Cooling and Ventilation System (PDEC)*, University of Seville, Seville

Argiriou, A. (1995) *Natural Cooling Techniques*, CIENE, Athens, Greece

Arya, P. (1988) *Introduction to Micrometeorology*, Academic Press, San Diego, CA

ASHRAE (American Society of Heating, Refrigerating and Air-Conditioning Engineers) (2005) *ASHRAE Handbook: Fundamentals*, American Society of Heating, Refrigerating and Air Conditioning Engineers, Inc, Atlanta, GA

Bahadori, M. (1985) 'An improved design of wind towers for natural ventilation and passive cooling', *Solar Energy*, vol 35, no 2, pp119–129

Brundrett, G. (2002) 'Controlling legionnaire's disease', *Indoor and Built Environment*, vol 12, pp19–23

Cook, M., Robinson, D., Lomas, K., Bowman, N. and Eppel, H. (2000) 'Passive down-draft evaporative cooling: Airflow modelling', *Indoor and Built Environment*, vol 9, pp325–334

Costelloe, B. and Finn, D. (2003) 'Indirect evaporative cooling potential in air-water systems in temperate climates', *Energy and Buildings*, vol 35, pp573–591

Cunningham, W. and Thompson, T. (1986) 'Passive cooling with natural downdraft cooling towers in combination with solar chimneys', in *Proceedings of the 5th International PLEA Conference: Passive and Low Energy in Housing*, Pecs, Hungary

Erell, E., Etzion, Y., Pearlmutter, D., Guetta, R., Pecornik, D. and Krutzler, F. (2005a) 'A novel multi-stage evaporative cool tower for space cooling. Part 1: Aerodynamic design', in *Passive and Low Energy Cooling for the Built Environment*, Santorini, Greece, 19–21 May

Erell, E., Etzion, Y., Pearlmutter, D., Guetta, R., Pecornik, D. and Krutzler, F. (2005b) 'A novel multi-stage evaporative cool tower for space cooling. Part 2: Preliminary experiments with a water spraying system', *Passive and Low Energy Cooling for the Built Environment*, Santorini, Greece, 19–21 May

Etzion, Y., Pearlmutter, D., Erell, E. and Meir, I. (1997) 'Adaptive architecture: Integrating low-energy technologies for climate control in the desert', *Automation in Construction*, vol 6, pp417–425

Fiala, D., Lomas, K., Martinez, D. and Cook, J. (1999) 'Dynamic thermal sensation in PDEC buildings', in *Proceedings of the 16th PLEA International Conference: Sustaining the Future: Energy – Ecology – Architecture*, Brisbane, Australia

Ford, B. and Diaz, C. (2003) 'Passive downdraft cooling: Hybrid cooling in the Malta Stock Exchange', in *Proceedings of the 20th Plea International Conference: Rethinking Development – Are We Producing a People Oriented Habitat?*, Santiago, Chile, 9–12 November

Ford, B., Patel, N., Zaveri, P. and Hewitt, M. (1998) 'Cooling without airconditioning', *Renewable Energy*, vol 15, pp177–182

Ford, B. and Schiano-Phan, R. (2003) 'Evaporative cooling using porous ceramic evaporators – product development and generic building integration', in *Proceedings of the 20th Plea International Conference: Rethinking Development – Are We Producing a People Oriented Habitat?*, Santiago, Chile, 9–12 November

Francis, E. (2000) 'The application of passive downdraft evaporative cooling (PDEC) to non-domestic buildings', in *Proceedings of the 17th PLEA International Conference: Architecture, City, Environment*, Cambridge

Givoni, B. (1992) 'Comfort, climate analysis and building design guidelines', *Energy and Buildings*, vol 18, pp11–23

Givoni, B. (1993) 'Semi-empirical model of a building with a passive evaporative cool tower', *Solar Energy*, vol 50, no 5, pp425–434

Givoni, B. (1994) *Passive and Low Energy Cooling of Buildings*, John Wiley and Sons, New York

Givoni, B. (1997) 'Performance of the "shower" cooling tower in different climates', *Renewable Energy*, vol 10, no 2/3, pp173–178

Guetta, R. (1993) *Energy from Dry Air: A Mathematical Model Describing Airflow and Evaporation of Water Drops in Vertical Tubes*, PhD thesis, Technion, Israel Institute of Technology, Haifa

Hay, H. and Yellot, J. (1969) 'Natural cooling with roofpond and moveable insulation', *ASHRAE Transactions*, vol 75, no 1, pp165–177

Ibrahim, E., Shao, L. and Riffat, S. (2003) 'Performance of porous ceramic evaporators for building cooling application', *Energy and Buildings*, vol 35, pp941–949

Incropera, F. and De Witt, D. (1990) *Fundamentals of Heat and Mass Transfer*, John Wiley and Sons, New York

Karakatsanis, C., Bahadori, M. and Vickery, B. (1986) 'Evaluation of pressure coefficients and estimation of air flow rates in buildings employing wind towers', *Solar Energy*, vol 37, no 5, pp363–374

Lomas, K., Fiala, D., Cook, M. and Cropper, P. (2004) 'Building bioclimatic charts for non-domestic buildings and passive downdraft evaporative cooling', *Building and Environment*, vol 39, pp661–676

Martinez, D., Fiala, D., Cook, J. and Lomas, K. (2000) 'Predicted comfort envelopes for office buildings with passive downdraft evaporative cooling', in *ROOMVENT 2000*, Reading, UK

Papagiannopoulos, G. and Ford, B. (2003) 'Evaporative cooling using porous ceramic bricks: Experimental results from Greece', in *Proceedings of the 20th Plea International Conference: Rethinking Development – Are We Producing a People Oriented Habitat?*, Santiago, Chile, 9–12 November

Pearlmutter, D., Erell, E., Etzion, Y., Meir, I. and Di, H. (1996) 'Refining the use of evaporation in an experimental down-draft cool tower', *Energy and Buildings*, vol 23, pp191–197

Pearlmutter, D., Hongfa, D., Etzion, Y., Erell, E. and Meir, I. (1994) 'The development of an evaporative cooling tower for semi-enclosed spaces', in *Architecture of the Extremes: 11th PLEA International Conference*, Dead Sea, Israel

Robinson, D., Lomas, K., Cook, M. and Eppel, H. (2004) 'Passive down-draft evaporative cooling: Thermal modelling of an office building', *Indoor and Built Environment*, vol 13, pp205–221

Rodriguez, E., Alvarez, S. and Martin, R. (1991a) 'Direct air cooling from water drop evaporation', in *Architecture and Urban Space: 9th PLEA International Conference*, Seville, Spain, 24–27 September

Rodriguez, E., Alvarez, S. and Martin, R. (1991b) 'Water drops as a natural cooling resource – physical principles', *Architecture and Urban Space: 9th PLEA International Conference*, Seville, Spain, 24–27 September

Rodriguez, E., Molina, J. L., Guerra, J. J. and Esteban, C. J. (1998) *Detailed Modelling of Roof Ponds: Final Report of Task 2, ROOFSOL Project*, European Commission, Brussels

Salmeron, J., Sanchez, F., Gordillo, M., Molina, J. L. and Alvarez, S. (2003) 'RSPT: A tool for simulating natural cooling techniques based on the roofs of buildings', in *Proceedings of the 20th Plea International Conference: Rethinking Development – Are We Producing a People Oriented Habitat?*, Santiago, Chile, 9–12 November

Tang, R., Etzion, Y. and Erell, E. (2003) 'Experimental studies on a novel roof pond configuration for the cooling of buildings', *Renewable Energy*, vol 28, pp1513–1522

Velazquez, R., Guerra, J., Alvarez, S. and Cejudo, J. (1991) 'Case study of outdoor climatic comfort: The Palenque at Expo '92', in *Architecture and Urban Space: 9th PLEA International Conference*, Seville, Spain, 24–27 September

Webster-Mannison, M. (2003) 'Cooling rural Australia: Passive downdraft evaporative cooling, Dubbo Campus, Charles Sturt University', *The Official Journal of the Australian Institue of Refrigeration, Air Conditioning and Heating*, vol 2, no 1, pp22–26

Woods, A., Short, A. and Gladstone, C. (2003) 'Stack driven natural ventilation with pre-cooled inflow from a central atrium', in *Proceedings of the 20th Plea International Conference: Rethinking development – Are We Producing a People Oriented Habitat?*, Santiago, Chile, 9–12 November

Yajima, S. and Givoni, B. (1997) 'Experimental performance of the shower cooling tower in Japan', *Renewable Energy*, vol 10, no 2/3, pp179–183

Yannas, S., Erell, E. and Molina, J. L. (2006) *Roof Cooling Techniques: A Design Handbook*, Earthscan, London

Yellot, J. (1989) 'Evaporative cooling', in Cook, J. (ed) *Passive Cooling*, MIT Press, Cambridge and London

# 7

# Radiative Cooling

*Evyatar Erell*

## INTRODUCTION

Radiative cooling of buildings has attracted considerable research over the years, much of it focused on evaluating the magnitude of the resource and the variations in cooling potential among different locations. However, radiative cooling is still not applied, in practice, in conventional buildings, partly because there has been relatively little research by engineers and architects dedicated to resolving problems of practical application.

This chapter will attempt to address this shortcoming. It begins with a brief overview of the principles governing radiant heat transfer as a foundation for the following sections dealing with recent research on application of radiative cooling to buildings. Readers interested in a more comprehensive discussion of the underlying mechanisms may refer to previous publications on this subject, such as Martin (1989), or to general texts on heat transfer, such as Incropera and De Witt (1990) or Duffie and Beckman (1991). The second part of the chapter deals with ways to assess the degree to which climatic conditions affect the applicability of radiative cooling in a specific location. The third section introduces the main concepts underlying the implementation of radiative cooling, in practice. The final section, which constitutes the main part of this chapter, includes a detailed discussion of specific issues relating to the construction and operation of radiative cooling systems for buildings, including coupling with thermal mass; the question of wind screens; the use of selective coatings; the effect of the radiator's angle of tilt; the detailed design of radiators for cooling water, including a discussion of geometry, materials and operating parameters; and general issues regarding the integration of radiative cooling components in buildings.

# RADIANT HEAT TRANSFER

The relationship between the total hemispherical radiation emitted by a black body and its surface temperature is given by the Stefan-Boltzmann Law:

$$R° = \sigma T^4 \tag{1}$$

where $R°$ is the total radiant energy in watts per square metre ($W/m^2$), $T$ is the absolute temperature in K and $\sigma$, known as the Stefan-Boltzmann constant, is $5.67 \times 10^{-8} W/m^2/K^4$.

The radiation given off by a real object may be calculated by introducing its total hemispherical emissivity, $\varepsilon$, which is, by definition, less than unity:

$$R = \varepsilon \sigma T^4 \tag{2}$$

The wavelength distribution of radiation emitted by a black body is given by Planck's Law:

$$R°_\lambda (T) = \frac{C_1}{\lambda^5 [e^{(C_2/\lambda T)} - 1]} \tag{3}$$

where $R°_\lambda (T)$ is the emittance at temperature $T$ (K) for the wavelength $\lambda (\mu m)$, and the constants are $C_1 = 3.741 \times 10^{-16} m^2 W$ and $C_2 = 1.4388 \times 10^{-2} mK$.

This function approaches zero at very small and very large wavelengths. The wavelength at which a black body emits radiation with the highest intensity depends only upon the temperature of the emitting surface, and may be calculated from Wien's Displacement Law, as follows:

$$\lambda_{max} \cdot T = 2897.8 \tag{4}$$

where $\lambda_{max}$ is the wavelength of maximum emission, in $\mu m$, and $T$ is the absolute temperature of the radiating surface (K).

The radiation incident on a body may be absorbed, reflected or transmitted through it. The fractions of the absorbed, reflected and transmitted radiation are called absorptivity ($\alpha$), reflectivity ($\rho$) and transmissivity ($\tau$), respectively. The sum of these fractions is, by definition, unity:

$$\alpha + \rho + \tau = 1 \tag{5}$$

It should be noted that these are total hemispherical values and characterize the overall interaction between an object and the radiation impinging on it.

The relation between the emitting and absorbing properties of a body is given by Kirchoff's Law, which states that for every wavelength and for every direction of propagation, the directional spectral emissivity is equal to its directional spectral absorptivity:

$$\alpha_\lambda (T, \varphi, \theta) = \varepsilon_\lambda (T, \varphi, \theta) \tag{6}$$

## APPLICABILITY: THE RADIATIVE COOLING RESOURCE

The previous discussion has dealt with the ideal process of radiative heat loss from a single surface. All terrestrial objects, however, not only emit long-wave radiation, but also receive it from their surroundings. Radiation must thus be viewed as a process involving the exchange of energy between at least two objects (radiative heat exchange occurs even between identical surfaces at the same temperature).

Building-integrated cooling radiators are installed on the roof, and radiant exchange therefore occurs primarily with the sky, which is typically colder than most terrestrial surfaces. The net outgoing radiative flux from a terrestrial object exposed to the open sky is equal to its emitted flux minus the incident flux absorbed from the atmosphere. While both the outgoing terrestrial radiation and the incoming sky radiation are relatively large fluxes, the net radiative flux, which is the difference between them, may be quite small. This is an extremely important observation because relatively small changes in either the environmental conditions (which affect the sky radiation) or in the temperature of the radiator may result in substantial changes to the net radiative flux.

## Atmospheric long-wave radiation

The long-wave radiant flux received at the Earth's surface varies with time and location as a result of differences in the properties of the atmosphere. The main components of the Earth's atmosphere – oxygen and nitrogen – are nearly transparent to infrared radiation. However, three relatively minor constituents – carbon dioxide, ozone and water vapour – are responsible for its high absorption in the infrared spectrum and, therefore (according to Kirchoff's Law), also for the emission of atmospheric long-wave radiation:

- Carbon dioxide comprises only 0.03 per cent of the atmosphere by volume; yet the increase in atmospheric carbon dioxide is one of the major reasons for global climate change. However, because it is present in similar concentrations throughout the world, its relevance to the design of radiative cooling applications is limited.
- Ozone, which is present mostly in the stratosphere, contributes about 1 per cent of the sky radiation and has a nearly constant peak of emission at $9.6\mu m$.
- Water vapour, unlike carbon dioxide, is found in the atmosphere in varying concentrations. Atmospheric humidity and the presence of clouds are the major environmental factors affecting the net radiative flux at a given location and specified time because water has a very high emissivity (close to unity).

More than 90 per cent of the total sky radiation, to which water vapour contributes over 95 per cent, is emitted by the lowest 5km of the atmosphere

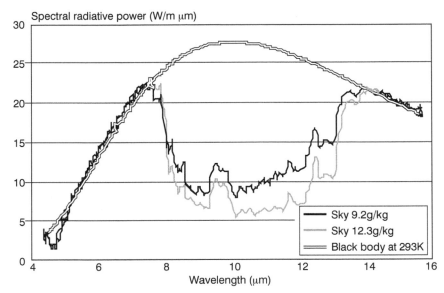

Spectral radiative power (W/m μm)

Sky 9.2g/kg
Sky 12.3g/kg
Black body at 293K

Wavelength (μm)

*Source:* adapted from Brunold (1989)

**Figure 7.1** *Measured spectra illustrating the effect of water vapour on atmospheric long-wave radiation received at the surface of the Earth*

(Argiriou, 1996). The spectral distribution of the sky radiation (Figure 7.1) is very similar to that of a black body at a temperature equal to the dry bulb temperature of the air near the ground, except in the spectral region of 8μm to 13μm. This part of the spectrum, known as the 'atmospheric window', is nearly transparent to infrared radiation – if the atmosphere is very dry. The presence of moisture in the atmosphere increases the sky radiation significantly, reducing the potential for radiative cooling at the Earth's surface accordingly. The importance of the spectral location of the atmospheric window is great because it corresponds to the spectrum in which terrestrial surfaces (most of which have a temperature of 270K to 330K) radiate with maximum intensity.

The incoming atmospheric infrared radiation can be expressed in two ways:

1  The sky may be assumed to behave like a black body, with an emissivity of 1.0. In this case, the long-wave radiation received from the sky is given by the expression:

$$L\downarrow = \sigma T_{sky}^{4} \qquad [7]$$

which requires a means of calculating $T_{sky}$, the apparent temperature of the sky (the apparent sky temperature at a particular location and time is defined as the temperature that a black body would have in order to radiate the same amount of energy as that received from the sky by an unobstructed horizontal radiator).

2    The sky may be assumed to have a temperature equal to the ambient dry bulb temperature near the ground $(T_a)$, in which case the differences in radiation emitted due to variations in atmospheric moisture content are accounted for by modifying the sky emissivity, $\varepsilon_{sky}$:

$$L\!\downarrow \; = \sigma\varepsilon_{sky}T_a^4 \tag{8}$$

where $T_a$ is the dry bulb temperature of the air near the ground in Kelvin.

Both of these methods deal only with the total flux of radiant energy, assumed to have a continuous spectrum, irrespective of the actual spectral distribution of the incoming sky radiation. They are applicable when the terrestrial radiating surface approximates a black body or is grey body – that is, it absorbs all wavelengths indiscriminately (most natural materials are grey bodies; but several building materials, particularly polished metals, are spectrally selective and have different emissivities in the long-wave part of the spectrum and in the solar spectrum).

Incoming long-wave radiation $(L\!\downarrow)$ is not generally measured at meteorological stations. Various statistical correlations have therefore been proposed between $L\!\downarrow$ and meteorological parameters that are measured on a widespread basis and which may be used as surrogates for atmospheric emissivity. Some of these models of atmospheric emissivity are summarized in Table 7.1 (Oke, 1987).

Arnfield (1979) evaluated the accuracy of some of these expressions on the basis of independent empirical data. He suggested that the Swinbank (1963) and Idso and Jackson (1969) relations performed best, and since they require no calibration to local conditions, they possess a high degree of spatial and temporal stability. The Brunt (1932) relation, on the other hand, requires *a priori* knowledge of local conditions for selection of appropriate regression coefficients. All expressions required correction for cloudy conditions (see below). However, the Idso and Jackson (1969) formula, with appropriate adjustment for cloud type and amount, gave good predictions of daily totals (albeit with variable accuracy on an hourly basis). An evaluation of eight parameterization models, including all of the ones in Table 7.1 except number 5, suggests that all of these schemes tend to underestimate incoming long-wave radiation in very clear, cold weather by $20W/m^2$ to $60W/m^2$ (Niemela et al, 2001). The study, based on measurements carried out in Finland, suggested different coefficients for two discrete ranges of $e_0$; but as its authors suggest, it is probably best suited to high-latitude locations.

In the models in Table 7.1, the effect of altitude on long-wave radiation at the surface is incorporated indirectly through its effect on air temperature and atmospheric moisture content. A recent model (Iziomon et al, 2003) proposes a parameterization scheme in which the effects of altitude are incorporated directly within two empirical coefficients:

**Table 7.1** *Empirical parameterizations for calculating the atmospheric emissivity of clear skies,* $\varepsilon_a$

| Source | Equation | Remarks |
|---|---|---|
| 1  Brunt (1932) | $\varepsilon_{a(0)}=0.51+0.066e_a^{1/2}$ | Coefficients vary with geographic location |
| 2  Brutsaert (1975) | $\varepsilon_{a(0)}=0.575e_a^{1/7}$ | Coefficient modified by Idso (1981) |
| 3  Idso (1981) | $\varepsilon_{a(0)}=0.70+5.95*10^{-5}T_a\exp(1500/T_a)$ | |
| 4  Swinbank (1963) | $\varepsilon_{a(0)}=0.92*10^{-5}T_a^2$ | For $T_a > 0°C$ |
| 5  Idso and Jackson (1969) | $\varepsilon_{a(0)}=1-0.261\exp\{-7.77*10^{-4}(273-T_a)^2\}$ | |

*Note:* All of the above equations use $T_a$ in Kelvin and $e_a$ in millibars.
*Source:* Oke (1987, p374)

$$L\downarrow = \sigma T_a^4 \{1 - X_s \exp(-Y_s e / T_a)\} \qquad [9]$$

The empirical parameters $X_s$ and $Y_s$ have values of 0.35 and 10K/hPa, respectively, for lowland sites and values of 0.43 and 11.5K/hPa for mountain sites. Vapour pressure $e$ is measured in hectopascals (hPa).

Clouds have a strong influence upon long-wave exchange in the atmosphere. Dense low stratus formations with relatively high base temperatures are the most intense emitters, while thin, high cirrus clouds, which are much colder, contribute much less to the downward-directed atmospheric infrared radiation.

Oke (1987) proposed the following modifications to the values for incoming sky radiation ($L\downarrow$) and for the net long-wave radiation ($L^*$) to account for the effect of clouds:

$$L\downarrow = L\downarrow_{(0)} (1 + an^2) \qquad [10]$$

$$L^* = L^*_{(0)} (1 - bn^2) \qquad [11]$$

where the subscript (0) refers to the clear sky radiation, the constants $a$ and $b$ are a function of the cloud type (see Table 7.2), and $n$ is the fraction of the sky covered in cloud, expressed in tenths on a scale from zero to unity.

A simpler cloud correction function that does not require input of cloud characteristics was proposed by Martin (1989):

$$L\downarrow = (1 + 0.0224n - 0.0035n^2 + 0.00028n^3)L\downarrow_{clear} \qquad [12]$$

If the difference between the absolute temperatures of the surface and the sky is not large, the following linear form of the Stefan-Boltzmann Law (Martin,

**Table 7.2** *Values of the coefficients used in Equations 10 and 11 to compensate for the effect of clouds on long-wave radiation in the atmosphere*

| Cloud type | Typical cloud height (km) | Coefficients | |
|---|---|---|---|
| | | a | b |
| Cirrus | 12.20 | 0.04 | 0.16 |
| Cirrostratus | 8.39 | 0.08 | 0.32 |
| Altocumulus | 3.66 | 0.17 | 0.66 |
| Altostratus | 2.14 | 0.20 | 0.80 |
| Cumulus | | 0.20 | 0.80 |
| Stratocumulus | 1.22 | 0.22 | 0.88 |
| Stratus | 0.46 | 0.24 | 0.96 |
| Fog | 0 | 0.25 | 1.00 |

*Source:* Oke (1987, p374)

1989) may provide a useful approximation of the net long-wave radiative heat transfer at the surface:

$$L^* = 4 \, \varepsilon \, \sigma T_a^3 \, (T_r - T_s) \qquad [13]$$

where $L^*$ (W/m²) is the net radiative heat loss, $T_a$ (K) is the ambient air temperature, $T_r$ (K) is the radiator temperature, $T_s$ (K) is the apparent sky temperature, and $\varepsilon$ and $\sigma$ are the emissivity and Stefan-Boltzmann constant, respectively.

In clear sky conditions, net radiative heat transfer may also be approximated by introducing a coefficient of radiant heat transfer at the surface. Setting this coefficient, $h_r$, to equal $4\varepsilon\sigma T_{air}^3$, Equation 13 may be rewritten as:

$$R_{net} = h_r \, (T_r - T_s). \qquad [14]$$

For the range of air temperatures normally encountered where radiative cooling may be expected to be applied – say 15°C to 30°C (288K to 303K) – the expression $4\varepsilon\sigma T_{air}^3$ has a value of approximately 5W/m²/K. The value of $h_r$ was also determined experimentally (Oliveti et al, 2003) for a range of temperatures and was found to vary within a fairly narrow range – about 3.9W/m²/K to 6.5W/m²/K. The typical diurnal range is about 0.5W/m²/K.

## Spatial and temporal variations

As the preceding section illustrates, the long-wave radiation received at the Earth's surface from the atmosphere depends strongly upon atmospheric moisture and upon cloud cover, both of which vary substantially in time and space. In order to assess the potential for radiative cooling at a given location, this flux must be evaluated for the duration of the period in which cooling is required. Unfortunately, infrared sky radiation is not measured in standard weather stations. Furthermore, pyrgeometers require very careful calibration and frequent maintenance; therefore, measured data may sometimes be inaccurate or unreliable. The only practical way of obtaining detailed time series reflecting the geographical and seasonal variations of the atmospheric long-wave radiation must rely on calculations based on measurements of air temperature and humidity near the surface, and of cloud cover, where available. To account for inter-annual variations, Typical Meteorological Years (TMYs) are assembled from composite data obtained over periods of many years. Such data have become available for an increasing number of locations and may be used to obtain reliable estimates of the potential for applying radiative cooling effectively.

The variation in the sky temperature depression over time and space may be presented by a number of means (the statistics are usually restricted to night-time hours only):

- tables giving the frequency distribution of the sky temperature depression for a list of locations over a given period of time, such as a month (Argiriou, 1996);
- monthly or seasonal histograms plotting the frequency of occurrence of a given sky temperature depression, either as an absolute number of hours or as a percentage of the time (Martin and Berdahl, 1984);
- maps with contour lines indicating the average sky temperature depression for a given period, such as monthly or seasonal data (Martin and Berdahl, 1984);

*Source:* Molina (1998)

**Figure 7.2** *Spatial variations in the potential for radiative cooling in Europe, represented by the sky temperature depression: total degree hours between 10 pm and 6 am during the months of June to September*

- colour-coded maps showing the spatial distribution of degree hours of the sky temperature depression (see Figure 7.2).

The above methods display only the climatic data characteristic of a given location or region, but give no indication of the actual cooling output that may be obtained from a specific radiator at a given location. The data used to generate these representations may also be used to calculate the cooling output of a hypothetical radiator if an appropriate model is available to describe the energy exchange of a specific design. Although such calculations necessarily represent the performance of only one radiator, they have the advantage of providing the user with an indication of the useful cooling power that may be obtained by the system. If the simulation is carried out for a large number of locations, the resulting data may be used to evaluate the applicability of the design in question at different locations. As in the case of the sky temperature depression, the output may be presented in tabular form (Argiriou, 1996) or in a map (Martin, 1989).

A further advantage of assessing the suitability of a location for radiant cooling on the basis of a full cooling simulation is that such a calculation takes into account not only the sky temperature depression, but also the difference between the temperature of the radiator surface and the adjacent ambient air (the air temperature depression). The importance of assessing the combined effect of both environmental parameters is illustrated in Figure 7.3. The graph shows the measured (instantaneous) cooling power of a radiator on the left-hand scale, and the sky temperature depression and the air temperature depression on the right, for the duration of one night. During the first part of the night, until about midnight, the cooling output increased continuously, although the sky temperature depression was almost constant or even declined a little. This is because ambient air temperature decreased much more rapidly than the temperature of the water reservoir that was being cooled. Much of the cooling is the result of convection rather than radiation. After midnight, the sky became overcast, and the sky temperature depression dropped substantially. Once the air temperature depression stabilized, the total cooling output of the radiator declined by almost 40 per cent as the radiative loss declined. At about 3 am, the sky temperature depression increased abruptly – and the total cooling power of the radiator increased substantially as a result. The increase in the sky temperature depression also resulted in a drop in the ambient air temperature, lagging by about half an hour, as indicated by the increase in the air temperature depression that began at about 3.30 am. This further reinforced the cooling output of the radiator until about 4 am.

The example in Figure 7.3 demonstrates that assessing the potential for radiative cooling at a given location requires a detailed description of the climate. While the sky temperature depression and air temperature near the ground are related, the relationship is often complex. As the discussion in the rest of this chapter will show, the cooling output of a radiator depends upon a number of environmental factors, in addition to the characteristics of the radiator itself. The sky temperature depression, taken in isolation, may be a

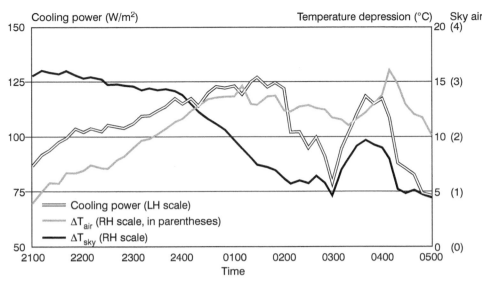

*Source:* Erell and Etzion, unpublished experimental data

**Figure 7.3** *Cooling output of a radiator exposed to changing environmental conditions; measured data from an experiment in Sde Boqer, Israel (24 July 1997)*

useful indicator of the potential for radiative cooling of buildings – but it is not always sufficient.

Each of the methods listed above has advantages and drawbacks. The data that they present should preferably be used only as a first approximation of the actual cooling potential at a given location. A more accurate assessment may be obtained by using the computer code, ROOFSOL and PDEC Tool (RSPT), available as part of the book *Roof Cooling Techniques* (Yannas et al, 2006). This software allows the user to simulate the performance of any type of radiator and to map the calculated cooling output for any location in Europe. The accuracy of the simulation depends upon the availability of a detailed climate file for the location in question. However, the output allows interpolation of the results to locations not represented in the database, as well as a facility for adding climate data for new locations.

## SYSTEM CONCEPTS

The main obstacle to implementing radiative cooling as a means of cooling buildings is the imbalance between incoming solar radiation during the daytime and the net long-wave radiation balance throughout the 24-hour daily cycle. Around noon in low latitude locations, incoming solar radiation (between the wavelengths of about $0.3\mu m$ to $3\mu m$) may reach an intensity of over $1000W/m^2$ on a horizontal surface. The balance between long-wave radiation emitted by building surfaces and long-wave radiation received from the sky depends upon local climatic conditions, but rarely produces a net cooling effect in excess of 150 W/m².

Harnessing radiative cooling as a means of providing thermal comfort must therefore resolve these conflicting requirements: how to allow dissipation of energy from a building element (typically the roof) by long-wave radiation, yet prevent undesirable solar gains at the same surface. The traditional solutions to the prevention of energy penetration through the roof deal with two heat transfer mechanisms: conduction and radiation. The aim is first to achieve the lowest possible temperature at the external surface of the roof and then to resist the flow of energy to the building interior by means of thermal insulation. The incorporation of radiative cooling systems on the roof does not obviate the need for these measures: *any cooling system incorporated within a building should not compromise the effectiveness of measures designed to reduce heat gain.*

The application of radiative cooling systems in buildings must recognize that high cooling rates are possible only if the radiator is relatively warm. While this may seem counter-intuitive, it should be kept in mind that in order to cool the structural mass of a building, the aim is to dissipate as much heat as possible to the relevant environmental heat sink, and this aim is not necessarily served by cooling the radiator itself as much as possible. *If the radiating surface is allowed to become substantially cooler than the internal mass of the building, its efficiency is greatly reduced.*

There are two reasons for this. First, although the Stefan-Boltzmann Law states that the rate at which energy is emitted by a radiator is proportional to its absolute temperature raised to the fourth power, net radiant loss from the radiator to the sky is, in practice, approximately proportional to the sky temperature depression (see Equations 13 and 14) because the difference between the absolute temperatures of the radiator and the sky is not large. The net cooling output of a radiator is thus approximately proportional to the difference between the temperature of the radiator and the sky temperature. The relative reduction in the net cooling output of a radiator as its surface temperature is reduced is illustrated in Figure 7.4.

Second, the effects of convection are directly proportional to the temperature difference between the radiator and the ambient air:

$$Q_c = h_c (T_r - T_a) \qquad [15]$$

where $Q_c$ is the rate of convective heat exchange (W/m²), $h_c$ the convective heat-exchange coefficient (W/m²/°C), and $T_r$ and $T_a$ the temperature of the radiator and ambient air (°C), respectively. The value of $h_c$ depends upon the wind speed and on the temperatures of the radiator and the air.

If the radiator is warmer than the air, convection will assist in the removal of energy from the radiator. As the radiator surface cools down relative to the air temperature, the beneficial effects of convection are reduced. If the radiator is colder than the surrounding air, then convection results in heat transfer from

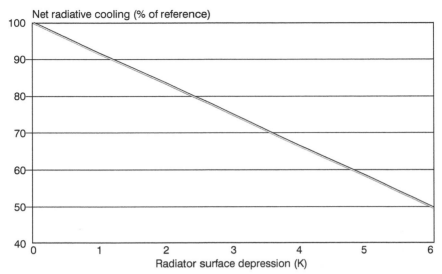

**Figure 7.4** *The effect of a reduction in the sky temperature depression from 15K to 7K on the net radiative heat loss of a surface with ε = 0.9, at ambient air temperature (with no net energy transfer by convection)*

the air to the radiator – obviously undesirable if the intent is to produce the greatest possible cooling effect.

From the discussion above, it is clear that in a cooling system designed to provide structural cooling, it is advantageous if the radiator is as warm as possible. This means that, ideally, it should be almost as warm as the warmest part of the building to be cooled. It should only be allowed to cool down in line with the reduction in temperature of the building interior.

Direct cooling of the building by long-wave radiation may be effective only if the radiating surface is not insulated during the period when cooling takes place. This may be done either through the installation of operable insulation, or by employing a radiator constructed above a roof with fixed insulation and transferring energy to it from the building interior by means of a suitable heat-exchange medium. Each of these approaches will be discussed, in turn, in the following sections.

## Moveable insulation

Thermal insulation installed on the roof of a building during the daytime may be removed at night (during the summer) to expose a massive roof or a roof pond to radiative cooling. In order to enable such a procedure, the thermal insulation must be held in rigid frames that can be moved by mechanical means for temporary storage over a building element that is not cooled, such as a garage roof. A small number of single-storey buildings incorporating this concept (the 'Skytherm' system), first proposed by Harold Hay (Hay and Yellot, 1969), were constructed in the US between the years 1973 to 1982. In these buildings, water enclosed in polyethylene bags and supported on a thin steel deck is exposed when the insulating panels are removed by means of a motor-operated system of cables and pulleys, and cooled by long-wave radiation. The first of these houses was built at Atascadero, California, and additional examples include the Bruder house (in Phoenix, Arizona), the Pala Passive Solar Facility (in San Diego, California), the Trinity University Test Facility (in San Antonio, Texas) and the Camelback School (in Paradise Valley, Arizona). Photographs of these buildings and construction details may be found in Yannas et al (2006).

The performance of some of these buildings was monitored extensively, particularly the test facility at Trinity University, and their thermal performance was claimed to be very satisfactory. However, Givoni (1994) questioned the accuracy and the validity of some of the published reports, and there is now no means of evaluating the results. The main drawback of the concept apparently lies with the mechanism installed to move the unwieldy insulation panels. Clark (1989) noted that 'the conventional horizontally moving panels have also been expensive and mechanically unreliable', a judgement supported by Givoni (1994) upon visiting the Atascadero building. A further drawback of the Skytherm system is the difficulty of creating a sufficiently good seal between the insulation and the parapet of the roof.

Since the 1980s, there has been only one more attempt to test and evaluate a radiative cooling system with moveable insulation. This was carried out at the PASSYS test cell in Almeria, Spain, as part of the ROOFSOL project (Roof Solutions for Natural Cooling, European Commission Joule programme, JOR3CT960074). However, this system only operated for a very short period, and there was no attempt to design a satisfactory mechanism for removing and replacing the insulation panels.

## Cooling a heat-exchange fluid

### Air-based systems

Early experiments with radiative cooling, carried out by Givoni in Israel (at Eilat and at Sde Boqer), were based on radiative cooling of air flowing in a narrow channel constructed above the roof.

At Eilat (Givoni and Hoffman, 1970), a 10cm channel was created by stretching a polyethylene film above a pitched asbestos-sheet roof insulated on its lower surface. At night, the channel was open at both ends to the interior of a test cell. During the daytime, the channel was closed by operable insulation panels. The authors presented data showing a substantial reduction in the air temperature of the test cell, air circulation being generated by thermosyphon. However, the performance of the system deteriorated rapidly, a process they attributed mainly to deposition of dust on the film.

At Sde Boqer (Givoni, 1977), the radiator panel was a corrugated metal sheet fixed above an insulated panel constructed above the flat concrete roof of a small test cell. Unlike the previous experiment, which was purely passive, the so-called 'roof radiation trap' relied on a small fan to draw external air into the test cell through the channel. A model developed later on the basis of the test data obtained at Sde Boqer (Givoni, 1994) predicts cooling rates (in good environmental conditions) of between 22.3W/m² to 46.9W/m² in excess of what could be obtained by night ventilation, depending upon the flow rate generated by the fan. The model predicts a decrease in cooling output of about 25 per cent if the sky is partly covered by cloud (three-tenths), and a further decrease if there is a light wind blowing (2.25m/s) to about 60 per cent of the cooling obtained in optimal conditions.

A system combining a solar chimney for passive heating during the winter with a radiator for nocturnal cooling during the summer was investigated by Brunold (1989). The system comprised a continuous channel 20cm thick extending from a vent at the bottom of the south wall of a room-sized test cell, covering the entire south wall, an upwardly sloping roof and then down to a vent near the ceiling on the north wall of the test cell. The primary radiator was a thin aluminium sheet fixed at the middle of the channel so that air could circulate freely both above and below it. The channel was insulated from the main structure of the test cell and sealed from the environment by means of a glass pane. In winter, the vents were opened only during the daytime, permitting thermosyphonic circulation of air entering the channel at the bottom of the south wall, heated by solar radiation and reintroduced into the test cell at the north wall. During summer, the vents were opened only at night, and relatively warm air was drawn into the channel through the vent near the ceiling on the north wall. It was then cooled by long-wave radiation and re-entered the test cell interior at the bottom of the south wall. An interesting feature of this system is that although the primary radiator was metallic, the air channel was glazed. Heat emitted by the radiator was absorbed by the glass cover sheet and re-emitted to the atmosphere. The maximum cooling output of the system in favourable environmental conditions was about 17W/m² (Erell et al, 1992). Although the air circulated in the channel was cooled by up to 6°C, it was never colder than the ambient air. The air speed in the channel resulting from this temperature difference was quite slow – less than 0.2m/s. This suggests that the cooling potential of any air-based radiative cooling that relies only on natural buoyancy to generate airflow might be fairly limited.

More recently, an analytical model was used to simulate the dynamic thermal performance of a lightweight metallic radiator covered by a single polyethylene wind screen (Mihalakakou et al, 1998). The metallic radiator was designed as part of a retrofit package for a rehabilitated historical building in the town of Legnano in Italy. The study predicts that the system, comprising a radiator 14m long and 7m wide, could deliver between 29.7Wh/m² to 55.8Wh/m² per day in clear sky conditions, depending upon wind speed. This cooling output is rather modest, and there is no indication if the system was, in fact, implemented.

## Water-based systems

Roof-mounted solar collectors have been used to heat water for many decades. The hot water has generally been delivered for domestic use, but it has also been used as a means of providing energy for space heating. One application of the concept is to circulate the water in a heat exchanger embedded in a concrete floor slab. Juchau (1981) proposed that the same system could be used to cool water in the storage tank by circulating it through a standard flat-plate collector at night. The chilled water would then be circulated through a radiant floor (originally installed for heating) to cool the building during the daytime. The concept was tested in an occupied house as part of a comparative study involving four different systems; but the focus of the study was on the performance of the radiant floor, and various monitoring difficulties prevented evaluation of the roof radiator itself.

A systematic evaluation of the potential of flat-plate collectors as radiative cooling devices was carried out at Sde Boqer in an extended research project beginning in 1987. The system tested was based on circulating water from a shallow roof pond through flat-plate collectors at night, to be cooled by long-wave radiation and convection (see Figure 7.5). The collectors were standard collectors designed for domestic solar water heating, with the glass top-cover removed. The roof pond was insulated from the environment from above and, in addition, was shaded by the collectors installed on a frame above it. It therefore absorbed heat mainly through the roof of the test building. The energy stored in the roof pond was dissipated gradually by circulating it through the collectors by means of an electric pump, typically after about 10 pm at night. The net cooling power of the radiator (not accounting for pumping energy) averaged $81 W/m^2$ for seven hours of operation each night over a three-week period in the summer of 1990 (Erell and Etzion, 1992). The lowest average cooling power for a single night during this period was $59 W/m^2$.

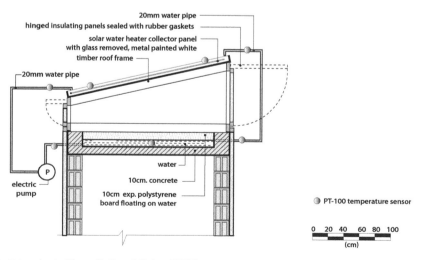

*Source:* adapted from Erell and Etzion (1992)

**Figure 7.5** *Schematic section of the radiative cooling system tested at Sde Boqer, Israel*

The advantage of this system, in particular, and of water-based radiative cooling systems, in general, compared with air-based systems, is that the close coupling between the thermal mass, embodied in the water, and the radiator surface means that the radiator is typically relatively warm. In dry desert climates where ambient air temperature drops quickly after sunset, the radiator may be warmer than the surroundings even fairly early at night. This makes wind screens redundant, and convection assists in cooling the radiator rather than impeding the cooling process.

## DESIGN ISSUES

The previous sections have given an overview of the effect of environmental conditions on the magnitude of the radiative cooling potential, and introduced conceptual solutions for harnessing this resource to cool buildings. The following discussion deals with the detailed design of the radiator, its mode of operation and its integration with the building.

### Coupling the radiator with building thermal mass

Thermal coupling of the radiator with the building has two roles. First, the radiator itself has insufficient mass to act as a heat sink for the entire building during daytime hours when the radiative cooling system is inoperative. Second, if the thermal coupling between the radiator and the building is poor, the radiator temperature decreases rapidly, its cooling rate is reduced and the building, as a whole, loses less energy to the environmental sink. The importance of coupling the radiator surface with the thermal mass was illustrated by Etzion and Erell (1991) in a study using reduced-scale models exposed to the environment. In the study, the radiator surface and the amount and location of thermal mass were varied. Although the surface temperature of a concrete radiator remained higher than that of lightweight metal radiators throughout the night, its average cooling rate was higher than that of any of the metallic radiators, irrespective of the location of the thermal mass (walls or floor) in the test cell. As a result, the internal air temperature of a test cell with a concrete radiator was lower than of any of the test cells with lightweight metallic radiators.

Two different approaches to incorporating thermal mass in a radiative cooling system have been explored in recent years:

The first uses water to absorb heat from the building during the daytime, releasing it at night. The water may be contained in polyethylene bags on the roof, which are exposed at night by removing a layer of thermal insulation (Hay and Yellot, 1969); in a shallow roof pond, from which it is circulated through radiators (Erell and Etzion, 1992, 2000; Etzion and Erell, 1999); or it may be stored in a purpose-built tank and circulated through heat exchangers in the building interior (Meir et al, 2002). Al-Nimr et al (1999) have suggested that the storage tank might contain a packed rock bed through which water would circulate. However, this design appears to offer few advantages over a storage volume containing water only because water has a lower density and a higher volumetric heat capacity ($4.18J/m^3/K$) than natural stone ($1.9J/m^3/K$ to $2.4J/m^3/K$), while turbulent mixing in the fluid ensures a more efficient heat distribution.

The second approach has been to use ceiling elements as the primary thermal storage mass. These elements, sometimes referred to as cooling panels, are typically made of concrete with embedded pipes through which a cooling fluid, such as water, is circulated (Dimoudi and Androutsopoulos, 2006). Heat absorbed by the fluid from the cooling panels is emitted to the environment by circulating it through radiators installed on the roof. A detailed sensitivity analysis of the parameters affecting the thermal performance of concrete cooling panels is included in the *Roof Cooling Techniques* handbook (Yannas et al, 2006). The software provided with the handbook allows calculation of the relative energy saving in a typical residential or office building resulting from the use of cooling panels in conjunction with a source of cooled water (or the improvement in thermal comfort, for a free-running building).

Roof ponds can cool the space immediately beneath them if the ceiling has a very high thermal conductance. While monolithic concrete ceilings may be adequate for small spans (where structural considerations allow the construction of a relatively thin roof), a thin metal deck supported on steel beams provides much better thermal contact. In this case, the entire interior surface of the ceiling has practically the same temperature as the water on the roof at all times. An attached roof pond also requires no distribution system to bring the cooled fluid into contact with the space to be cooled.

However, roof ponds have several drawbacks:

- They require very careful waterproofing.
- They need additional structural strength to support the added weight of the water.
- They have a large surface area exposed to the environment. Clark (1989) suggested that as much as one third of the heat absorbed in the Skytherm roof pond systems installed by Hay resulted from heat transferred through the closed insulation panels.
- They require a distribution system to cool interior spaces that are not immediately beneath them.

## Interaction with convection (wind screens)

In favourable environmental conditions, long-wave radiation to the night sky can lower a radiator's surface below ambient air temperature. Once this happens, heat exchange by convection tends to counteract radiant loss. The point at which radiative cooling and convective heat gain exactly cancel each other out is known as a radiator's stagnation temperature.

A wind screen may be added to a radiator to restrict convective heat exchange, especially in windy conditions. The stagnation temperature of such a radiator may be lowered by creating a layer of still air above the radiator that is prevented from mixing with the surroundings. Two types of wind screens have been proposed:

1  An open honeycomb covering, preferably made of a material that is highly reflective in the infrared spectrum (Martin, 1989): such a covering would allow radiation to be emitted from the radiator since it would be open at the top, while restricting the motion of air near the radiator surface.
2  Covering the radiator with a glazing that is transparent in the infrared spectrum, particularly between 8µm and 13µm: such a glazing would transmit infrared radiation emitted by the radiator, but prevent air sealed beneath it from mixing with the surrounding air. Glass and other materials with adequate mechanical strength and chemical stability are opaque over this spectral range. Polyethylene film manufactured without ultraviolet (UV) inhibitors is probably the only material currently available that is sufficiently transparent in the atmospheric window: a very thin (about 0.05mm) low-density polyethylene (LDPE) film has infrared transmissivity of about 75 per cent (Clark and Berdahl, 1980).

Early research work focused on means of obtaining the lowest possible stagnation temperature, and on evaluating the contribution of wind screens in the pursuit of such a goal. Mostrel and Givoni (1982) reported that in favourable environmental conditions – a clear, dry atmosphere with no wind – the temperature of a thin metallic radiator exposed to the environment could drop to as much as 6°C to 8°C below ambient air temperature. Similar radiators covered with a polyethylene wind screen were cooled to 9°C to 11°C below ambient air temperature under the same environmental conditions. In more typical conditions, where wind speed was not negligible and the contribution of a wind screen might therefore be expected to be more important, the respective temperature depressions were 4°C to 5°C for an exposed radiator and 6°C to 8°C for a screened one.

As noted above, polyethylene is the only material suitable for use in wind screens. However, while polyethylene is quite cheap, it has a low tensile strength and deteriorates rapidly when exposed to solar radiation. In addition, lacking rigidity, it must be stretched tightly over a supporting grid or reinforced by a web of nylon or another similar fibre to provide tensile strength: if it is not held in tension, it flutters in the wind, producing air motion in the space between the wind screen and the radiator, and, as a result, losing most of its insulating effect.

As Mostrel and Givoni (1982) noted, all wind screens suffer from two additional drawbacks. First, the surface of the wind screen is often cooled to the dew point temperature of the surrounding air. As the stagnant air trapped beneath the wind screen is gradually cooled as well, it begins to cool the lower surface of the wind screen. Since the wind screen is generally very thin, its upper surface is soon cooled as well. If it reaches the dew point and moisture begins to condense on the surface of the wind screen, the process is self-accelerating, as water has a very high emissivity. The wind screen then loses its transparency and becomes, in effect, the primary radiator, albeit one exposed to the very effects of convection that it was designed to prevent. This is a common occurrence even in very dry locations since the clear atmosphere of such regions promotes nocturnal radiative cooling.

Second, the accumulation of dust or airborne pollutants on the exposed surface of a wind screen requires frequent cleaning. In the absence of such maintenance, the transmissivity of the glazing to infrared radiation may be reduced significantly. The problem is exacerbated by the periodic formation of dew on the exposed surface since dust particles deposited on it are retained, rather than being blown away by the wind.

The practical problems associated with wind screens tend to obscure an important fact: when the fluid entering the radiator is well above the ambient temperature, most of the radiator surface is also likely to be above that temperature and convection will contribute to heat loss from the radiator. In such cases, wind improves radiator performance, and the addition of a wind screen is counterproductive. Any decision regarding the implementation of a wind screen should therefore first establish whether the radiator surface is likely to be below air temperature in typical operating conditions.

## Selective surfaces

The solar absorptivity of the radiator plate is only important where the radiator is also used as a solar collector (e.g. to provide domestic hot water, swimming pool heating, etc.), and is of little importance when the system is designed for nocturnal operation only. Where the heat exchange fluid is not circulated during daytime, it may be preferable to paint the radiator in a light colour, thus providing it with a low solar absorptance to reduce solar gains. If the radiator is not in direct thermal contact with the building, the colour is irrelevant since even a black radiator cools down very rapidly near sunset, and cooling operation is not delayed (Erell and Etzion, 1992).

Unlike solar absorptivity, the long-wave emissivity of the radiator has an important effect on its performance. The maximum possible radiation is emitted from a body at any given temperature if it has an all-wave emissivity of 1.0 (i.e. it is a so-called black body). However, net radiant exchange is also affected by the properties of the surrounding surfaces. Since the clear sky does not radiate as a black body, it is possible for a spectrally selective surface to have a higher net outgoing radiation balance than a black body at the same temperature. Such a radiator should have a high emissivity in the atmospheric window (wavelengths of between 8µm and 13µm), and be highly reflective in all other parts of the spectrum. It is useful in this context to note that according to Wien's Law, most of the energy emitted from buildings is, in fact, within this range of wavelengths because the temperature of building surfaces in hot climates is typically between 288K to 333K (about 15°C to 60°C).

Some materials with the desired properties are applied as thin films deposited on highly reflective aluminium foil. Berdahl (1984), for example, experimented with magnesium oxide (MgO) and lithium fluoride (LiF) backed with aluminium foil, and reported favourable spectral properties. Solid polished MgO surfaces also have a very high solar reflectivity, opening the possibility of using the material as a radiator capable of producing net cooling even during the daytime.

In the pursuit of daytime radiative cooling, several researchers have experimented with thin coatings applied to polyethylene film, which is then stretched above the primary reflector. In order to fulfil this role, a polyethylene sheet must be coated with a reflective pigment designed to reduce solar gains and to increase the longevity of the material, while retaining its high transmittance to long-wave radiation. The concept was tested as a means of cooling an experimental facility in Naples, Italy, which was covered with a double polyethylene film having the following properties: solar reflectance: upper face = 0.6; lower face = 0.2; solar transmittance: 0.1; infrared transmittance: 0.55 (Addeo et al, 1980). The air temperature inside the test shed was reported to be lower than in an adjacent open-air weather screen at all times, including during sunny daytime conditions with a solar flux of up to $700W/m^2$. However, the exact chemical properties of the film were not disclosed, and the result has not been reproduced by other researchers.

Nilsson and Niklasson (1995) experimented with several pigments to provide the desired wavelength selective transmissivity to polyethylene film. The most appropriate pigments (from an optical point of view) are zinc sulphide (ZnS) and zinc selenide (ZnSe), although their long-term chemical stability is poor. Results of a field experiment in Tanzania indicated that the temperature of a black body radiator could be restricted by a pigmented polyethylene foil to as little as 1.5K above ambient air temperature in the daytime, even when exposed to a solar flux of about $1000W/m^2$. Nilsson and Niklasson (1995) recommended that an optimized cover foil for practical use should include a pigment volume fraction of 0.15 on a foil 400μm thick. They also suggested that the night-time cooling performance of a radiator covered with such a film would not be affected negatively, although no experimental data was provided in support of this conclusion. Furthermore, although simulation suggested that long-wave radiative cooling during daytime was possible, this was only achieved in practice for 19 hours of the day, when incident solar flux was relatively low.

More recently, Dobson et al (2003) reported that chemical solution deposition of thin semiconductor films of lead sulphide (PbS) or lead selenide (PbSe) onto a polyethylene substrate pigmented with zinc oxide (ZnO) or ZnS resulted in very low solar transmission, coupled with a transmissivity in the atmospheric window of about 40 to 50 per cent.

The progress in research on wavelength-selective pigments is promising; but further development is needed before the concept can be applied in radiative cooling systems for buildings. However, a number of practical issues suggest that even if the technological problems can be overcome and stable high-quality selective coatings can be produced at a reasonable cost, they may not be used except in unusual circumstances. First, the application of selective-emitting materials to radiative cooling on a commercial scale requires a means of maintaining the optical properties of the surface after prolonged exposure to the environment. Like simple infrared transparent wind screens, films with selective coatings need to be kept clean to preserve their advantage over conventional radiators. In addition, they must be kept dry – the formation of dew on the radiator eliminates any spectral selectivity that the surface may have.

## Tilt angle and sky view factor

As the previous discussion shows, installing a radiator horizontally maximizes the net long-wave radiant heat loss. Nevertheless, it may still be desirable to install the radiator at a small angle to the horizon for a variety of reasons, such as integration with an existing (pitched) roof, ease of drainage, or possible use in a heating mode in addition to its primary function of cooling (Erell and Etzion, 1996). In these circumstances, the benefits anticipated from tilting the radiator should be assessed carefully to establish whether they outweigh the loss in cooling performance.

It may be assumed that the difference in temperature between the radiator and other terrestrial surfaces is likely to be small, so net radiant cooling results almost exclusively from exchange with the sky. Tilting the radiator increases the incoming long-wave radiation at the radiator surface for two reasons:

1   it reduces the sky view factor from the radiator surface (the view factor to the unobstructed sky is given by $f_{sky}(\alpha) = \frac{1}{2}(1+\cos\alpha)$, where $\alpha$ is the angle of tilt with respect to the horizon);
2   if the radiator is tilted, it is more exposed to the warmer regions of the sky near the horizon.

The combined effect of these mechanisms would appear to be relatively small. Assuming the temperature of the ground surface is approximately the same as air temperature, tilting the radiator by 45° only increases long-wave radiation received by about 4 to 6 per cent, depending upon atmospheric conditions (Martin, 1989). However, even such a small difference may lead to a reduction in the *net* cooling power of the radiator of up to 20 per cent.

## Design and operation of water-based radiative cooling systems

As shown above, a flat-plate solar collector may also be used for radiative cooling if its glazing cover is removed. A typical flat-plate collector (see Figure 7.6) consists of parallel riser pipes connected to main distributors (headers) at the inlet and outlet. In order to achieve high fluid temperatures, each of the pipes is, in turn, attached to a fin, designed to concentrate solar energy from a relatively large surface area onto a small volume of water (the fins are sometimes joined together to form a continuous plate). The mass flow rate in

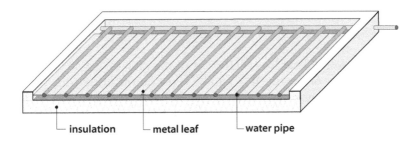

insulation      metal leaf      water pipe

**Figure 7.6** *Schematic design of a typical tube-and-fin flat-plate solar collector, with no glass cover plate*

a solar collector is generally very low – about 0.003kg/m²/s to 0.05kg/m²/s, corresponding to a water velocity of about 0.5cm/s to 10cm/s.

### Radiator design

The overall performance of a solar collector may be characterized by a collector efficiency factor, which relates the actual useful energy gain of a collector to the useful gain that would result if the collector absorbing surface had been at the local fluid temperature (Duffie and Beckman, 1991) and by a fin efficiency factor. The efficiency of the fins and collector are affected by a number of parameters, such as the diameter and spacing of the risers, the thickness and conductivity of the pipes and the fins, and the quality of the thermal bond between them.

Duffie and Beckman (1991) give the following expression for the collector efficiency factor $F'$:

$$F' = \frac{\dfrac{1}{U_L}}{W\left[\dfrac{1}{U_L[D + (W - D)F]} + \dfrac{1}{C_b} + \dfrac{1}{\pi D_i h_{fi}}\right]} \quad [16]$$

where $C_b$ is the bond conductance between pipes and absorber plate (W/m/K), $h_{fi}$ is the heat transfer coefficient between the fluid and the tube interior (W/m²/K), $D$ is the external diameter of the pipes (m) and $D_i$ is the internal diameter of the tubes (m). $F$, the fin efficiency factor, is a convenient means of expressing the overall geometric characteristics of the radiator, and is given by:

$$F = \frac{tanh[m(W - D) / 2]}{m(W - D) / 2} \quad [17]$$

where $W$ is the riser spacing (m), $D$ the pipe diameter (m), and $m$ is given by:

$$m = \sqrt{\frac{U_L}{k\delta}} \qquad\qquad [18]$$

where $U_L$ is the overall heat loss coefficient of the radiator (W/m²/K), $k$ is the thermal conductivity of the fin (W/K) and $\delta$ is the fin thickness (m).

The fluid temperature at any given point along the riser in a solar collector may then be derived from the following expression:

$$\frac{T_f - T_a - \dfrac{S}{U_L}}{T_{fi} - T_a - \dfrac{S}{U_L}} = exp\left( -\frac{U_L n W F' y}{\dot{m}\ C_p} \right) \qquad\qquad [19]$$

where $T_f$ is the fluid temperature (°C) at a distance $y$ (m) from the collector inlet; $T_a$ is the ambient air temperature (°C); $T_{fi}$ is the fluid temperature at the collector inlet (°C); $S$ is the absorbed solar energy (W/m²); $U_L$ is the overall heat loss coefficient of the collector (W/m²/C); $n$ is the number of pipes; $W$ is the distance between the pipes (m); $F'$ is the collector efficiency factor; $\dot{m}$ is the mass flow rate through the collector (kg/s); and $C_p$ is the specific heat of the fluid (J/g/°C).

The adaptation of the Duffie and Beckman (1991) model (see Equation 16) to describe a nocturnal cooling radiator, proposed by Erell and Etzion (2000), requires the replacement of the solar heat gain, $S$, with an expression for the net long-wave radiant exchange between the radiator and the sky (the model is applicable for night-time only). Figure 7.7 shows the predicted temperature of a fluid (water) at the outlet of a given radiator, compared with the measured temperature during an experiment at Sde Boqer, Israel.

The efficiency of a typical flat-plate solar collector with pipes and an absorber plate is determined by its geometry, the materials used in its construction and the quality of the glazing cover, which affects both the solar heat gain and the heat loss from the absorber plate. The relative importance of some of these factors is somewhat changed in a cooling application.

The primary geometric parameters describing flat-plate collectors are the diameter of the riser tubes and their spacing. In heating collectors, the use of an absorber plate allows a concentration of solar energy resulting in higher water temperatures. This is because sufficiently thick copper plates (up to 1mm thick) in good thermal contact with the pipes lose little energy to the surroundings even if the collector cover consists of a single glazing.

The effects of convection on the fin efficiency factor are expressed in Equations 17 and 18. Figure 7.8, derived from these equations, illustrates the effect of increasing convective heat exchange and, thus, the overall heat loss coefficient ($U_L$), on the fin efficiency of a typical rectangular fin (the calculations were carried out for a typical collector design, with a steel absorber plate 0.5mm thick and 10mm risers in good thermal contact with it,

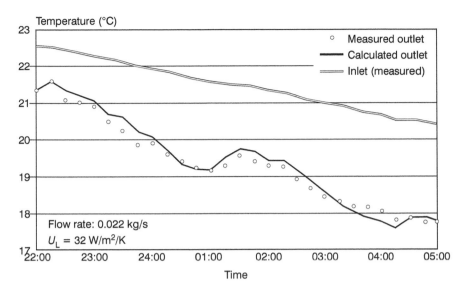

*Source:* adapted from Erell and Etzion (2000)

**Figure 7.7** *Time series comparing predicted and measured temperature of cooling fluid (water) at the outlet of a radiator in Sde Boqer, Israel*

for several values of *W*, the spacing of the risers).

It is evident from Figure 7.8 that an increase in the convective heat loss coefficient of the radiator (which may result, for instance, from increased wind speed) causes a reduction of the fin efficiency (*F*). This effect is noticeable for the whole range of fin widths, but becomes dominant when the tube spacing is large. When the tube spacing is small – 5cm or less in this case – the reduction in fin efficiency is less than 10 per cent even for fairly high heat loss coefficients.

The effect of changes in the fin efficiency on the overall performance of a cooling radiator are complex and depend upon the relationship between the ambient air temperature ($T_a$) and the temperature of the radiator surface ($T_r$):

- If $T_a < T_r$, convection increases the rate of heat dissipation, reinforcing the effects of radiation. In this case, a higher fin efficiency (*F*), which in turn leads to a higher collector efficiency (*F'*), results in improved radiator performance both in radiation and in convection.
- If $T_a > T_r$, convection leads to heat transfer from the air to the radiator, counteracting the effects of radiation. In this case, higher fin efficiency still leads to improved heat dissipation through radiation, but also results in increased convective gains. The net balance of these opposite forces depends upon the properties of the specific radiator, on the environmental conditions and on the fluid inlet temperature and flow rate.

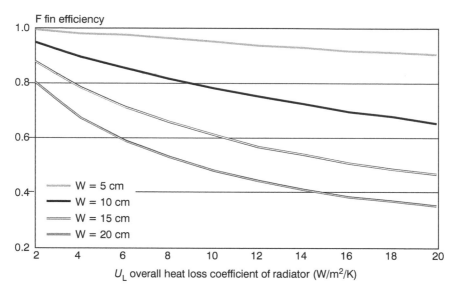

*Source:* adapted from Erell and Etzion (2000)

**Figure 7.8** *The effect of the overall heat loss coefficient (U$_L$) of an exposed, non-glazed tube-and-fin radiator on fin efficiency (F) for several widths of fin*

Figure 7.9 shows the effect of the fin efficiency factor ($F'$) on changes in the fluid temperature of a hypothetical radiator exposed to different ambient air temperatures. The radiator is 1m wide by 2.5m long; fluid inlet temperature is 23°C; mass flow rate is 0.03kg/s; $U_L$ = 15W/m$^2$/K; and the sky temperature depression is 15°C. Under these conditions, there is a net cooling effect (indicated by a negative change in the fluid temperature) until the ambient air temperature rises to 3°C above the radiator temperature. At this point, radiators with different fin efficiencies (but identical in other respects) would have a similar neutral energy balance.

The optimal design of a cooling radiator therefore depends upon an understanding of the environmental conditions in which it is likely to operate. In general, if air temperatures are very high, the net cooling output of a radiator would probably be too low to justify installation, assuming that convection cannot be suppressed entirely. If ambient air temperatures are equal to or lower than the fluid inlet temperature for most of the operating period, then high fin efficiency is preferred. On the other hand, if the ambient air is warmer than the fluid at the radiator inlet, high fin efficiency increases convective heat gains and thus *reduces* the net amount of energy given off by the radiator.

Fin efficiency may be improved (assuming the radiator is warmer than the air) by using thicker, more conductive fins and by reducing the distance from the edge of the fin to the tube with the fluid. Unlike a heating collector, fins might ultimately be dispensed with altogether, leaving a continuous emitting

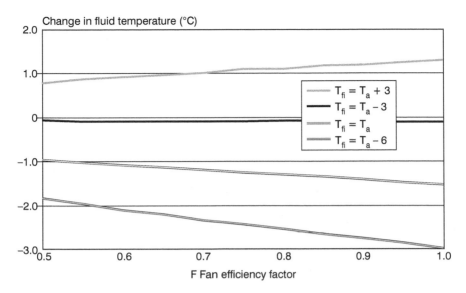

Change in fluid temperature (°C)

Legend:
$T_{fi} = T_a + 3$
$T_{fi} = T_a - 3$
$T_{fi} = T_a$
$T_{fi} = T_a - 6$

F Fan efficiency factor

*Note:* The radiator is 1m wide by 2.5m long; fluid inlet temperature is 23°C; mass flow rate is 0.03kg/s; UL = 15W/m²/K; and the sky temperature depression is 15K.
*Source:* adapted from Erell and Etzion (2000)

**Figure 7.9** *The effect of the fin efficiency factor (F') on changes in the fluid outlet temperature of a hypothetical radiator exposed to different ambient air temperatures*

surface. If fins are retained, the minimum thickness required for structural reasons is also sufficient to satisfy thermal requirements (for metal plates, this is about 1mm to 2mm). The effect of fin width (W) on fin efficiency is shown in Figure 7.10. For any given heat loss coefficient of the radiator ($U_L$), itself affected by convective exchange, a narrower fin is more efficient.

The effect of collector efficiency on the cooling output of a hypothetical radiator is illustrated in Figure 7.11. For the given radiator design and environmental conditions, an increase in collector efficiency from 0.7 to 0.9 may result in an increase in cooling output of approximately 15 per cent at low flow speeds (less than 1cm/s) and up to 25 per cent at high flow velocities (3cm/s or more). The difference in absolute terms is even greater since the cooling rate is slightly higher at high flow rates. Optimization of the collector design should therefore focus on operation at relatively high flow velocities of the fluid (3cm/s or more).

**Materials**
Early radiative cooling systems utilizing water as a heat transfer medium (Erell and Etzion, 1992) were based on metallic radiators, similar to those used as flat-plate solar collectors. Metallic radiators, whether of copper or galvanized iron, have excellent thermal characteristics and good mechanical strength. They are also fairly durable (especially copper collectors). However, such

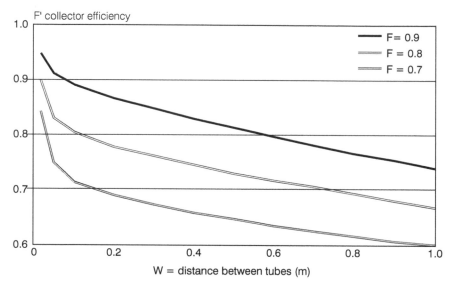

*Source:* adapted from Erell and Etzion (2000)

**Figure 7.10** *The effect of fin width on the collector efficiency (F') for different values of the fin efficiency factor (F)*

collectors are fairly expensive, and a number of studies have been carried out to investigate the suitability of other materials. O'Brien-Bernini and McGowan (1984) performed a detailed thermal analysis of solar collectors made of a variety of plastics. They noted that although the thermal conductivity of polypropylene, for example ($k$ = 0.36W/m/K), was much lower than that of copper ($k$ = 384W/m/K), little increase in performance could be expected through the use of copper – if the transport fluid is in contact with the entire surface of the absorbing plate (i.e. a fully wetted configuration). This is because the thickness of the material is quite small – typically less than 1mm.

The performance of two radiators made of non-metallic materials was evaluated concurrently in an experiment at Sde Boqer, Israel (Erell and Etzion, 1999; Etzion and Erell, 1999):

- a mat-type solar collector for low- to medium-temperature applications, such as heating swimming pools: it was made of polypropylene tubes with a diameter of 6mm (4.5mm internal), about 10mm between centres, attached to a manifold 40mm in diameter; and
- a double-glazed polycarbonate (PC) sheet with a total thickness of 10mm, comprised of two plates about 0.5mm thick connected by vertical ribs at 10mm spacing: the sheet was attached to proprietary manifolds, 25mm in diameter, at either end.

Both designs were evaluated during two week-long monitoring periods characterized by different environmental conditions. The polycarbonate sheet

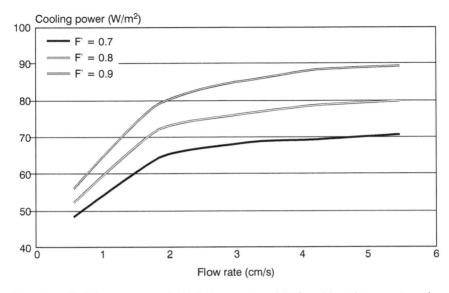

*Note:* The calculation assumes a fluid inlet temperature of 23°C, ambient air temperature of 21°C, sky temperature depression of 12K and a collector heat loss coefficient of 15W/m²/K.
*Source:* adapted from Erell and Etzion (2000)

**Figure 7.11** *The effect of collector efficiency (F') for different values of the fin efficiency factor (F')*

radiator proved to be a more efficient design, giving a mean cooling output for eight hours of operation per night of 105.9W/m² compared to 91.4W/m² for the polypropylene mat when the sky temperature depression was, on average, 12.2K; and 91.1W/m² compared to 65.1W/m² when the sky temperature depression was only 7.9K, on average.

The efficiency of both types of radiator is higher than that of conventional metallic tube-and-fin flat-plate collectors converted to cooling. The analysis of the thermal performance of cooling radiators described above presents the theoretical justification: if the definition of the collector efficiency factor $(F')$ is applied to a cooling radiator, then the radiator efficiency factor is the ratio of the energy dissipated, in practice, to the heat loss that would result if the radiator surface had been at the local fluid temperature. The two designs demonstrate this logic, doing away with fins entirely. Both thus have a theoretical 'fin efficiency' of unity and an overall radiator efficiency, according to this definition, which approaches unity as well. Since both polycarbonates and polypropylene have a thermal emittance ($\lambda$ = 3μm to 40μm) in excess of 0.95 and the thin pipes have a similar thermal conductance, the difference in the net cooling output of the radiators stems from differences in the density of the pipes. In the PC sheet radiator, the whole sheet is in contact with the water, whereas in the polypropylene mat radiator, the pipes comprise approximately 80 per cent of the gross area of the radiator.

The radiators investigated were much cheaper than conventional metallic flat-plate collectors (Etzion and Erell, 1999). A polypropylene mat radiator costs the equivalent of about US$100 for a 2.77 square metre unit (2007 prices) and is extremely durable. The polypropylene pipes do not corrode, do not deteriorate upon exposure to sunlight and are not clogged by minerals deposited from the water. The useful service life of such a unit is many years, barring mechanical damage or exposure to sub-zero temperatures when full of water. The PC sheet radiator is based on an ordinary polycarbonate sheet with a UV protective coating, attached to proprietary manifolds made of the same material. The polycarbonate sheet costs the equivalent of approximately US$30 per square metre (2007 prices), so a 2 square metre unit could be expected to cost less than US$100 if it were produced commercially. Both radiators are available at almost any length, are very easy to transport and require little expertise to install.

A similar experiment was carried out in Oslo, Norway, using a double-glazed sheet made of polyphenyleneoxide (PPO) as the radiator (Meir et al, 2002). Like polycarbonates, PPO degrades when exposed to UV radiation and therefore requires a UV-resistant coating. The internal channels of the sheet, formed by the rib structure, were filled with lightweight expanded clay-aggregate granules to increase turbulence and thus enhance heat transfer between the fluid and the radiator surface. However, no comparison was made of the performance of the radiator with and without the granules, nor was a calculation provided of the additional pumping energy required to overcome the extra flow resistance.

## Operating parameters

The effect of altering the operating parameters – the inlet temperature and the mass flow rate – is illustrated for a hypothetical radiator 1m wide, with an overall efficiency ($F'$) of 0.8. The environmental conditions assumed were an ambient air temperature of 22°C, a sky temperature depression of 15K and little wind. Under these conditions, an unglazed, exposed radiator would have a convective heat loss coefficient ($U_c$) of about 10W/m/K.

Figure 7.12 shows the effect of altering the flow rate on the longitudinal temperature profile. For a specific radiator length, a desired outlet temperature may be attained, within the constraints of the environmental conditions, by manipulating the flow rate.

Figure 7.13 shows the effect of altering the flow rate on the cooling output at several inlet temperatures. Increasing the mass flow rate reduces the temperature difference between radiator inlet and outlet, resulting in an increase in the mean surface temperature of the radiator. This, in turn, results in increased radiative cooling, under all environmental conditions. When the radiator is warmer than the ambient air, increasing the flow rate also increases the cooling output due to convection; if the radiator is cooler than the ambient air, convective heat exchange tends to counteract the effects of radiation. Increasing the fluid flow does, however, have diminishing returns. As the flow rate increases, the surface temperature of the radiator approaches that of the

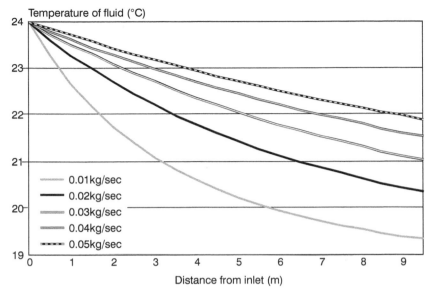

*Source:* adapted from Erell and Etzion (2000)

**Figure 7.12** *The effect of changes in the mass flow rate on the longitudinal temperature profile of a hypothetical cooling radiator*

fluid at the inlet, which in a cooling radiator is the theoretical limit. The practical limit takes account of the power required to operate the pump, as well as the heat transfer processes occurring at the radiator surface.

In conclusion, it may be useful to note the following observations regarding the design of the radiator:

- Heat exchange between the tube walls and the fluid is improved by a turbulent flow regime. However, in most collector designs, the flow speeds noted above result in laminar flow. An increase in flow rate to force a turbulent regime would not significantly increase the thermal performance and would substantially increase the energy needed to pump the water.
- A round pipe section is not essential; other considerations, such as the sky view factor of each pipe or simplicity of construction, may recommend a radiator with rectangular pipes.
- The final design requires optimization of the hydraulic parameters in order to attain the flow rate calculated to maximize cooling output, with a minimum investment of electricity to operate the pump.
- The length of the radiator may be adapted to the geometry of the roof on which it is installed; the flow velocity and mass flow rate are limited by hydraulic constraints, such as the diameter of the manifold pipe or the operating parameters of the pump. Thus, there is no advantage to be gained, for instance, from dividing a long radiator into several shorter ones of equal cross-section.

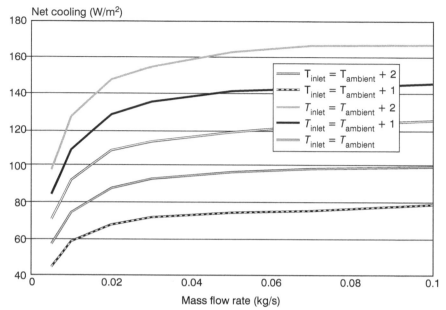

Net cooling (W/m²)

$T_{inlet} = T_{ambient} + 2$
$T_{inlet} = T_{ambient} + 1$
$T_{inlet} = T_{ambient} + 2$
$T_{inlet} = T_{ambient} + 1$
$T_{inlet} = T_{ambient}$

Mass flow rate (kg/s)

*Source:* adapted from Erell and Etzion (2000)

**Figure 7.13** *The effect of changes in the mass flow rate on the net cooling power of a hypothetical radiator at different temperatures of the fluid at the radiator inlet*

## Building integration and compatibility with conventional cooling systems and other passive methods

Most passive cooling or heating strategies have a dominant effect on building form. Unlike conventional HVAC systems, they cannot be applied at a late stage in the design process, and few may be added as a retrofit to existing buildings. This remains one of the factors preventing the widespread adoption of passive heating and cooling. Radiative cooling also suffers from these drawbacks to a certain extent:

- Radiative cooling panels must be placed where they are fully exposed to the sky. This means, in practice, that their installation is typically restricted to the building roof.
- Since the cooling output per unit surface area of the cooling panel is fairly small, compared with the demands of buildings in hot climates, they cannot meet the cooling requirements of multi-storey buildings where the ratio of roof area to floor area is small.
- The efficiency of radiative cooling panels is reduced if they are not mounted horizontally or at a relatively low tilt angle. The roof must therefore be horizontal (e.g. concrete) or moderately pitched if the cooling panels are to be integrated with it.

- If the radiators are employed to cool water in a roof pond, the structure must be able to support the additional weight. This load may not be excessive as a proportion of the overall weight of concrete-frame buildings with concrete roofs; but it might be very large in proportion to the weight of many low-rise timber-framed houses.
- Since radiative cooling cannot produce a large temperature depression (the fluid cannot be cooled to the temperature typically found in other modes of chilling), the surface area of the heat exchangers in the building interior must be large. This is why roof ponds and ceiling-integrated cooling panels have been the focus of most of the research on the application of radiative cooling systems.
- Air-based systems do not provide a sufficiently high flow rate if they rely only on buoyancy flow. However, the geometry of the roof-mounted radiators is not conducive to efficient operation of a fan-assisted system. A cooling radiator must have a narrow air channel and a large surface area to maximize contact with the air, creating substantial internal friction and resistance to the airflow.

Purely passive methods of providing thermal comfort in buildings while conserving non-renewable energy are often incapable of producing the same level of comfort obtained by conventional HVAC systems. Building occupants in passively heated or cooled buildings cannot control indoor air temperature to a desired range, and can only modify conditions by relatively crude methods if existing conditions are unsatisfactory. They are also unable to deal with peak loads that are substantially higher than background levels of demand and that are imposed suddenly because most heat transfer occurring in nature is characterized by gradual changes in magnitude. Finally, alternative energy resources are frequently not available when they are in greatest demand: solar energy is available during the daytime, while minimum air temperature occurs at night or on overcast days when direct solar radiation is limited. Likewise, most modes of passive cooling are most effective during the night – while demand is typically greatest during the daytime.

The application of radiative cooling can succeed only if it can respond to at least some of the difficulties outlined above. It would appear that a successful radiative cooling system must therefore be a water-based one. Insulated roof ponds with radiators installed above them as shading devices can provide continuous cooling throughout the day and night, even though heat dissipation from the pond to the environment occurs only at night. The cooling effect is almost independent of interior conditions and the application of other cooling strategies, such as night ventilation. An experiment carried out over a period of about four weeks at a demonstration building at Sde Boqer, Israel, showed that a room in contact with a roof pond cooled by nocturnal radiation was substantially cooler than an otherwise similar reference room in each of five operating modes (Etzion and Erell, 1993); with windows closed and shaded for the entire 24 hours; with windows shaded but allowing natural ventilation at night, during the daytime or throughout the

24-hour cycle; and with all windows closed, but with the south-facing windows exposed to the sun during the daytime. The mean daily temperature in the cooled room was, on average, 2.1°C lower than in the reference room. Even when both rooms were ventilated at night, the difference was 1.6°C, and even the minimum temperature difference was 1.2°C.

An alternative to a roof pond as a means of applying radiative cooling is to supply the chilled water to a distribution system that circulates water through heat exchangers installed in the building interior. Such a system would probably be a hybrid one, and would comprise a storage tank and a conventional cooling apparatus to address peak loads in order to provide additional cooling if the radiative system was incapable of responding to demand on a regular basis, and as a back-up for days of unfavourable environmental conditions. The added cost of roof-mounted radiators to such a system would be moderate; but its cost-effectiveness must be evaluated in the context of the specific project.

## CONCLUSION

The application of radiative cooling to buildings requires a thorough understanding of local climatic conditions. If these are suitable, careful design can result in a system capable of providing a net cooling output of over 100W/m$^2$ of installed radiator surface for most of the night. However, the decision on whether to install a radiative cooling system should be taken at an early design stage so that the system may be integrated with the overall design of the building.

## ACKNOWLEDGEMENTS

The experimental work that underpins much of this chapter was carried out at Sde Boqer, Israel, over many years. Professor Yair Etzion played a leading role in all of this research, which would not have been possible without his contribution.

## REFERENCES

Addeo, A., Nicolais, L., Romeo, G., Bartoli, B., Coluzzi, B. and Silvestrini, V. (1980) 'Light selective structures for large scale natural air conditioning', *Solar Energy*, vol 24, no 1, pp93–98

Al-Nimr, M., Tahat, M. and Al-Rashdan, M. (1999) 'A night cold storage system by radiative cooling – a modified Australian system', *Applied Thermal Engineering*, vol 19, pp1013–1026

Argiriou, A. (1996) 'Radiative cooling', in Santamouris, M. and Asimakopoulos, D. (eds) *Passive Cooling of Buildings*, James and James, London

Arnfield, A. J. (1979) 'Evaluation of empirical expressions for the estimation of hourly

and daily totals of atmospheric longwave emission under all sky conditions',
*Quarterly Journal of the Royal Meteorological Society*, vol 105, no 446,
pp1041–1052

Berdahl, P. (1984) 'Radiative cooling with MgO and LiF layers', *Applied Optics*, vol
23, no 3, pp370–372

Brunold, S. (1989) *Untersuchungen zum Potential der Strahlungskhlung in Ariden
Klimazonen*, PhD thesis, University of Freiburg, Freiburg, Germany

Brunt, D. (1932) 'Notes on radiation in the atmosphere', *Quarterly Journal of the
Royal Meteorological Society*, vol 58, pp389–420

Brutsaert, W. (1975) 'On a derivable formula for longwave radiation from clear skies',
*Water Resources Research*, vol 11, pp742–744

Clark, G. (1989) 'Passive cooling systems', in Cook, J. (ed) *Passive Cooling*, MIT
Press, Cambridge and London

Clark, G. and Berdahl, P. (1980) 'Radiative cooling: Resource and applications', in
*Proceedings of the Fifth National Passive Solar Conference*, Amherst, MA, 20–22
October

Dimoudi, A. and Androutsopoulos, A. (2006) 'The cooling performance of a radiator
based roof component', *Solar Energy*, vol 80, pp1039–1047

Dobson, K. D., Hodes, G. and Mastai, Y. (2003) 'Thin semiconductor films for
radiative cooling applications', *Solar Energy Materials and Solar Cells*, vol 80, no
3, pp283–296

Duffie, J. A. and Beckman, W. A. (1991) *Solar Engineering of Thermal Processes*,
John Wiley and Sons, New York

Erell, E. and Etzion, Y. (1992) 'A radiative cooling system using water as a heat
exchange medium', *Architectural Science Review*, vol 35, pp39–49

Erell, E. and Etzion, Y. (1996) 'Heating experiments with a radiative cooling system',
*Building and Environment*, vol 31, no 6, pp509–517

Erell, E. and Etzion, Y. (1999) 'Analysis and experimental verification of an improved
cooling radiator', *Renewable Energy*, vol 16, pp700–703

Erell, E. and Etzion, Y. (2000) 'Radiative cooling of buildings with flat-plate solar
collectors', *Building and Environment*, vol 35, pp297–305

Erell, E., Etzion, Y., Brunold, S., Rommel, M. and Wittwer, V. (1992) 'A passive
cooling laboratory building for hot-arid zones', in *Proceedings of the Third
International Conference – Energy and Building in Mediterranean Area*,
Thessaloniki, 8–10 April

Etzion, Y. and Erell, E. (1991) 'Thermal storage mass in radiative cooling systems',
*Building and Environment*, vol 26, no 4, pp389–394

Etzion, Y. and Erell, E. (1993) *Radiative Cooling of Buildings*, Research report
submitted to the Bundesministerium für Forschung und Technologie (BMFT) and
to the Israel Ministry of Science and Technology (MOST), Midreshet Ben-Gurion

Etzion, Y. and Erell, E. (1999) 'Low-cost long wave radiators for passive cooling of
buildings', *Architectural Science Review*, vol 42, no 2, pp79–86

Givoni, B. (1977) 'Solar heating and night radiation cooling by a roof radiation trap',
*Energy and Buildings*, vol 1, pp141–145

Givoni, B. (1994) *Passive and Low Energy Cooling of Buildings*, John Wiley and
Sons, New York

Givoni, B. and Hoffman, M. (1970) *Preliminary Study of Cooling of Houses in Desert
Regions by Utilizing Outgoing Radiation*, Research report submitted to the Israel
Ministry of Energy, Haifa

Hay, H. and Yellot, J. (1969) 'Natural cooling with roofpond and moveable insulation', *ASHRAE Transactions*, vol 75, no 1, pp165–177

Idso, S. (1969) 'Thermal radiation from the atmosphere', *Journal of Geophysical Research*, vol 74, pp397–403

Idso, S. (1981) 'A set of equations for full-spectrum and 8–18μm and 10.5–12.5μm thermal radiation from cloudless skies', *Water Resources Research*, vol 74, pp5397–5403

Idso, S. and Jackson, R. (1969) 'Thermal radiation from the atmosphere', *Journal of Geophysical Research*, vol 74, pp397–403

Incropera, F. and De Witt, D. (1990) *Fundamentals of Heat and Mass Transfer*, John Wiley and Sons, New York

Iziomon, M. G., Mayer, H. and Matzarakis, A. (2003) 'Downward atmospheric longwave irradiance under clear and cloudy skies: Measurement and parameterization', *Journal of Atmospheric and Solar-Terrestrial Physics*, vol 65, pp1107–1116

Juchau, B. (1981) 'Nocturnal and conventional space cooling via radiant floors', in *Proceedings of the International Passive and Hybrid Cooling Conference*, Miami Beach, FL

Martin, M. (1989) 'Radiative cooling', in Cook, J. (ed) *Passive Cooling*, MIT Press, Cambridge and London

Martin, M. and Berdahl, P. (1984) 'Summary of results for the spectral and angular sky radiation measurement program', *Solar Energy*, vol 33, no 3/4, pp241–252

Meir, M., Rekstad, J. and Lovvik, O. (2002) 'A study of a polymer-based radiative cooling system', *Solar Energy*, vol 73, no 6, pp403–417

Mihalakakou, G., Ferrante, A. and Lewis, J. (1998) 'The cooling potential of a metallic nocturnal radiator', *Energy and Buildings*, vol 28, pp251–256

Molina, J. L. (1998) *Visual Atlas – Final Research Report of Task 4*, ROOFSOL Project, JOR3CT9600074, European Commission, Brussels

Mostrel, M. and Givoni, B. (1982) 'Windscreens in radiant cooling', *Applied Research*, vol 1, no 4, pp229–238

Niemela, S., Raisanen, P. and Savijarvi, H. (2001) 'Comparison of surface radiative flux parameterizations. Part I: Longwave radiation', *Atmospheric Research*, vol 58, pp1–18

Nilsson, T. and Niklasson, G. (1995) 'Radiative cooling during the day: simulations and experiments on pigmented polyethylene cover foils', *Solar Energy Materials and Solar Cells*, vol 37, pp93–118

O'Brien-Bernini, F. and McGowan, J. (1984) 'Performance modeling of non-metallic flat plate solar collectors', *Solar Energy*, vol 33, no 3/4, pp305–319

Oke, T. R. (1987) *Boundary Layer Climates*, Methuen, London and New York

Oliveti, G., Arcuri, N. and Ruffolo, S. (2003) 'Experimental investigation on thermal radiation exchange of horizontal outdoor surfaces', *Building and Environment*, vol 38, pp83–89

Swinbank, W. C. (1963) 'Long wave radiation from clear skies', *Quarterly Journal of the Royal Meteorological Society*, vol 89, pp339–348

Yannas, S., Erell, E. and Molina, J. L. (2006) *Roof Cooling Techniques: A Design Handbook*, Earthscan, London

# Index